HPE ATP – Data Center and Cloud V2
HPE ASE – Data Center and Cloud Architect V3
OFFICIAL CERTIFICATION STUDY GUIDE
(EXAMS HPE0-D33, HPE0-D34 AND HPE0-D35)

First Edition

Ken Radford

HPE Press
660 4th Street, #802
San Francisco, CA 94107

HPE ATP – Data Center and Cloud V2
HPE ASE – Data Center and Cloud Architect V3
Official Certification Study Guide
(Exams HPE0-D33, HPE0-D34 and HPE0-D35)
Ken Radford

Published by:

Hewlett Packard Enterprise Press
660 4th Street, #802
San Francisco, CA 94107

ISBN: 978-1-942741-33-6

Printed in Mexico

WARNING AND DISCLAIMER
This book provides information about the topics covered in the HPE ATP—Data Center and Cloud V2 (HPE0-D33) and the HPE ASE— Data Center and Cloud Architect V3 (HPE0-D34) certification exams. Every effort has been made to make this book as complete and as accurate as possible, but no warranty or fitness is implied.

The information is provided on an "as is" basis. The author, and Hewlett Packard Enterprise Press, shall have neither liability nor responsibility to any person or entity with respect to any loss or damages arising from the information contained in this book or from the use of the discs or programs that may accompany it.

The opinions expressed in this book belong to the author and are not necessarily those of Hewlett Packard Enterprise Press.

TRADEMARK ACKNOWLEDGEMENTS
All third-party trademarks contained herein are the property of their respective owner(s).

GOVERNMENT AND EDUCATION SALES
This publisher offers discounts on this book when ordered in quantity for bulk purchases, which may include electronic versions. For more information, please contact U.S. Government and Education Sales 1-855-447-2665 or email sales@hpepressbooks.com.

Feedback Information

At HPE Press, our goal is to create in-depth reference books of the best quality and value. Each book is crafted with care and precision, undergoing rigorous development that involves the expertise of members from the professional technical community.

Readers' feedback is a continuation of the process. If you have any comments regarding how we could improve the quality of this book, or otherwise alter it to better suit your needs, you can contact us through email at hpepress@epac.com. Please make sure to include the book title and ISBN in your message.

We appreciate your feedback.

Publisher: Hewlett Packard Enterprise Press

HPE Contributors: Wim Groeneveld, Richard Hykel, Jeroen Kleen, Chris Powell, and Colin Taylor

HPE Press Program Manager: Michael Bishop

About the Author

Ken Radford is a certified HP ASE—Cloud Architect, HP ASE—Converged Infrastructure Architect, HP ATP—Cloud Administrator, and HP Certified Instructor. He was author of the HP ASE Cloud Architect V1 and V2 official certification study guides, and courseware developer for the Designing HPE Data Center and Cloud Solutions, Rev. 16.11 course (ID number: 01045403).

Introduction

This study guide helps you prepare for the HPE ATP—Data Center and Cloud V2 (HPE0-D33) certification exam and the HPE ASE—Data Center and Cloud Architect V3 (HPE0-D34) certification exam. Organized in two parts along the lines of exam topics and objectives, chapters can be studied independently when preparing for each level of certification. If you already hold an HP Data Center and Cloud certification and want to acquire the HPE ASE—Data Center and Cloud Architect V3 certification, this guide also covers the topics in the Designing HPE Data Center and Cloud Solutions (HPE0-D35) delta exam (see Preparing for the HPE0-D35 Delta Exam on page vii for details).

This guide provides a solid foundation for recommending optimal solutions when moving critical IT operations and applications to the cloud. Topics discussed include creating an on-demand hybrid infrastructure using HPE CloudSystem and HPE Helion OpenStack technologies. With a focus on meeting the customer's business and technical needs, the guide serves as a practical on-the-job reference tool to describe, position, plan and design HPE data center and cloud solutions.

Interactive embedded links and Videos

Hewlett Packard Enterprise Partner Ready Certification and Learning is driving innovation across our portfolio of technologies to enhance the way you learn. We have formed a partnership with Aurasma to provide embedded digital content as an additional benefit for customers who purchase this study guide.

Within this study guide you will see the My Learning HPE icon , which indicates that additional digital content is available to you. To view this content, you have to download the My Learning HPE app, from the Play Store/App Store, open the app, and then point the viewfinder at the page wherever you see the icon .

For additional learning content and practical demonstrations, you will also find URLs to YouTube videos throughout the guide.

QR Codes

To expand your print to digital reading experience, Quick Response (QR) codes are also used throughout this book. These two-dimensional bar codes can be read by a smart phone or tablet that has a QR reader installed. Here is an example:

 Note

> If you do not have a smart phone or tablet with a QR reader, enter the URL underneath the QR code into a browser to access the content.

www.hpe.com

Scan this with a Smart Phone app that reads QR Codes

Certification and Learning

Hewlett Packard Enterprise Partner Ready Certification and Learning provides end-to-end continuous learning programs and professional certifications that can help you open doors and succeed in the idea economy. We provide continuous learning activities and job-role based learning plans to help you keep pace with the demands of the dynamic, fast-paced IT industry; professional sales and technical training and certifications to give you the critical skills needed to design, manage, and implement the most sought-after IT disciplines; and training to help you navigate and seize opportunities within the top IT transformation areas that enable business advantage today.

As a Partner Ready Certification and Learning certified member, your skills, knowledge, and real-world experience are recognized and valued in the marketplace. To continue your professional and career growth, you have access to our large HPE community of world-class IT professionals, trend-makers and decision-makers. Share ideas, best practices, business insights, and challenges as you gain professional connections globally.

To learn more about HPE Partner Ready Certification and Learning certifications and continuous learning programs, please visit **http://certification-learning.hpe.com**

Audience

This guide is suitable for anyone seeking a deeper understanding of how to recognize converged infrastructure and cloud opportunities and apply the right solutions to the right environments. It is ideal for solution architects and technical presales professionals who need to describe, position, and recommend HPE Converged Infrastructure and HPE Helion cloud solution offerings and how they can be applied to meet a customer's business and technical needs.

Assumed Knowledge

It is assumed that you have a good understanding of the HP Converged Infrastructure and its components, including HP servers, storage, networking, IT operations, software, security, and services.

Minimum Qualifications

There are no prerequisites for the HPE ATP – Data Center and Cloud V2 certification. However, it is strongly recommended that you have foundational HPE server, storage, networking, and management product knowledge and have taken the recommended trainings.

The prerequisite for the HPE ASE – Data Center and Cloud Architect V3 certification is the HPE ATP – Data Center and Cloud V2 certification or HP ATP – Data Center and Cloud V1 certification. If eligible, you can take the Designing HPE Data Center and Cloud Solutions Delta exam (HPE0-D35).

The specific requirements for each certification are available on the HPE Partner Ready Certification and Learning website.

Relevant Certifications

After you pass these exams, your achievement may be applicable toward more than one certification. To determine which certifications can be credited with this achievement, log in to the Learning Center and view the certifications listed on the exam's More Details tab. You might be on your way to achieving additional certifications.

Preparing for the Certification Exams (HPE0-D33 and HPE0-D34)

This self-study guide does not guarantee that you will have all the knowledge you need to pass the HPE ATP—Data Center and Cloud V2 (HPE0-D33) exam and the HPE ASE—Data Center and Cloud Architect V3 (HPE0-D34) exam. It is expected that you will also need to draw on real-world experience and would benefit from completing the hands-on activities delivered in the instructor-led training.

Preparing for the Delta Exam (HPE0-D35)

The Designing HPE Data Center and Cloud Solutions delta exam (HPE0-D35) tests your skills to identify, describe, position, demonstrate, and specify the correct cloud solution based on customer needs. Knowledge of how to architect, install, manage, and administer the HPE Helion Cloud solution, based on CloudSystem 9.0 Foundation or Enterprise, is required to pass the exam. To prepare for the delta exam, it is recommended that you study Chapters 7–11 in this guide. It may be helpful to review Chapter 4 because in-depth OpenStack knowledge is also required.

Recommended HPE Training

Recommended training to prepare for each exam is accessible from the exam's page in The Learning Center. See the exam attachment, "Supporting courses," to view and register for the courses.

Obtain Hands-on Experience

You are not required to take the recommended, supported courses, and completion of training does not guarantee that you will pass the exams. Hewlett Packard Enterprise strongly recommends a combination of training, thorough review of courseware and additional study references, and sufficient on-the-job experience prior to taking an exam.

Exam Registration

To register for an exam, go to http://certification-learning.hpe.com/tr/certification/learn_more_about_exams.html

CONTENTS

HPE ATP — Data Center and Cloud V2

1 IT Transformation for the Idea Economy

WHAT'S IN THIS CHAPTER FOR YOU?

After completing this chapter, you should be able to:

✓ Describe the major trends changing the IT industry today

✓ Name the four transformation areas that form the centerpiece of Hewlett Packard Enterprise and explain what they mean to customers

The idea economy is here

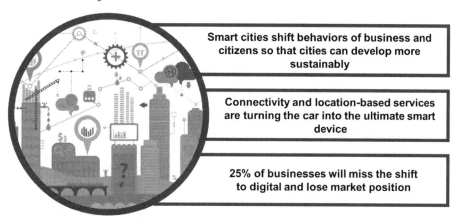

Smart cities shift behaviors of business and citizens so that cities can develop more sustainably

Connectivity and location-based services are turning the car into the ultimate smart device

25% of businesses will miss the shift to digital and lose market position

Figure 1-1 The idea economy

Ideas have always fueled business success. Ideas have built companies, markets, and industries. However, there is a difference today.

Figure 1-1 shows how businesses operate in the idea economy, which is also called the digital, application, or mobile economy. Doing business in the idea economy means turning an idea into a new product, a capability, a business, or an industry that has never been easier or more accessible—for customers and for their competitors.

Today, an entrepreneur with a good idea has access to the infrastructure and resources that a traditional Fortune 1000 company would have. That entrepreneur can rent compute capacity on demand, implement a software-as-a-service (SaaS) enterprise resource planning system, use PayPal or Square for transactions, market products and services using Facebook or Google, and have FedEx or UPS run the supply chain.

Companies such as Vimeo, One Kings Lane, Dock to Dish, Uber, Pandora, Salesforce, and Airbnb used their ideas to change the world with very little start-up capital. Uber had a dramatic impact after launching its application connecting customers and drivers in 2009. Without owning a single car, Uber now serves more than 300 cities in 58 countries (as of May 28, 2015). The company has completely disrupted the taxi industry. San Francisco Municipal Transportation Agency reported that cab use in San Francisco has dropped 65% in two years.

In a technology-driven world, it takes more than just ideas to be successful. Success is defined by how quickly ideas can be turned into value.

Creating disruptive waves of new demands and opportunities

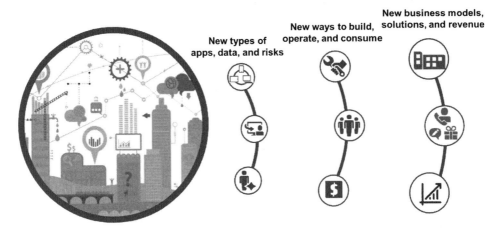

Figure 1-2 New demands and opportunities

The idea economy presents an opportunity and a challenge for most enterprises. On the one hand, cloud, mobile, big data, and analytics give businesses the tools to accelerate time to value. This increased speed allows organizations to combine applications and data to create dramatically new experiences, even new markets.

On the other hand, most organizations were built with rigid IT infrastructures that are costly to maintain. This rigidity makes it difficult, if not impossible, to implement new ideas quickly.

Figure 1-2 shows how creating and delivering new business models, solutions, and experiences requires harnessing new types of applications, data, and risks. It also requires implementing new ways to build, operate, and consume technology. This new way of doing business no longer just supports the company—it becomes the core of the company.

Faster application development enables accelerated innovation

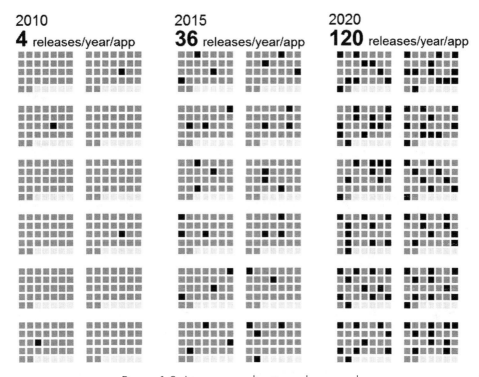

Figure 1-3 Average application release cycle

From 2010 to 2015, much changed from an application development perspective. In 2010, the average application release cycle was four releases per year, per application. In 2015, this number went up to 36 releases per year, per application. It is projected that by 2020, there will be 120 releases per year, per application (30 times more releases than in 2010) as shown in Figure 1-3.

Considerations from the Forrester Thought Leader Paper commissioned by Hewlett Packard Enterprise, *Better Outcomes, Faster Results: Continuous Delivery and the Race for Better Business Performance*, help summarize this trend:

- Agility is paramount.
- "Even when delivering at cadences of less than a week, 20% of organizations want to go even faster."
- Developers need flexibility.
- Organizations expect to deploy 50%–70% of code to cloud environments by 2015.
- Companies want open, flexible architectures for application portability and lock-in prevention.

Long time to value is costly

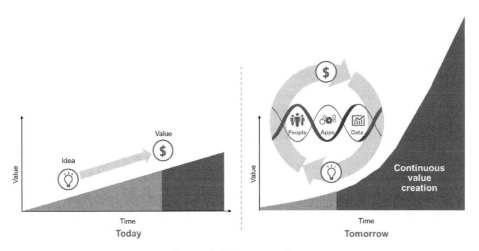

Figure 1-4 Time to value

As shown in Figure 1-4, success today is defined by how quickly an enterprise can turn ideas into value—how quickly a business can experiment, learn, test, tune, and make things better. Speed is a key differentiator in all industries.

Uber did not invent a new technology. Instead, the company took advantage of the explosion of smartphones and mobile applications to design a compelling customer experience, ultimately creating a new way of doing business.

This example is not only about Uber executing a good idea. It is also about the taxicab industry's inability to act quickly to transform its business models to compete. Examples such as Uber serve as a warning. Every Fortune 1000 company is at risk of missing a market opportunity, failing to secure its enterprise, and being disrupted by a new idea or business model.

Timelines for IT projects and new applications used to be planned over years and months. Today, these projects take weeks or days. Increasingly, it is shrinking to hours. Now is the time for a company to ask:

- How quickly can the company capitalize on a new idea?

- How rapidly can the company seize a new opportunity?

- How fast can the company respond to a new competitor that threatens the business?

The good news is that any company can use these technologies in order to adapt quickly to changing business models and achieve faster time to value. However, many established companies are working with a rigid IT infrastructure, which may present significant challenges. Changing a data center strategy is a costly, lengthy, and complex process.

Being successful in the idea economy

Thriving in the idea economy requires enterprises to adopt a new style of business that:

- **Is experience and outcome driven**—Rapidly compose new services from any source to meet the evolving needs of customers and citizens.

- **Proactively manages risks**—Remain safe and compliant in a world of rapidly changing threat landscape.

- **Is contextually aware and predictive**—Harness 100% of data to generate real-time instant insights for continuous improvement, innovation, and learning.

- **Is hyper-connected to customers, employees, and the ecosystem**—Deliver experiences that enable employees and engage customers in a persistent, personalized way.

In the idea economy, applications and information are the products.

IT must become a value creator that bridges the old and the new

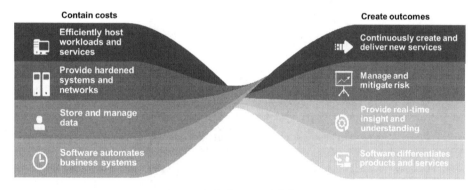

Figure 1-5 IT must shift focus

To respond to the disruptions created by the idea economy, IT must transform from a cost center to a value creator as shown in Figure 1-5. In order to evolve, IT must shift focus:

- From efficiently hosting workloads and services to continuously creating and delivering new services

- From simply providing hardened systems and networks to proactively managing and mitigating risks

- From just storing and managing data to providing real-time insight and understanding

- From using software to automate business systems to differentiating products and services

Customers need to make IT environments more efficient, productive, and secure. They need to enable their organizations to act rapidly on ideas by creating, consuming, and reconfiguring new solutions, experiences, and business models.

One of the first steps in achieving this kind of agility is to break down the old infrastructure silos that make enterprises resistant to new ideas internally and vulnerable to new ideas externally. Designing compelling new experiences and services does not work if the infrastructure cannot support them.

The right compute platform can make a significant impact on business outcomes and performance. Examples include storage that "thinks" as much as it stores; networking that moves information faster and more securely than ever before; and orchestration and management software that provides predictive capabilities.

Each company is on a unique journey to the cloud, custom-made for the way it consumes and allocates resources, transforms to the changing landscape, implements financial models, and achieves desired outcomes.

This unique journey starts with four transformation areas

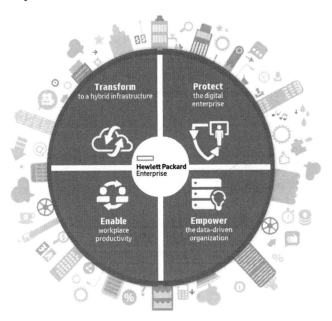

Figure 1-6 The four HPE transformation areas

The HPE transformation areas are designed to:

- Generate revenue and profitable growth

- Increase agility and flexibility

- Deliver remarkable customer experience

- Amplify employee productivity

- Reduce cost and risk

These transformation areas, shown in Figure 1-6, reflect what customers consider most important:

- **Protecting the digital enterprise**—Customers consider it a matter of when, not if, their digital walls will be attacked. The threat landscape is wider and more diverse than ever before. A complete risk management strategy involves security threats, backup and recovery, high availability, and disaster recovery.

- **Empowering the data-driven organization**—Customers are overwhelmed with data; the solution is to obtain value from information that exists. Data-driven organizations generate real-time, actionable insights.

- **Enabling workplace productivity**—Many customers are increasingly focused on enabling workplace productivity. Delivering a great digital workplace experience to employees and customers is a critical step.

- **Transforming to a hybrid infrastructure**—A hybrid infrastructure enables customers to get better value from the existing infrastructure and delivers new value quickly and continuously from all applications. This infrastructure should be agile, workload optimized, simple, and intuitive.

Learning check

1. Describe today's idea economy.

_____

_____

_____

_____

2. In order to make a significant impact on business outcomes, what should IT organizations focus on? (Select three.)

   a. Continuously creating and delivering new services

   b. Hosting workloads and services

   c. Software differentiating products and services

   d. Providing real-time insight and understanding

   e. Storing and managing data

Learning check answers

1. Describe today's idea economy.

   In today's world, the ability to turn an idea into a new product, capability, business, or industry has never been easier or more accessible

This presents an opportunity and a challenge for most enterprises

Creating and delivering new business models, solutions, and experiences requires new types of applications, data, and risks as well as implementing new ways to build, operate, and consume technology

2. In order to make a significant impact on business outcomes, what should IT organizations focus on? (Select three.)

   a. **Continuously creating and delivering new services**

   b. Hosting workloads and services

   c. **Software differentiating products and services**

   d. **Providing real-time insight and understanding**

   e. Storing and managing data

Protect the digital enterprise

Protecting a digital enterprise requires alignment with key IT and business decision-makers for a business-aligned, integrated, and proactive strategy to protect the hybrid IT infrastructure and data-driven operations, as well as enable workplace productivity. By focusing on security as a business enabler, HPE brings new perspectives on how an organization can transform from traditional static security practices to intelligent, adaptive security models to keep pace with business dynamics.

Protecting the digital enterprise: Video activity

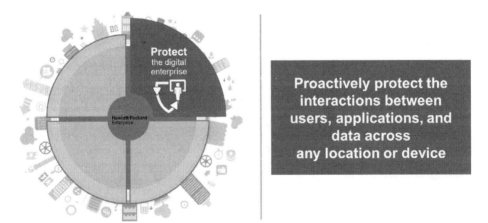

Figure 1-7 Protecting the digital enterprise

Watch a five-minute video about this transformation area by scanning Figure 1-7 into the My Learning HPE app on your mobile device. Think about the main ideas outlined in the video. What steps are needed to protect the digital enterprise? Take notes in the space here:

_____

_____

_____

_____

_____

_____

The following three pages detail the challenges, action plan, and enabling technologies to address this issue. Your notes should reflect the ideas presented on these pages.

What problems can be solved?

Figure 1-8 Security problems solved by IT transformation

All businesses must manage the emerging risks created by the proliferation of apps, new consumption models, and the shift to mobile and cloud capabilities. With the right strategy, organizations can access all the benefits of an app-centric, hybrid world, and proactively protect their networks.

As shown in Figure 1-8, many customers struggle with:

- **Growing threats and vulnerabilities**—Lack of integrated protection mechanisms and inadequate technology maintenance and testing

- **Reactive strategy**—Uncoordinated spending, compliance issues, and underinvestment to handle emerging threats and data protection gaps

- **Rigid operations**—Manual and siloed allocation of backup jobs to target devices, managed separately from business applications with fragmented security controls

- **Overreliance on silver bullets**—Limited impact of tools due to insufficient integration, inadequately trained staff, and suboptimal security processes

To achieve critical business outcomes, customers must focus on:

- **Built-in resilience**—Automated and integrated data protection and security controls, robust security governance, and high-availability infrastructure

- **Planned ecosystem**—Strategic planning and investment in sophisticated enterprise security, latest protection topologies, and tools for compliance

- **Adaptive and federated systems**—Integrated tools, elastic pools of protection capacity, and analytics-based optimization to balance performance

- **Integrated solutions**—Regular assessments of capabilities to ensure people, technology, and processes are aligned to deliver better business outcomes

Action plan and enabling technologies

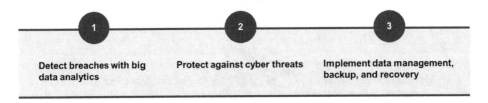

Figure 1-9 Protecting digital assets

Security and risk protection should be integrated when the infrastructure is set up. Enhancing security after the infrastructure is in place can cost 10 times more than initial prevention. Businesses need a single solution that balances regulatory requirements, cyber threats, asset protection, and business change. The key is to protect the most important part of the business and understand how people access it, and then create policies and tools for those users.

Figure 1-9 shows how HPE products, solutions, and services align with each step of protecting enterprise digital assets:

1. **Detect breaches with big data analytics**—ArcSight and managed security services identify potential and successful security and compliance breaches.

2. **Protect against cyber threats**—ClearPass from Aruba, HPE Networking, Security Voltage, Atalla data security and encryption, Fortify, and Security Research deploy next-generation vulnerability analysis, encryption, and intrusion protection using the latest threat intelligence.

3. **Implement data management, backup, and recovery**—HPE StoreServ, StoreOnce Backup, and Data Protector ensure business continuity during a crisis and simplify regulatory compliance.

Empower the data-driven organization

A data-driven organization leverages valuable feedback that is available consistently from both internal and external sources. By harnessing insights from data in the form of information, organizations can determine the best strategies to pave the way for seamless integration of agile capabilities into an existing environment. Because both technical and organizational needs must be considered, HPE helps organizations define the right ways to help ensure that processes, security, tools, and overall collaboration are addressed properly for successful outcomes.

Empowering the data-driven organization: Video activity

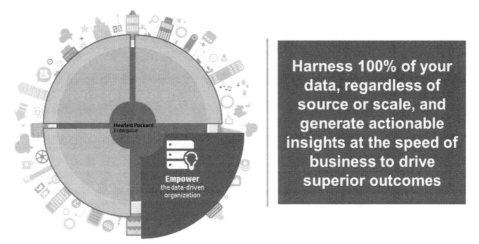

Figure 1-10 Empowering the data-driven organization

Watch a five-minute video about this transformation area by scanning Figure 1-10 into the My Learning HPE app on your mobile device. Think about the main ideas outlined in the video. What steps are needed to empower the data-driven organization? Take notes in the space here:

_____

_____

_____

_____

_____

_____

_____

The following four pages detail the challenges, action plan, and enabling technologies to address this issue. Your notes should reflect the ideas presented on these pages.

What problems can be solved?

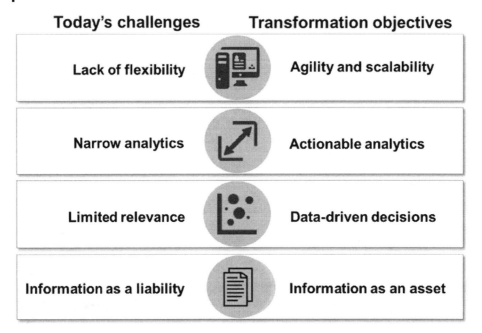

Figure 1-11 Data problems solved by IT transformation

Rapid evolution in technology has created a distributed digital world—data is everywhere. It presents new opportunities to capture value, as well as new sources of risk. To compete, businesses must generate actionable insights that can drive better business outcomes.

As shown in Figure 1-11, many customers struggle with:

- **Lack of flexibility**—Inadequate IT investment planning and expensive proprietary systems constrain ability to scale out or extend to new data types.

- **Narrow analytics**—Insights are backward-looking and generated in silos, with limited relevance to future business decisions.

- **Limited relevance**—Analytics output is not always useful because search queries are too slow and draw from only a fraction of available data.

- **Information as a liability**—Inadequate tracking and indexing of information creates compliance and business risks.

To achieve critical business outcomes, customers must focus on:

- **Agility and scalability**—An investment road map enables the rapid deployment of powerful open hardware and software at a lower cost with more flexibility to scale.

- **Actionable analytics**—Predictive insights should be constantly refined and highly relevant to multiple facets of the business.

- **Data-driven decisions**—Powerful analytics solutions (traditional or cloud based) connect to virtually any data source quickly and easily.

- **Information as an asset**—Information is governed in a secure end-to-end life cycle, balancing value, cost, and risk.

Action plan and enabling technologies

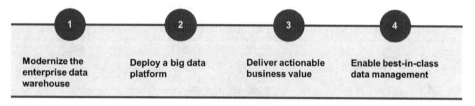

Figure 1-12 Empowering a data-driven enterprise

According to a survey of HPE customers, companies realize only 10%–15% of the expected value on their big data investments. There are three main lessons from past HPE customer engagements:

- First, customers must optimize for their existing data. Optimizing the core infrastructure and hardware allows for evolving data sources such as media and text.

- Second, customers need to drive continuous analytics into business processes. Insights must happen in real-time and be embedded into the decision flow, not created and processed as separate events.

- And third, modernized business intelligence (BI) sources are not properly monetized for two reasons: they do not integrate all new data sources and they do not properly combine data from existing data warehouses.

Figure 1-12 shows how HPE products, solutions, and services align with each step of empowering a data-driven enterprise:

1. **Modernize the enterprise data warehouse**—Vertica, ProLiant servers, Converged Systems, and ISVs improve scalable performance and responsiveness by adopting a more effective cost model.

2. **Deploy a big data platform**—Haven Big Data platform powered by Vertica, IDOL, Distributed R, and Analytics and Data Management Services help develop analytics apps and services on premise and in the cloud.

3. **Deliver actionable business value**—Haven Enterprise, Haven OnHadoop, Haven OnDemand, Haven ISVs, Helion, and HPE ConvergedSystems deliver simple insight that is responsive to business needs.

4. **Enable best-in-class data management**—Connected MX, ControlPoint, HPE Archiving, and HPE Storage integrate data management and collaboration tools to maximize efficiency and effectiveness.

 Note

To get more information about Haven Enterprise, Haven OnHadoop, and Haven OnDemand, scan the QR code or enter the URL into your browser.

http://www8.hp.com/us/en/software-solutions/big-data-platform-haven/

Enable workplace productivity

Organizations seeking to improve their efficiency and speed place a premium on creating a desirable work environment for their employees, including offering technology that employees want and need. They believe they must enable employees to work how, where, and whenever they want.

HPE solutions for the workplace provide secure, easy, mobile collaboration, and anywhere, anytime access to data and applications for better productivity and responsiveness. To achieve the expected potential, spending growth for mobile resources is expected to be twice the level of IT spending growth in general, according to 2014 IDC survey results.

Enabling workplace productivity: Video activity

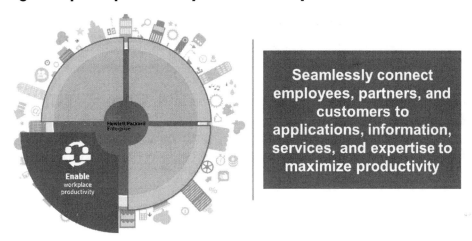

Figure 1-13 Enabling workplace productivity

Watch a five-minute video about this transformation area by scanning Figure 1-13 into the My Learning HPE app on your mobile device. Think about the main ideas outlined in the video. What steps are needed to enable workplace productivity? Take notes in the space here:

_____

_____

_____

_____

_____

The following three pages detail the challenges, action plan, and enabling technologies to address this issue. Your notes should reflect the ideas presented on these pages.

What problems can be solved?

Figure 1-14 User experience problems solved by IT transformation

Delivering a superior user experience to customers, employees, and partners is a major driver of productivity. To be competitive, the modern enterprise needs to support ubiquitous access, seamless communication, and high-performing applications—without jeopardizing data security and corporate assets.

As shown in Figure 1-14, many customers struggle with:

- **Increasing costs**—Meeting user expectations is more costly and time-consuming due to an aging, rigid infrastructure.

- **Desk-bound workers**—Wired networks with separate voice and data capabilities make desktop devices a necessity and limit opportunities for creative collaboration.

- **Constrained mobility**—Inadequate support for mobile devices limits productivity and prompts employee workarounds that create risk.

- **Legacy investment limitations**—Options for technology refresh are constrained by past purchasing decisions.

To achieve critical business outcomes, customers must focus on:

- **Greater efficiency**—Software-defined infrastructure and user-based management reduce costs and improve user experience.

- **Universal accessibility**—High-performance wireless devices, new working practices, and cross-device collaboration improve communication.

- **Anywhere workforce**—The flexibility to work anywhere means accessing resources on any device with secure, tested, and monitored apps.

- **Adaptable investment strategy**—Evolve capabilities, devices, and applications in line with business needs.

Action plan and enabling technologies

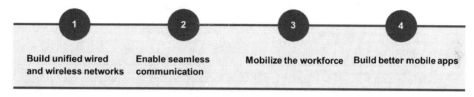

Figure 1-15 Enabling a more connected community

From an infrastructure perspective, plan the full technology stack from end to end. From the initial touch point to core data, the infrastructure must be performance-ready for more devices, more features, and smarter environments. It is not just a matter of adding more switches and Wi-Fi nodes. It is about providing real-time access to information.

Additionally, think about how to constantly improve the user experience. That means optimizing across mobile apps and infrastructure (designing, testing, and securing every aspect) and exploiting the full power of analytics into the feedback loop.

Figure 1-15 shows how HPE products, solutions, and services align with each step of enabling workplace productivity:

1. **Build unified wired and wireless networks**—HPE Aruba AirWave (previously known as the Intelligent Management Center) and HPE campus and data center switches reduce costs and improve the user's experience.

2. **Enable seamless communication**—HPE Technology Services, Software-Defined Networking (SDN) Network Optimizer application, and HPE WorkSite/HPE LinkSite deploy the latest productivity applications.

3. **Mobilize the workforce**—AirWave, Network Protector SDN Application, and ClearPass from Aruba enable bring your own device (BYOD) without compromising security.

4. **Build better mobile apps**—HPE application development and delivery services, StormRunner Load, Network Virtualization, Mobile Center, and AppPulse Mobile can build, test, and monitor mobile apps for optimal user experience.

Transform to a hybrid infrastructure

An organization might see cloud services as a key component of its ability to access the IT services they need, at the right time and at the right cost. A hybrid infrastructure is based on standards, is built on a common architecture with unified management and security, and enables service portability across deployment models. Getting the most out of hybrid infrastructure requires planning performance, security, control, and availability strategies. For this reason, organizations must understand where and how a hybrid infrastructure strategy can most effectively be applied to their portfolio of services.

Transforming to a hybrid infrastructure: Video activity

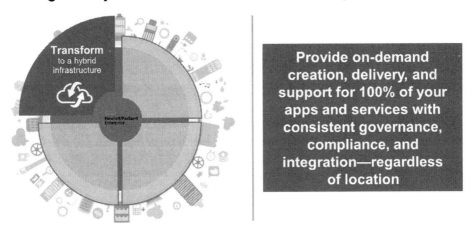

Figure 1-16 Transforming to a hybrid infrastructure

Watch a five-minute video about this transformation area by scanning Figure 1-16 into the My Learning HPE app on your mobile device. Think about the main ideas outlined in the video. What steps are needed to transform the IT environment? Take notes in the space here:

_____

_____

_____

_____

_____

The following three pages detail the challenges, action plan, and enabling technologies to address this issue. Your notes should reflect the ideas presented on these pages.

What problems can be solved?

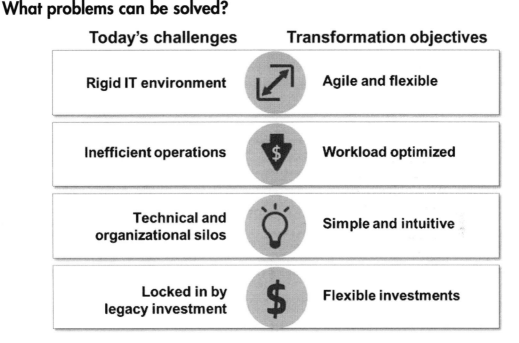

Figure 1-17 Infrastructure problems solved by IT transformation

Hybrid is a reality. In IT operations, extracting optimum performance and efficiency from applications is essential. The best environment for applications, whether traditional, mobile, or cloud native, is unique to each business.

As shown in Figure 1-17, many customers struggle with:

- **Rigid IT environment**—Legacy hardware scales poorly and slows deployment of apps and workloads.

- **Inefficient operations**—Data center has high operating costs and overhead, slow IT services, and poor utilization with inadequate availability and performance.

- **Technical and organizational silos**—Inefficiencies and lack of collaboration mean IT is dedicated to "keeping the lights on."

- **Being locked in by legacy investment**—Proprietary systems and depreciation schedules limit upgrade opportunities.

To achieve critical business outcomes, customers must focus on:

- **Agility and flexibility**—A converged and virtualized hybrid infrastructure scales easily and delivers continuous value to make IT a service provider.

- **Workload optimization**—Modern infrastructure offers better utilization, adjusting performance, and availability dynamically.

- **Simplicity and intuitiveness**—Software-defined controls, along with automation and converged management, free up IT resources.

- **Flexible investments**—Open-standards-based systems and new IT consumption models enable continuous business innovation.

Action plan and enabling technologies

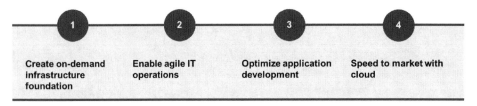

Figure 1-18 Transforming to on-demand IT infrastructure

The journey to hybrid is difficult, nonlinear, and different for every customer. More than 90% of enterprises say their implementation of hybrid is behind company goals and lagging competitors (based on Enterprise Services interviews with customers).

The most successful journeys often require bold moves, such as shifting to new platforms and SaaS. Sometimes, it is best to move straight to cloud, even from old, non-virtualized environments.

Figure 1-18 shows how HPE products, solutions, and services align with each step of transforming to a hybrid infrastructure:

1. **Create an on-demand infrastructure foundation**—ConvergedSystems, ProLiant Gen9 servers, and StoreServ improve efficiency and create agility for the next generation of applications and services.

2. **Enable agile IT operations**—OneView and Operations Analytics transform management of infrastructure and clouds with analytics and automation.

3. **Optimize application development**—HPE application development and delivery services, Application Lifecycle Management, and Codar deliver high-quality applications across legacy, cloud, and mobile environments.

4. **Speed to market with cloud**—HPE Helion CloudSystem and Helion OpenStack and Development Platform access unlimited scale and speed in a secure way.

Learning check

1. Which solutions should you consider when protecting against cyber threats? (Select two.)

   a. ClearPass from Aruba

   b. Fortify

   c. ArcSight

   d. Data Protector

2. Complete the action plan to empower a data-driven enterprise:

   - Modernize the _____ _____ warehouse

   - Deploy a _____ _____ platform

   - Deliver actionable _____ _____

   - Enable best-in-class data _____

3. Random acts of mobility will deliver a major shift in productivity.

   ☐ True

   ☐ False

4. What matters when transforming to a hybrid infrastructure? (Select three.)

   a. Open-standards-based solutions

   b. IT dedicated to "keeping the lights on"

   c. Rigid IT environment

   d. Easy control of infrastructure and apps

   e. The right capabilities across people, processes, and governance

   f. Technical and organizational silos

Learning check answers

1. Which solutions should you consider when protecting against cyber threats? (Select two.)

   a. ClearPass from Aruba

   b. Fortify

   c. ArcSight

   d. Data Protector

2. Complete the action plan to empower a data-driven enterprise:

   - Modernize the **enterprise data** warehouse
   - Deploy a **Big Data** platform
   - Deliver actionable **business value**
   - Enable best-in-class data **management**

3. Random acts of mobility will deliver a major shift in productivity.

   ☐ True

   ☐ **False**

4. What matters when transforming to a hybrid infrastructure? (Select three.)

   a. **Open-standards-based solutions**

   b. IT dedicated to "keeping the lights on"

   c. Rigid IT environment

   d. **Easy control of infrastructure and apps**

   e. **The right capabilities across people, processes, and governance**

   f. Technical and organizational silos

Summary

Today's idea economy means enhanced access, data, and connections are driving exponential innovation, which creates disruptive new challenges and opportunities for IT.

HPE has identified four transformation areas that are keys to success:

- Protecting the digital enterprise
- Empowering the data-driven organization
- Enabling workplace productivity
- Transforming to a hybrid infrastructure

These four transformation areas enable organizations to:

- Generate revenue and profitable growth
- Increase agility and flexibility
- Deliver remarkable customer experience
- Amplify employee productivity
- Reduce cost and risk

2 Creating On-Demand Hybrid IT Infrastructure

ON-DEMAND IT INFRASTRUCTURE FUNDAMENTALS

Before proceeding with this section, assess your existing knowledge by attempting to answer the following questions.

Assessment questions

1. What is the correct definition of a converged infrastructure?

   a. An IT environment where the server assets are standardized and consolidated

   b. An IT environment where server, storage, and networking resources are standardized, virtualized, organized, and managed as resource pools

   c. An IT environment that provides on-demand and self-service provisioning of infrastructure services

   d. An IT environment that provides a development platform for consumer-developed or acquired applications created with programming languages or tools from the provider

Make additional notes explaining the correct answer:

_____

_____

_____

_____

_____

2. The following graphic represents a customer's transition from a legacy data center built with traditional, siloed IT toward infrastructure convergence and cloud. A converged infrastructure provides a foundation for which phases?

Phase 1:
Converge and
virtualize

Traditional IT

Phase 2:
Automate and
orchestrate

Phase 3:
Transform app and
service delivery

Phase 4:
Deploy and
manage private
cloud with
traditional apps

Phase 5:
Broker cloud
services across
multiple clouds

Phase 6:
Extend and
modernize
with new cloud-native
applications

a. Phases 1 and 2

b. Phases 2 and 3

c. Phases 1, 2, and 3

d. Phases 4, 5, and 6

e. Phases 1 through 6

Make additional notes explaining the correct answer:

_____

_____

_____

_____

_____

3. HPE OneView and HPE Operations Analytics for HPE OneView provide a foundation for which phase of the transformation to hybrid infrastructure and on-demand IT?

   a. Converge and virtualize

   b. Automate and orchestrate

   c. Transform application and service delivery

   d. Deploy and manage private cloud with traditional applications

   e. Broker cloud services across multiple clouds

   f. Extend and modernize with new cloud-native applications

   Make additional notes explaining the correct answer:

   _____

   _____

   _____

   _____

   _____

Assessment question answers

1. What is the correct definition of a converged infrastructure?

   a. An IT environment where the server assets are standardized and consolidated

   b. An IT environment where server, storage, and networking resources are standardized, virtualized, organized, and managed as resource pools

   c. An IT environment that provides on-demand and self-service provisioning of infrastructure services

   d. An IT environment that provides a development platform for consumer-developed or acquired applications created with programming languages or tools from the provider

2. A converged infrastructure provides a foundation for which phases?

   a. Phases 1 and 2

   b. Phases 2 and 3

   c. Phases 1, 2, and 3

   d. Phases 4, 5, and 6

   e. **Phases 1 through 6**

3. HPE OneView and HPE Operations Analytics for HPE OneView provide a foundation for which phase of the transformation to hybrid infrastructure and on-demand IT?

   a. Converge and virtualize

   b. **Automate and orchestrate**

   c. Transform application and service delivery

   d. Deploy and manage private cloud with traditional applications

   e. Broker cloud services across multiple clouds

The journey to on-demand IT infrastructure

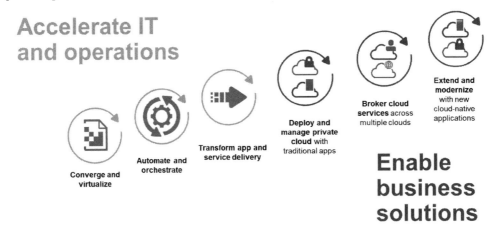

Figure 2-1 The journey to on-demand IT infrastructure

The journey to hybrid infrastructure and on-demand IT has some common issues:

- The journey to hybrid infrastructure and on-demand IT is hard.

- The journey is nonlinear and different for every customer.

- Many customers report their implementation of hybrid is behind schedule.

Chief information officers (CIOs) around the world are being challenged to add more value to the business by rapidly developing and deploying new solutions—solutions that enable new products, new services, new business models, and new customer experiences, all to help compete. Business leaders are seeking a flexible platform that embraces cloud in all of its forms.

The ultimate goal for these CIOs is an IT environment that:

- Deploys and manages private cloud

- Manages cloud services across multiple clouds

- Extends and modernizes new applications and workloads

- Enables business solutions

To complete the journey, the traditional, siloed IT must undergo a number of phases as shown in Figure 2-1. Companies must converge silos, including servers, storage, networking, and management software. Enterprises must virtualize their resources into pools of assets that can be automated and orchestrated. Application and service delivery should be driven by the business, not by IT constraints. Some customers undergo several phases at once, some deal with them one by one, and others move straight to cloud.

Laying the foundation — Converge and virtualize

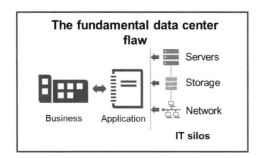

- Sluggish service delivery: **10 weeks**
- Admin time on maintenance: **75%**
- Data center utilization: **25%**

- Service delivery: **minutes**
- Admin time on maintenance: **25%**
- Data center utilization: **70%+**

Figure 2-2 Converge and virtualize

Why are businesses interested in moving to infrastructure convergence and cloud solutions? Most data centers are built on the traditional, siloed IT framework. The underlying technologies of the existing infrastructures are not aligned to the way applications are built, managed, and consumed.

For example, companies no longer have a year to plan and nine months to develop version 1.0 of an application (and another nine months to develop version 2.0). They can no longer take 10 weeks to deliver a new service. Today, things happen in weeks or days. Sometimes, even in hours or minutes. Companies can no longer build overprovisioned and underutilized technology silos that waste time, money, and assets, and cannot be easily repurposed for different workloads. Companies can no longer spend 75% of their IT budgets on maintenance of the infrastructure and 25% on data center innovation.

Aligning IT to the applications, workloads, and business is the primary goal of a converged infrastructure and cloud. Figure 2-2 shows how the hybrid infrastructure and on-demand IT dramatically simplify the users' experience to accelerate time to service and application value for any workload at any scale.

Laying the foundation—Automate and orchestrate for continuous delivery

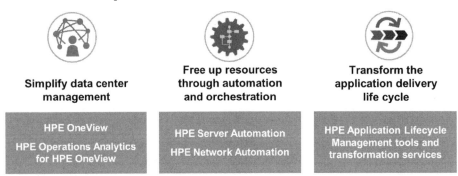

Figure 2-3 Automate and orchestrate for continuous delivery

The demand for continuous delivery makes it critical to bring automation, orchestration, and advanced analytics to IT operations. Figure 2-3 shows the HPE solutions to these challenges. Most customers start with solutions such as HPE OneView to simplify data center management and enable a more programmable infrastructure, and they use HPE Operations Analytics for HPE OneView to predict issues and avoid outages.

HPE automation tools and services such as HPE Server Automation, HPE Network Automation, and the HPE Application Lifecycle Management portfolio free up resources by automating, provisioning, patching, and orchestrating end-to-end processes including change management and disaster recovery.

What a converged infrastructure requires

Converged technologies	Converged management	HPE ConvergedSystem	Converged expertise
• Modern technologies and leadership across servers, storage, and networking • Unique IP at convergence points • Built to meet unique requirements and any workload or scale	• Integrated app-aware infrastructure with life cycle management capabilities • Modernized IT so it is simpler, faster, and more powerful and efficient • Role-specific views for business operations	• Workload-optimized engineered systems • Easy to procure, manage, and support • Simplified user experience	• Smooth transition to an app-centric, converged infrastructure • Consulting, support, education, and financial services across people, processes, facilities, and infrastructure • The most powerful partner community

Figure 2-4 Converged infrastructure requirements

As shown in Figure 2-4, in order for hybrid infrastructure and on-demand IT to be effective, purposeful innovation is required in these areas:

- **Converged technologies**—HPE offers a complete portfolio of modern technologies based on common architectures. Designed for convergence, these technologies can integrate into shared pools of resources and be managed though one platform.

- **Converged management**—Converged management is quickly becoming the central nervous system of the data center. HPE offers several management solutions for simple data center management that is aligned with applications.

- **HPE Converged System**—For customers looking to gain the value of convergence fast, HPE offers a complete Converged System portfolio. These workload-optimized, engineered systems are built for the way people work. They provide a new level of seamlessness from virtualization to cloud and from physical to virtual.

  Each system shortens the procurement lifecycle with a set of core platforms that are easy to configure, quote, order, and deploy. Customers can go from ordering to implementing operations in 15–30 days. Each system includes management at the systems level, not the component level. These integrated systems improve standardizing and automation capabilities without losing application performance.

- **Converged expertise**—HPE Converged Infrastructure consulting, support, education, and financial services work with a customer's IT teams, processes, facilities, and infrastructure on transitioning to hybrid. These resources help create a smooth transition from a traditional, siloed, product-focused IT infrastructure to an app-centric, converged infrastructure.

  This transition also requires partners with capabilities that go well beyond product-specific consulting and implementation. Today, the HPE partner community can implement complex solutions based on customer requirements. For example, HPE channel partners designated as Platinum partners are considered Converged Infrastructure Specialists.

Assessing the customer's journey

Customers who want to embark on the journey to a hybrid infrastructure and on-demand IT have different starting points, destinations, budgets, and time frames. HPE and various channel partners offer a range of tools and services to help customers assess their journey. These tools and services evaluate the customer's IT infrastructure, business processes, people readiness, and so forth.

You might have experience with assessment tools, services, and methodologies from channel partners. Or you might be familiar with assessment tools from industry analysts and advisors, such as Gartner. Gartner's Five-Stage Demand-Driven Maturity Model that provides a methodology for assessing a customer's IT infrastructure, including advice for these five stages: react, anticipate, integrate, collaborate, and orchestrate.

Gartner also developed another tool, the ITScore, which evaluates IT organizations in terms of how IT services are provided and how the enterprise is affected.

Note

For more information about ITScore, scan this QR code or enter the URL into your browser.

http://www.gartner.com/technology/research/methodologies/it-score.jsp

Paths to the destination

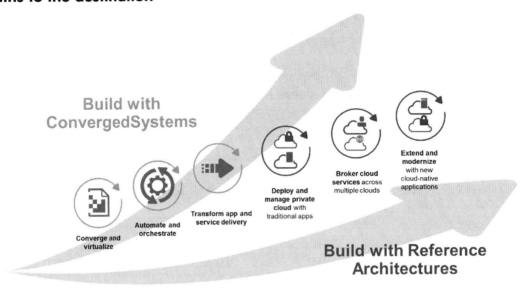

Figure 2-5 Two infrastructure building options

Each customer on the journey to a hybrid infrastructure can decide how they want to proceed. To build the infrastructure themselves, they have two options as shown in Figure 2-5: construct it piece by piece using RAs from HPE, or use prebuilt, preintegrated ConvergedSystem solutions.

Building with RAs
What are HPE RAs?

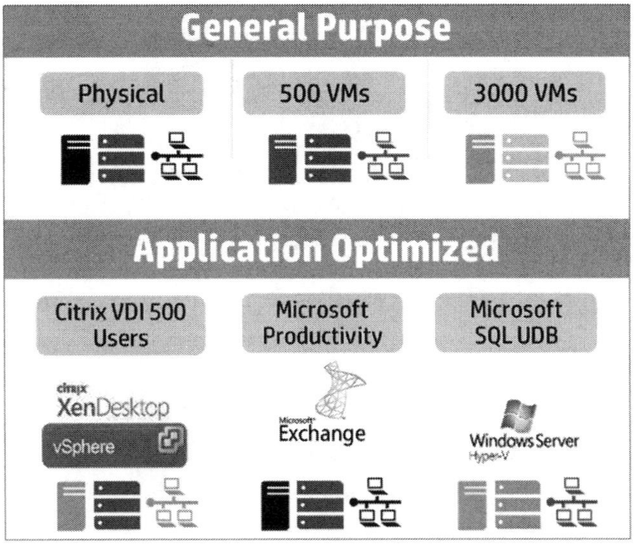

Figure 2-6 HPE RAs

Building a hybrid infrastructure and on-demand IT piece by piece is a complex process that requires many components, people, decisions, and diverse skills. If businesses make any mistakes, they may have to start over or repeat certain steps. This approach involves:

- Researching many individual products and components

- Developing a proposed complex design

- Ordering a large number of individual parts

- Coordinating the delivery of the parts

- Assembling the individual parts when they arrive

- Installing and updating all firmware and software

- Testing and adjusting the design as required

HPE RAs are complete, open solutions designed, and optimized for specific workloads as shown in Figure 2-6. Each presized and tested configuration follows a proven deployment methodology to help reduce provisioning time, cost, risk, and errors.

Each workload-based Reference Architecture includes hardware and software components from HPE and other trusted partners. In addition, each one can be deployed on a wide-ranging set of platforms including HPE ConvergedSystem, HPE hyper-converged systems, and Converged Infrastructure. HPE solution architects and channel partners can use these RAs to quickly deliver customized solutions in a consistent, repeatable manner.

Business value of RAs

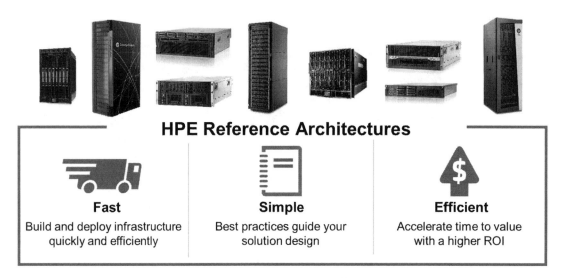

Figure 2-7 Business value of RAs

When customers choose to deploy a RA, they can quickly adapt to their market, meet service-level agreements, and achieve their business goals. These RAs have multiple benefits as shown in Figure 2-7:

- **Fast**—Enabling businesses to build and deploy infrastructure quickly and efficiently

- **Simple**—Guiding the solution design with best practices, making it easier to get optimal results

- **Efficient**—Accelerating time to value and delivering a higher return on investment (ROI)

Accessing RAs—Activity

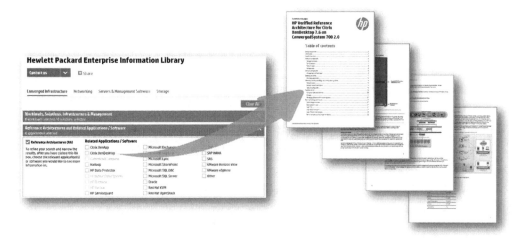

Figure 2-8 Reference Architectures

As shown in Figure 2-8, RAs are posted in the HPE Information Library under the RAs and Related Applications/Software menu.

To familiarize yourself with the RAs, access the related documents in the HPE Information Library.

 Note

To access these documents, scan this QR code or enter the hyperlink into your browser.

http://h17007.www1.hpe.com/us/en/enterprise/converged-infrastructure/info-library/index.aspx#.VmVBHXnruUk

---

1. Click the disclosure arrow to the right of RAs and Related Applications/Software and left-click the RAs box.

2. Select a reference architecture that interests you.

3. Find the reference architecture in the HPE Information Library and open it.

4. Answer these questions:

    – Which reference architecture did you choose?

    _____

    _____

    _____

    _____

    – What business problem does the reference architecture solve?

    _____

    _____

    _____

    _____

— Who is the target audience for this document?

_____

_____

_____

_____

— Which HPE products does your reference architecture use?

_____

_____

_____

_____

— Which third-party software does your reference architecture use?

_____

_____

_____

_____

— Does the reference architecture contain capacity and sizing information?

_____

— Does the reference architecture contain performance and high-availability expectations?

_____

— Does the reference architecture contain a bill of materials (BoM)?

_____

Applying RAs—Activity

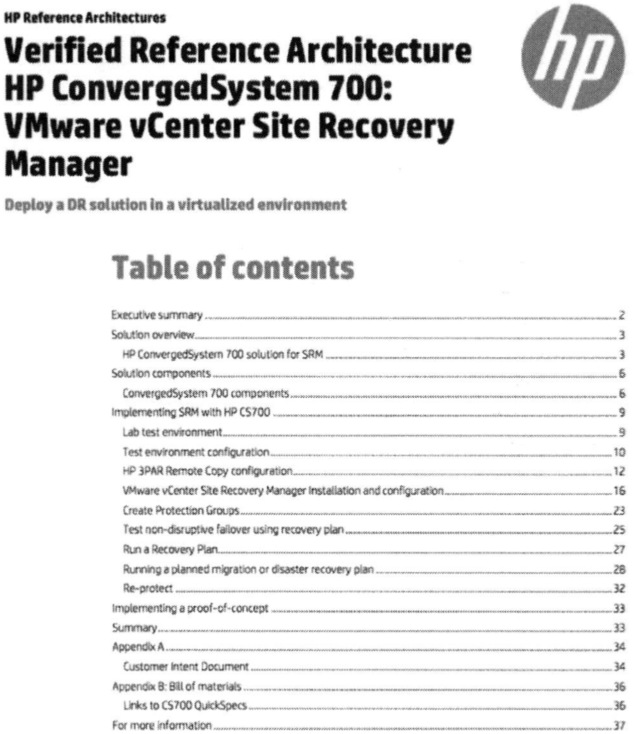

HP Reference Architectures

Verified Reference Architecture HP ConvergedSystem 700: VMware vCenter Site Recovery Manager

Deploy a DR solution in a virtualized environment

Table of contents

Figure 2-9 HP ConvergedSystem 700 with VMware vCenter reference architecture

For this activity, use the HPE Information Library to locate the *Verified Reference Architecture HP ConvergedSystem 700: VMware vCenter Site Recovery Manager* RA shown in Figure 2-9. It is located in the Converged Infrastructure section, under the RAs and Related Applications/Software menu.

1. What is the primary purpose of the solution presented in the RA?

    _____

    _____

2. What is the purpose of the ProLiant DL360 Gen9 servers and how many virtual machines are configured on each server?

    _____

    _____

3. How many and what model network switches are shown?

_____

_____

4. How many M6710 300 GB 6G SAS 15K 2.5 in disk drives are shown?

_____

_____

Applying RAs activity—Solution

1. What is the primary purpose of the solution presented in the reference architecture?

   This RA provides details on deploying a VMware®vCenter Site Recovery Manager (SRM) solution across two data centers or sites using HP ConvergedSystem 700 systems.

2. What is the purpose of the ProLiant DL360 Gen9 servers and how many virtual machines are configured on each server?

   The DL360 Gen9 servers are management servers running VMware 5.5 Update 2, and they are each deployed with eight management virtual machines.

3. How many and what model network switches are shown?

   Two HP SN6000B Fibre Channel, two HP 5900AF-48G-4XG-4QSFP+, and two HP 5900AF-48XG-4QSPF+ switches are shown.

4. How many M6710 300 GB 6G SAS 15K 2.5 in disk drives are shown in the HP 3PAR StoreServ 7440c?

   120 disk drives are shown.

Customizing RAs

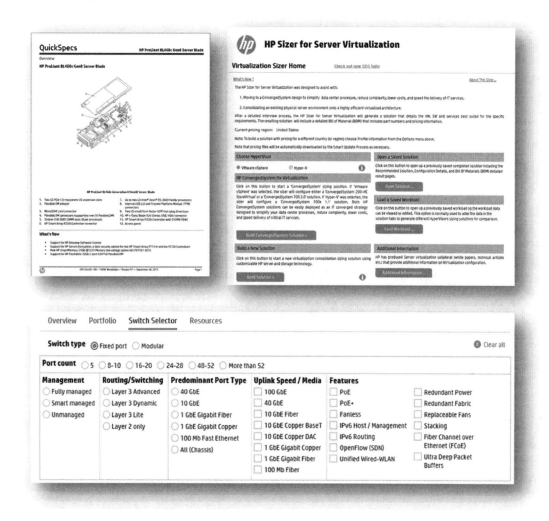

Figure 2-10 Customizing RAs using HPE tools

You can customize the solution presented by the RAs using HPE tools, shown in Figure 2-10, that include:

- **QuickSpecs**—Documents that provide technical overviews and specifications for HPE hardware and software

- **Sizers**—Automated tools that assist you with managing the size and scope of a solution and IT environment

- **Selectors and configurators**—Tools designed to help you choose the correct product for different environments and configure it appropriately

Most of these tools are publicly accessible and free to use.

Customizing RAs—Activity

The *HP ConvergedSystem 700: VMware vCenter Site Recovery Manager* RA calls for ProLiant BL460c Gen9 server blades to be used as workload servers. These server blades have specific hardware components and configuration already predetermined by HPE.

For the purposes of this activity, imagine that you want to customize these servers and add more memory.

Find the ProLiant BL460c Gen9 Server Blade QuickSpecs in the HPE Marketing Document Library.

 Note

To access the HPE Marketing Document Library, scan this QR code or enter the hyperlink into your browser.

http://h41370.www4.hp.com/quickspecs/overview.html

Use the ProLiant BL460c Gen9 Server Blade QuickSpecs to answer these questions.

1. How many DIMM slots are available in this server blade?

2. What is the maximum supported memory speed?

3. What options do you have if you want to increase the amount of RAM beyond 256 GB? What are the maximums?

Customizing RAs activity — solution

1. How many DIMM slots are available in this server blade?

   16 DDR4 DIMM slots (8 per processor) are available.

2. What is the maximum supported memory speed?

   The maximum supported memory speed of the BL460c Gen9 configured with Intel Xeon E5-2690 v3 processors is 2133 MHz.

3. What options do you have if you want to increase the amount of RAM beyond 256 GB? What are the maximums?

   The maximum supported LRDIMM memory is 1 TB and the maximum supported RDIMM memory is 512 GB.

HPE sizers

HPE sizers are automated tools that assist with managing the size and scope of a solution environment. The sizing information and algorithms have been developed using testing and performance data on a wide range of HPE solutions from partners such as Citrix, Microsoft, SAP, and VMware. These tools provide a consistent methodology to help determine the best solution for the customer's environment.

A partial list of HPE sizers includes:

- HPE Converged Infrastructure Solution Sizer Suite
- HPE Sizer for Citrix Mobile Workspace
- HPE Insight Management Sizer
- HPE Power Advisor
- HPE Sizer for Server Virtualization
- HPE Sizer for Microsoft Exchange Server
- HPE Sizer for Microsoft Hyper-V 2008 R2
- HPE Sizer for Microsoft SharePoint
- HPE Storage Sizing Tool
- HPE SAP Sizing Tool

Most sizers are downloadable and freely accessible through the HPE Information Library. Many of the sizers are presented as PDF documents and when the PDF is opened a download link will be shown.

Note

To access HPE sizers, scan this QR code or enter the hyperlink into your browser.

http://h17007.www1.hp.com/us/en/enterprise/converged-infrastructure/info-library/index.aspx?type=20#.Vlwlcf7ruUl

 Note

To access the Storage Sizer, scan this QR code or enter the hyperlink into your browser.

http://h30144.www3.hp.com/SWDSizerWeb

HPE selectors and configurators

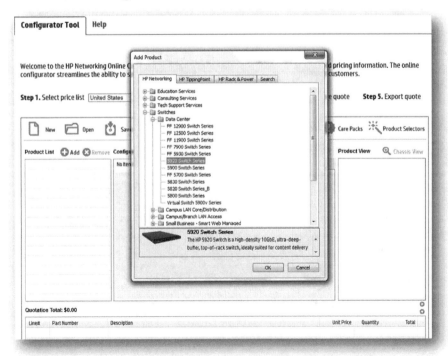

Figure 2-11 HPE selectors and configurators

Similar to HPE sizers, HPE selectors and configurators are designed to help you select the correct product for each customer and configure it appropriately. These selectors and configurators are accessible online and Figure 2-11 shows two examples.

HPE Sales Builder for Windows

HPE Sales Builder for Windows (SBW) is the premier HPE configuration and quotation tool for the sales force and channel partners. SBW is a downloadable tool designed to support the complete HPE product portfolio of servers and related service and support products.

The SBW Configurator displays the system diagram and modifications for HPE clusters, servers, and storage running HPE-UX, Windows, Linux, or mixed environments. You can use this tool to configure and customize technical solutions for new systems, upgrades, and add-ons. Components of the Configurator include:

- **Whiteboard**—The center of SBW that shows technical solutions and forms the unit of quoting and storing

- **Configurator worksheet**—Main configuration tool that enables you to configure complete technical solutions

- **System diagram**—A graphical view of your configuration that can be used for modifications

Other components of SBW include:

- **Quoter**—Prepares a budgetary quotation for the customer that shows numbers, descriptions, and prices of all the products in the solution

- **Price book**—Data files containing the latest product descriptions and prices that are updated every two weeks

- **Knowledge base**—Data file containing the rules and product modeling used by SBW to check configurations

SBW exports schematic diagrams in Microsoft Visio format. SBW is accessible via the HPE Partner Ready Portal.

Additional HPE selectors and configurators

Additional HPE selectors and configurators include:

- **HPE Networking Switch Selector**—This web-based tool helps you select the correct HPE networking product based on specific requirements, such as switch type, port count, management capabilities, routing and switching capabilities, predominant port type, and others.

Note

To access the Networking Switch Selector tool, scan this QR code or enter the hyperlink into your browser.

**http://www8.hp.com/us/en/mpc/networking-switches/index.
html#!view=grid&page=1**

- **HPE Networking Online Configurator**—This configurator enables you to create price quotations for HPE networking products quickly and easily using your web browser. You can save quotation files to your hard drive or export them in several formats including Microsoft Excel.

 Note

To access the Networking Online Configurator, scan this QR code or enter the hyperlink into your browser.

**http://h17007.www1.hp.com/th/en/networking/products/
configurator/index.aspx#.VmVCJXnruUk**

- **One Config Simple (OCS)**—This configurator tool allows users to configure HPE products by streamlining the selection process. It maps customer requirements to a set of products or service options. The guided selling capability directs users to an optimal solution for each customer's specific workloads or application needs.

 Note

To access the OCS, scan this QR code or enter the hyperlink into your browser.

http://h22174.www2.hp.com/SimplifiedConfig/Index

Customizing RAs—Activity

Figure 2-12 HPE Networking switch selector

Answer these questions using the HPE Networking switch selector shown in Figure 2-12.

 Note

To access the Networking switch selector, scan this QR code or enter the hyperlink into your browser.

http://www8.hp.com/us/en/mpc/networking-switches/index. html#!view=grid&page=1

1. Determine which HPE switches meet these requirements:

   – Modular configuration

   – Primary port type 100 GbE

   – FCoE capable

   – OpenFlow capable

2. Compare the switches based on the different descriptions and features.

Customizing RAs activity—solution

1. Determine which HPE switches meet these requirements:

   – Modular configuration

   – Primary port type 100 GbE

   – FCoE capable

   – OpenFlow capable

   Three switches meet these requirements:

   – **HPE FlexFabric 7900 Switch Series**

   – **HPE FlexFabric 12900 Switch Series**

   – **HPE 12500 Switch Series**

2. Compare the switches based on the different descriptions and features.

   – **HPE FlexFabric 7900 Switch Series: Up to 10 I/O module slots, up to 5.8 Bpps throughput**

   – **HPE FlexFabric 12900 Switch Series: Up to 16 I/O module slots, up to 28.8 Bpps throughput**

   – **HPE 12500 Switch Series: Up to 18 I/O module slots, up to 10.8 Bpps throughput**

Learning check

1. Which HPE automation tools and services free up resources through automation and orchestration? (Select two.)

   a. HPE OneView

   b. HPE Server Automation

   c. HPE Operations Analytics for HPE OneView

   d. HPE Application Lifecycle Management portfolio

   e. HPE Network Automation

2. What are HPE RAs?

   _____

   _____

   _____

   _____

3. Match each use case to the relevant RA:

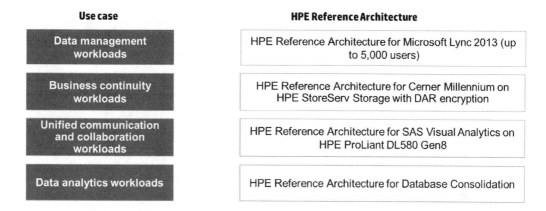

Use case	HPE Reference Architecture
Data management workloads	HPE Reference Architecture for Microsoft Lync 2013 (up to 5,000 users)
Business continuity workloads	HPE Reference Architecture for Cerner Millennium on HPE StoreServ Storage with DAR encryption
Unified communication and collaboration workloads	HPE Reference Architecture for SAS Visual Analytics on HPE ProLiant DL580 Gen8
Data analytics workloads	HPE Reference Architecture for Database Consolidation

Learning check answers

1. Which HPE automation tools and services free up resources through automation and orchestration? (Select two.)

    a. HPE OneView

    b. HPE Server Automation

    c. HPE Operations Analytics for HPE OneView

    d. HPE Application Lifecycle Management portfolio

    e. HPE Network Automation

2. What are HPE RAs?

    – **Complete, open solutions designed and optimized for specific workloads**

    – **Each presized and tested configuration follows a proven deployment methodology to help reduce provisioning time, cost, risk, and errors**

    – **Each workload-based RA includes hardware and software components from HPE and other trusted partners**

    – **HPE solution architects and channel partners can use these RAs to quickly deliver customized solutions in a consistent, repeatable manner**

3. Match each use case to the relevant Reference Architecture:

    a. **Data management workloads: HPE RA for Database Consolidation**

    b. **Business continuity workloads: HPE RA for Cerner Millennium on HPE StoreServ Storage with DAR encryption**

    c. **Unified communication and collaboration workloads: HPE RA for Microsoft Lync 2013 (up to 5000 users)**

    d. **Data analytics workloads: HPE RA for SAS Visual Analytics on HPE ProLiant DL580 Gen8**

Building with ConvergedSystems

HPE ConvergedSystem 700

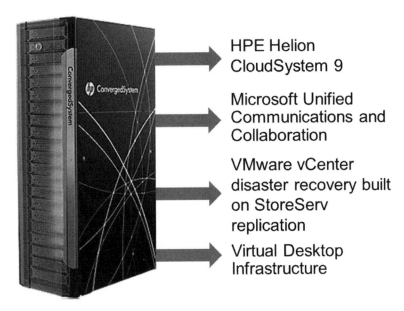

HPE Helion CloudSystem 9

Microsoft Unified Communications and Collaboration

VMware vCenter disaster recovery built on StoreServ replication

Virtual Desktop Infrastructure

Figure 2-13 HPE ConvergedSystem 700

HPE ConvergedSystem 700 for on-demand IT infrastructure is optimized for infrastructure as a service (IaaS), private cloud, and single and mixed workloads. Every system ships with factory-integrated server, storage, networking, and HPE OneView management. Systems also include HPE Proactive Care to help you avoid problems and resolve issues more quickly.

Designed to be open and flexible, HPE ConvergedSystem 700 can meet general purpose infrastructure needs and provide a foundation for private cloud. This powerful IaaS platform can run a single enterprise application or support a range of common workloads.

ConvergedSystem 700 includes:

- HPE OneView 2.0

- Increased flexibility of data center connectivity

- ProLiant Gen9 blades and servers

- Increased scale-out capabilities

- StoreServ 7xx0c series storage

As shown in Figure 2-13, ConvergedSystem 700 serves as a platform for:

- HPE Helion CloudSystem 9

- Microsoft Unified Communications and Collaboration

- VMware vCenter disaster recovery built on StoreServ replication

- Virtual Desktop Infrastructure (VDI) (HPE ConvergedSystem 700 for Citrix XenDesktop)

HPE Hyper Converged 250 family

Figure 2-14 HPE Hyper Converged 250 family

Hyper-converged systems integrate servers, storage, networking, and management into a small footprint as shown in Figure 2-14, with simplified installation and administration:

- **Fast deployment**—Deploy and expand in minutes with HPE OneView for all IP addressing, server and storage clustering, and system startup.

- **Simplified management**—Manage the entire infrastructure from a single interface without any specialized server, storage, or virtualization expertise.

- **Efficient operations**—Get four times the compute in 75% less space than traditional systems.

Note

> For a one-minute explanation of hyper-converged systems, scan this QR code or enter the hyperlink into your browser.

https://www.youtube.com/watch?v=EEFDlmHUb10

Each system arrives preconfigured with servers, storage, networking, and VMware vSphere. A complete virtualized environment can be deployed in under 15 minutes. All products comply with the open software-defined data center (SDDC) framework developed by HPE in conjunction with VMware and other partners. Ideal markets for hyper-converged systems include midsize businesses, remote offices, and line of business.

The current HPE hyper-converged portfolio includes:

- **HPE Hyper Converged 250**—Provides robust data services using best-in-class storage management software and powerful features typically offered by SANs, without the SAN complexity (for example, 99.999% availability).

- **HPE Hyper Converged 250 for Microsoft**—Provides a factory-integrated modular appliance that is easy to use, hybrid cloud-ready, scalable to four appliances in the same cluster, and managed from a single interface/console.

- **HPE ConvergedSystem 200-HC StoreVirtual**—Provides an entry-level solution with an easy, open path to hybrid cloud delivery. The HPE Helion CloudSystem is a smaller platform, built on the HPE ConvergedSystem 200-HC system.

 Note

> For an eight-minute video explaining the difference between hyper-converged and converged systems, scan this QR code or enter the hyperlink into your browser.

https://www.youtube.com/watch?v=WJyWSEeCvDg

Complementary benefits of "hyper-converged cloud"

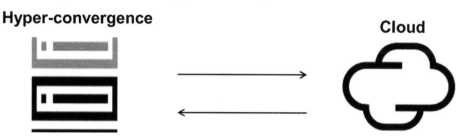

Figure 2-15 Hyper-converged cloud

The combination of cloud plus hyper-convergence as illustrated in Figure 2-15 yields maximum IT efficiency. In addition, this combination:

- Allows you to create and consume virtual resources quickly via self-service

- Provides elastic resource pools that scale up and down

- Offer services to users via Marketplace

- Delivers 4× the compute in 75% less space when compared with traditional servers

- Has no centralized SAN

- Provides straightforward linear scale-out architecture

HPE Hyper Converged 250 for Microsoft

Figure 2-16 HPE Hyper Converged 250 for Microsoft

HPE extends hyper-convergence into Microsoft environments with an integrated cloud solution powered by Microsoft Hyper-V and HPE Hyper Converged 250 for Microsoft, as shown in Figure 2-16.

The Microsoft software that makes up the offering is preinstalled and configured at the factory. These factory-installed and configured Microsoft software components include Microsoft Windows Server 2012 R2, System Center 2012 R2, and Windows Azure Pack.

HPE StoreVirtual built-in resiliency is paired with add-on cloud services from Microsoft: Azure Backup and Azure Site Recovery. The result is an enterprise-class virtualization solution, in a package that is easy to acquire, quick to set up, and simple to operate.

This fast, simple, and efficient Azure-consistent cloud in-a-box is optimized to handle a variety of virtualization workloads—including on-demand IT infrastructure, business-critical applications, and virtual desktop infrastructure (VDI). It is based on the latest HPE Apollo Gen9 technology, delivering Intel Xeon performance and economies of power and cooling in less space when compared with traditional servers. This compact 2U platform ships with either three or four identically configured nodes. Nodes can be configured with:

- Two Xeon E5-2640 v3 or E5-2680 v3 processors

- 128 GB, 256 GB, or 512 GB memory

- Hybrid configuration with SAS and SSD

- Capacity configuration with all SAS drives

- Two 10 GbE SFP+ module network connectivity

Features and benefits

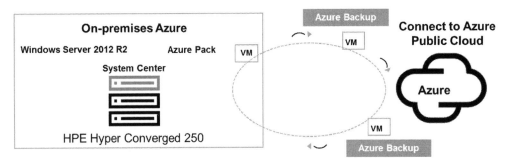

Figure 2-17 HPE Hyper Converged 250 for Microsoft features and benefits

As shown in Figure 2-17, HPE Hyper Converged 250 for Microsoft builds high availability, data protection, and back-up and recovery capabilities into a single solution. High availability is standard with 99.999% uptime out of the box.

Additional features and benefits of the HPE Hyper Converged 250 for Microsoft include:

- Reduced OPEX with compact 2U appliance able to be managed by IT generalist

- Built-in resiliency for lower CAPEX (no need to build your own cluster)

- Simple start up with factory-integrated hardware and Microsoft software

- Storage-only expansion to existing systems

- Self-service ability to create and consume virtual resources quickly

- Elastic resource pools for scalability up and down

- Security and availability

Integrated software in Hyper Converged 250 for Microsoft

| **Configure and deploy** | **Monitor** | **Patch and upgrade** |

Figure 2-18 Windows Azure Pack Management Cluster used by HPE Hyper Converged 250

Windows Azure Pack, shown in Figure 2-18, is a collection of Microsoft Azure technologies available to Microsoft customers at no additional cost. It integrates with Windows Server, System Center, and SQL Server to offer a self-service portal and cloud services, such as virtual machine hosting (IaaS), database as a service (DBaaS), scalable web app hosting (PaaS), and more. Add-on Azure services, such as Backup and Disaster Recovery facilitate business continuity.

The Management Cluster keeps all services and operations running. HPE Hyper Converged 250 ships with the infrastructure and management functions predeployed and preconfigured in a highly available and resilient manner. They are also automated to minimize manual involvement in the operation of the system.

HPE OneView for Microsoft System Center Virtual Machine Manager (SCVMM) uses a single console to manage both physical and virtual environments. SCVMM is used to deploy and configure additional racks when scale out is required. VMM Service templates are the basis of all other management services. When the system is preconfigured in the factory, VMM is the first management service implemented and all other management systems are deployed from it. Through this standardized approach, patches and updates to the management systems can be automated and integrated into the Orchestrated Update and Patching subsystem.

Using Microsoft System Center 2012 R2 Operations Manager capabilities, you can monitor the fabric compute, storage, and network components from centralized dashboards. The dashboards enable you to see component health at a glance and drill into granular health, performance, and capacity. Because the exact hardware and software configurations are known and have been operated by Microsoft, the management packs are optimized to eliminate alerting noise and help you focus on what is important. HPE has connected the System Center Operations Manager Management Pack and SCVMM add-in to enable you to see the System Center Operations Manager alerts in the SCVMM Add-in instead of toggling between two interfaces.

The Storage Management Pack provides insight into the health of underlying storage resources such as VM, volumes, and disks. It shows the storage provisioned to hosts and the VM capacity utilization Server Management Pack shows health status and alerts for the servers. This provides a close view of infrastructure health, enabling corrective action for alerts and warnings, and the flexibility to provision and manage virtual resources to meet business needs such as to expand volumes in case of more capacity required and to deploy StoreVirtual VSA if required.

 Note
The Patching and Upgrade functionality is available for servers only, and does not apply to storage in the HC 250.

Seamless integration into Microsoft System Center

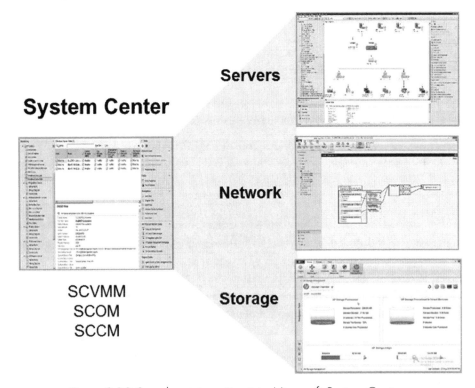

Figure 2-19 Seamless integration into Microsoft System Center

As shown in Figure 2-19, you can access the HPE deep management ecosystem from within the System Center consoles, eliminating the need for administrators to use multiple management tools. With HPE OneView for Microsoft System Center, you can:

- Simplify administration with single-console access to health, inventory, and configuration monitoring

- Reduce planned and unplanned downtime with detailed resolution information for health alerts

- Take control by launching trusted HPE management tools in context

- Proactively manage changes with detailed insight into the relationship between physical and virtual infrastructure from the VM to the network edge

- Maintain the stability and reliability of the environment with simplified driver and firmware updates

Reduced CAPEX with three-node configuration

Figure 2-20 Four-node configuration rear view

The HPE Hyper Converged 250 for Microsoft is ideal for remote offices and data centers because you can start small and scale at the speed of business using predictable, modular, hyper-converged building blocks. It starts with a three-node configuration for reduced investment at remote and branch offices and medium-sized businesses. Customers can expand existing installations of current and previous generation models to grow to four nodes, as shown in Figure 2-20, with a single-node expansion.

HPE OneView InstantOn start up and expansion wizard

The HPE OneView InstantOn startup and expansion wizard helps you deploy and configure a new HPE Hyper Converged 250 for Microsoft appliance in 15 minutes. It is as easy as entering an IP address range, credentials, and license information.

The HPE OneView InstantOn quick start wizard provides:

- IP addressing

- Cluster build

- System standup

Then log in to HPE OneView for Microsoft System Center. You are then ready to provision your first VM.

To add more capacity, use the InstantOn wizard to recognize new appliances automatically to add to existing clusters, with no interruption to services. The wizard provides the same experience, whether the first or the fourth appliance deployment.

Built-in business continuity with HPE StoreVirtual VSA

The Hyper Converged 250 is built on the enterprise software-defined storage solution HPE StoreVirtual VSA, providing integrated thin provisioning, full snapshot integration with Microsoft, site-to-site replication and multisite high availability. Based on seven years of experience in software-defined storage, HPE Hyper Converged 250 for Microsoft:

- Delivers enterprise storage features without a SAN. With tens of thousands of systems deployed, HPE StoreVirtual VSA is the solution trusted by Fortune 500 companies and small businesses.

- Provides the benefits of traditional SAN storage in the server infrastructure. Shared storage is created out of internal or external disk capacity and is accessible to all virtual and physical application servers on the network.

- Creates availability zones within the environment across racks, rooms, building, and cities for seamless application high availability with transparent failover and failback across zones automatically.

- Tightly integrates with storage management functions of Microsoft applications. Centralized backup and disaster recovery on a per-volume basis with application-integrated snapshots provide fast recovery.

Predictable scalability for compute and storage

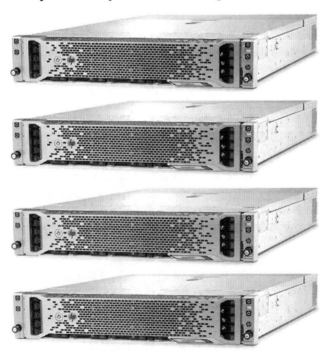

Figure 2-21 Predictable scalability

Planning ahead can ensure that there is enough storage and compute to prevent a slowdown in the business. Central to HPE Hyper Converged systems is the concept of scalable performance by simply adding nodes to the cluster, as shown in Figure 2-21, with no fear of disruption to performance. Scaling occurs linearly in both performance and capacity with HPE scale-out architecture.

Because many storage products become CPU-bound, adding a node to a cluster not only adds disk drive performance and capacity but also another CPU, cache, and IO ports. HPE Hyper Converged 250 for Microsoft maintains a consistent ratio of CPU/cache/IO to capacity. The maximum number of supported appliances in a cluster is four, for a total of 16 nodes. To add more storage, but not more compute, simply expand outside the system to any x86 server or HPE VSA Storage array.

Additional storage for hyper-converged environments

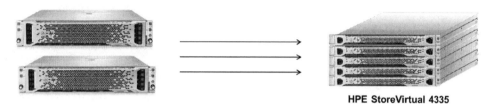

HPE StoreVirtual 4335

Figure 2-22 Additional storage

As shown in Figure 2-22, with the HPE Hyper Converged 250 for Microsoft, it is easy to add components such as StoreVirtual VSA running on HPE ProLiant servers, or StoreVirtual appliances to expand storage. This is done by creating a separate management group with the HPE OneView for Microsoft System Center plug-in or the Centralized Management Console for StoreVirtual. With built-in HPE Peer Motion, data volumes can nondisruptively replicate and migrate data volumes from one Hyper Converged 250 for Microsoft Cloud Platform System (CPS) appliance to another, or to any x86-based server running any major hypervisor and HPE StoreVirtual VSA software.

HPE Customer Intent Document

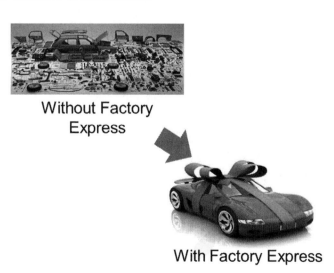

Without Factory
Express

With Factory Express

Figure 2-23 HPE Factory Express

The HPE Customer Intent Document (CID) is a key requirement that enables successful delivery of a solution. The CID details how the technical design will be built, integrated, and configured. The build and configuration requirements are collected by HPE Sales and HPE Presales consultants along with channel partners. These team members work with the HPE project manager to document the requirements within the CID so that the order can be processed, integrated, and deployed.

When buying a car, most people prefer to buy the car completely assembled and tested, rather than buying the components and building the car themselves. As illustrated in Figure 2-23, HPE Factory Express allows HPE to build and deliver a complex solution from order to operations in as little as 20 days, meaning that the customer takes delivery of the finished solution rather than the individual parts. However, it is possible to achieve this goal and meet customer expectations only if an accurate and complete CID is provided in a timely manner.

 Note

CID is applicable to and required for HPE ConvergedSystems.

The CID may contain:

- Customer contact details
- Visio diagrams containing logical and physical view of the hardware layout and connectivity
- Questionnaire containing software configuration details
- Site survey checklist
- Bill of materials

The associated benefits include the following:

- Reduced overall cost of the installation due to lower labor rates and better resource utilization
- Reduced onsite deployment time and disruptive activities at the customer site (systems are pre-configured and ready to deploy when delivered)
- Reduced time to utilization and faster time to production, ultimately helping the customer achieve a faster ROI

At the end of the engagement, after the system is built and configured, the project manager generates a Solution Information Manual. This document contains all information in the CID, installation documents, log files, and other documentation generated during the build process.

What is included in the process?

The CID process is a separate, complementary process from the system design. The design and sizing of the solution are not part of the CID process; they are owned by HPE Presales or the channel partner. These items are also excluded from the CID process:

- Creation of the budgetary or legal quote, which is part of the standard sales cycle.

- Performing solution design for integration with the customer environment, which is part of HPE presales and channel partner roles. If the customer requires a more complex design, appropriate consulting services can be purchased.

- The project manager's engagement in presales activities outside of the CID. The project manager only supports the validation of the quotes and CID details. The project manager does not support other presales activities.

Roles and responsibilities—How to engage

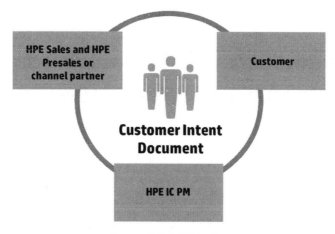

Figure 2-24 CID roles

The CID process involves the customer, HPE Sales and HPE Presales consultants, the channel partner, and the project manager, all working closely together as shown in Figure 2-24. HPE Sales or HPE Presales consultants and the channel partner collaboratively own the CID and need to work together with the customer to complete the CID. The project manager provides necessary support and assistance when required.

After being assigned, the project manager assists HPE Sales and HPE Presales consultants and channel partners in completing the CID. The assigned project manager also manages the solution end to end, from factory build to onsite deployment.

To get a project manager assigned, send an email to the regional HPE Factory Express engagement desk at:

- sol_eng_support@hpe.com (for EMEA)

- AICEngagementProjectMgmtTeam@hpe.com (for AMS)

- apic-sg.pdl@hpe.com (for APJ)

Examples

Figure 2-25 Hardware and software details

As shown in Figure 2-25, starting with the architectural and high-level drawings that are established at the presales stage, detailed assembly and cabling instructions are developed. These are needed by the factory and the on-site teams.

Details for software configuration must be defined with the customer. The software installation and configuration is then executed centrally at HPE.

Other HPE tools and resources

HPE Solution Demonstration Portal

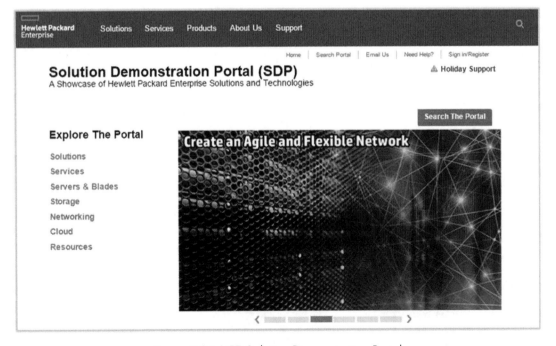

Figure 2-26 HPE Solution Demonstration Portal

The HPE Solution Demonstration Portal (SDP) shown in Figure 2-26, formerly, known as the *Virtual Resource Portal*, provides a central location for all demonstrations, webinars, and supporting collateral that showcase how HPE technologies lead, innovate, and transform enterprise businesses. Live and prerecorded demonstrations feature HPE hardware, software, services, and partnerships in an exciting multimedia format, illustrating how HPE can help solve business and IT problems.

 Note

To access the SDP, scan this QR code or enter the hyperlink into your browser.

https://vrp.glb.itcs.hpe.com/SDP/default.aspx

More HPE tools and resources

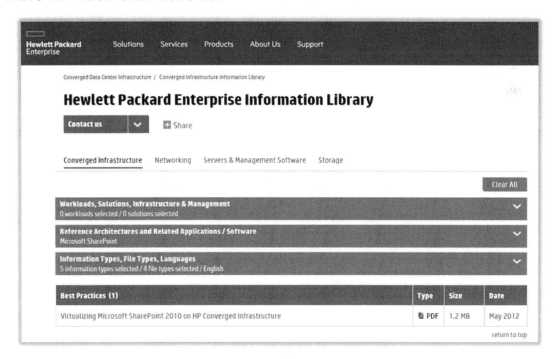

Figure 2-27 HPE Information Library

Shown in Figure 2-27, the **Hewlett Packard Enterprise Information Library** is a search-based tool used to access user guides, technical documentation, and other collateral for a variety of products and solutions.

 Note

To access the HPE Information Library, scan this QR code or enter the hyperlink into your browser.

http://www.hp.com/go/cloudsystem/docs

The **HPE Single Point of Connectivity Knowledge (SPOCK)** portal provides detailed information about supported HPE storage product configurations.

You need an HPE Passport account to enter the SPOCK website.

 Note

To access SPOCK, scan this QR code or enter the hyperlink into your browser.

https://h20272.www2.hp.com/spock/

HPE Live Network is a value-added service available to customers who purchase products enabled through HPE Live Network. Through this network, customers can access the most current content add-ons and extensions for their HPE IT Performance Suite products.

 Note

To access this network, scan this QR code or enter the hyperlink into your browser.

https://hpln.hpe.com/

Learning check

1. What are the ideal markets for hyper-converged systems?

_____

_____

_____

_____

2. During the HPE CID process, who is responsible for obtaining and completing the CID document? (Select three.)

   a. Customer

   b. HPE Sales consultant

   c. HPE Presales consultant

   d. Channel partner

   e. HPE project manager

3. What is contained within the HPE CID? (Select four.)

   a. Bill of materials

   b. Pricing

   c. Site survey checklist

   d. Software licenses

   e. Customer contact details

   f. Visio diagrams of the hardware layout and connectivity

4. Which resource provides detailed information about supported HPE storage product configurations and requires an HPE Passport account?

   a. HPE Single Point of Connectivity Knowledge

   b. HPE Information Library

   c. HPE Sales Builder for Windows

   d. HPE OneView

   e. Microsoft Unified Communications and Collaboration

Learning check answers

1. What are the ideal markets for hyper-converged systems?

   a. Midsize businesses

   b. Remote offices

   c. Line of business

2. During the HPE CID process, who is responsible for obtaining and completing the CID document? (Select three.)

   a. Customer

   b. HPE Sales consultant

   c. HPE Presales consultant

   d. Channel partner

   e. HPE project manager

3. What is contained within the HPE CID? (Select four.)

   a. Bill of materials

   b. Pricing

   c. Site survey checklist

   d. Software licenses

   e. Customer contact details

   f. Visio diagrams of the hardware layout and connectivity

4. Which resource provides detailed information about supported HPE storage product configurations and requires an HPE Passport account?

   a. HPE Single Point of Connectivity Knowledge

   b. HPE Information Library

   c. HPE Sales Builder for Windows

   d. HPE OneView

   e. Microsoft Unified Communications and Collaboration

Summary

- On the journey to hybrid infrastructure and on-demand IT, CIOs are seeking an IT environment that:

  - Deploys and manages private cloud

  - Manages cloud services across multiple clouds

  - Extends and modernizes new applications and workloads

  - Enables business solutions

- HPE RAs are complete, open solutions designed and optimized for specific workloads. Each presized and tested configuration follows a proven deployment methodology to help reduce provisioning time, cost, risk, and errors.

- HPE ConvergedSystems for on-demand IT infrastructure is optimized for IaaS, private cloud, and single and mixed workloads. Every system ships with factory-integrated server, storage, networking, and HPE OneView management. Systems also include HPE Proactive Care to help you avoid problems and resolve issues more quickly. Hyper-converged systems integrate servers, storage, networking, and management into a small footprint, with simplified installation and administration

- The HPE CID is a key requirement that enables successful delivery of a solution. The CID details how the technical design will be built, integrated, and configured. The build and configuration requirements are collected by HPE Sales and HPE Presales consultants along with channel partners.

- HPE offers various tools and resources to support customers and channel partners. These resources include the Solution Demonstration Portal, HPE Information Library, SPOCK, and HPE Live Network.

3 Moving to Cloud Computing

WHAT'S IN THIS CHAPTER FOR YOU?

After completing this chapter, you should be able to:

✓ Make the argument for customers with traditional IT to move toward converged infrastructure, cloud, and on-demand IT

✓ Describe the key benefits of private, public, and hybrid cloud deployment models

✓ Explain how the Hewlett Packard Enterprise (HPE) Helion portfolio can benefit customers

✓ Summarize the service and deployment models of Helion Cloud Services, including how they enable the new style of business and IT

CLOUD FUNDAMENTALS

HPE Helion brings together all the speed, agility, and cost benefits of cloud computing, with all the possibilities and interoperability of open source.

Assessment questions

Before proceeding with this section, assess your existing knowledge by attempting to answer the following questions.

1. What is the correct definition of a hybrid cloud?

    a. The cloud infrastructure is operated solely for a single organization; it can be owned or operated by that organization or by a third party; it can be on or off premise.

    b. The cloud infrastructure is made available from an external organization and is owned and operated by the organization selling cloud services.

    c. A hybrid cloud is a cloud computing environment in which an organization provides and manages some resources in-house and has others provided externally.

    d. The cloud infrastructure is shared by several organizations and supports a specific community that has shared interests.

Take additional notes explaining the correct answer:

_____

_____

_____

_____

2. What is orchestration?

   a. A method of increasing resources to handle peak loads by using either additional local and private resources or remote and public resources

   b. The process of making a service available based on multiple providers

   c. The execution of a predefined workflow used in the process of provisioning and managing infrastructure

   d. A method of providing access to services from different resource pools, such as from traditional IT or private and public clouds, often to minimize service downtime

Take additional notes explaining the correct answer:

_____

_____

_____

_____

3. When a user accesses applications, hosted on a provider's infrastructure, from various devices using a web browser, what cloud service model is being used?

   a. Infrastructure as-a-Service (IaaS)

   b. Platform-as-a-Service (PaaS)

   c. Software-as-a-Service (SaaS)

   d. Hybrid cloud

   e. Public cloud

   f. Private cloud

Take additional notes explaining the correct answer:

_____

_____

_____

_____

4. Which cloud service model is described by these definitions?

    The consumer can deploy their own applications onto the cloud infrastructure using programming languages and tools supported by the provider.

    The consumer does not manage or control the underlying cloud infrastructure, network, servers, operating systems, or storage. However, the consumer controls the deployed applications and (possibly) application-hosting environment configurations.

    a. PaaS

    b. IaaS

    c. SaaS

    d. Hybrid cloud

    e. Public cloud

    f. Private cloud

    Take additional notes explaining the correct answer:

    _____

    _____

    _____

    _____

Assessment question answers

1. What is the correct definition of a hybrid cloud?

    a. The cloud infrastructure is operated solely for a single organization; it can be owned or operated by that organization or by a third party; it can be on or off premise.

    b. The cloud infrastructure is made available from an external organization and is owned and operated by the organization selling cloud services.

    c. **A hybrid cloud is a cloud computing environment in which an organization provides and manages some resources in-house and has others provided externally**

    d. The cloud infrastructure is shared by several organizations and supports a specific community that has shared interests.

2. What is orchestration?

   a. The process of making a service available based on multiple providers.

   b. **The execution of a predefined workflow used in the process of provisioning and managing infrastructure.**

   c. A method of increasing resources to handle peak loads by using either additional local and private resources or remote and public resources.

   d. A method of providing access to services from different resource pools, such as from traditional IT or private and public clouds, often to minimize service downtime.

3. When a user accesses applications, hosted on a provider's infrastructure, from various devices using a web browser, what cloud service model is being used?

   a. IaaS

   b. PaaS

   c. **SaaS**

   d. Hybrid cloud

   e. Public cloud

   f. Private cloud

4. Which cloud service model is described by these definitions?

   The consumer can deploy their own applications onto the cloud infrastructure using programming languages and tools supported by the provider.

   The consumer does not manage or control the underlying cloud infrastructure, network, servers, operating systems, or storage. However, the consumer controls the deployed applications and (possibly) application-hosting environment configurations.

   a. **PaaS**

   b. IaaS

   c. SaaS

   d. Hybrid cloud

   e. Public cloud

   f. Private cloud

What does "cloud native" mean?

Table 3-1 Differences between traditional IT and "cloud native" IT

Traditional IT	Cloud native
Scale up/hardware-defined reliability	Scale out/applications designed without relying on infrastructure reliability
Virtualized	Lightweight runtimes (containers, processes, and PaaS platforms)
Three-tier architectures and stateful, tight couplings	Loosely coupled microservices connected through APIs to enable hybrid application deployment
Operating system, VM aware, and dependencies	Operating system and VM abstraction
Administrator controlled	System controlled (auto scale and recovery, self-configuring)
Waterfall → agile	Agile → CI/CD, DevOps

Typical deployment models force developers to configure and maintain servers and databases in addition to writing code. These developers need to be experts in many diverse technologies, and the high learning curves for different platforms slow down development. Also, the application landscape is heterogeneous—applications are written in multiple languages and frameworks; application services span many vendors; and standardization is costly.

Customers need to provide a complete set of development services to developers that allow them to rapidly develop cloud-native applications or applications designed specifically to run in the cloud. What does it mean for an application to be cloud native? Table 3-1 shows some of the differences. There are some key characteristics and definitions:

- Cloud-native applications are unique to each customer because of the different ways applications will support the business.

- Cloud-native applications are written to use open-source application program interfaces, tools, and languages and are built for hybrid application deployment.

- The intent of cloud-native applications is to allow users to change the underlying infrastructure without having to rewrite the application.

- Cloud-native applications are stateless, meaning they maintain their state internally, as opposed to relying on another layer, such as a database. This means that if a server fails, the application can be restarted on another server with minimal disruption to the user.

Primary, Helion customers have a DevOps focus. They hire developers with different skill sets and enable them to use development tools of their own choice to shrink the development cycle. As a result, applications are published faster, with continuous integration and continuous development (CI/CD) along with constant releases. Think about how often applications automatically update on your smartphone. Developers perform several updates and patches throughout a year, often as frequently as once per day.

PaaS becomes critical because it is delivered in the context of developers. It is important to provide them with the infrastructure and development environment needed to quickly and continuously write and release code. Examples of cloud-native applications include Gmail, Netflix, and Facebook.

The evolution of traditional IT—Activity

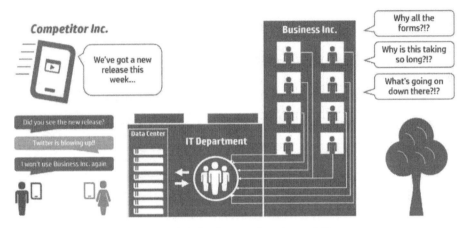

Figure 3-1 Evolution of traditional IT

Watch this 13-minute video by Petamber Pahuja, Helion Business Unit, explaining why and how a traditional IT organization needs to evolve to effectively support its company by scanning Figure 3-1 into the My Learning HPE app on your mobile device. After you watch the video, answer these questions.

1. What is the focus of the traditional IT model?

   _____

   _____

   _____

   _____

2. Explain the competitive threats to the traditional IT organizations.

   _____

   _____

   _____

   _____

3. When the traditional IT organization cannot quickly and efficiently support new requests from the lines of business (LOBs), where do these LOBs turn?

_____

_____

_____

_____

4. After the LOBs bypass the IT department, what is the resulting negative impact?

_____

_____

_____

_____

5. How does the channel partner or reseller conversation need to change to embrace the new style of business?

_____

_____

_____

_____

6. What benefits does a private cloud bring to traditional IT departments?

_____

_____

_____

_____

7. What does the traditional IT department evolve into after adopting cloud technologies?

_____

_____

_____

_____

The evolution of traditional IT activity—Answers

1. What is the focus of the traditional IT model?

   - **Adequate compute resources that are needed by the IT department to support the business**

2. Explain the competitive threats to the traditional IT organizations.

   - **Competitors to traditional IT organizations are offering the business new applications and services that provide better customer experiences at a lower cost, with a faster time to market, and better overall results**

3. When the traditional IT organization cannot quickly and efficiently support new requests from the LOBs, where do these LOBs turn?

   - **Many LOBs turn to public cloud service providers such as Google, Amazon Web Services, and Microsoft Azure for IT on demand, which are low cost and easy to use**

4. After the LOBs bypass the IT department, what is the resulting negative impact?

   - **IT double purchases, lost governance, and security concerns**

5. How does the channel partner or reseller conversation need to change to embrace the new style of business?

   - **The focus of the conversation with the customer needs to be placed on customer business and IT pain points**

6. What benefits does a private cloud bring to traditional IT departments?

   - **Automation of manual tasks, orchestration of services, and overall better agility of the data center**

7. What does the traditional IT department evolve into after adopting cloud technologies?

   - **An internal service provider**

What is a cloud?

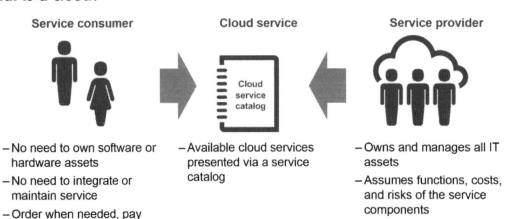

Figure 3-2 What is a cloud?

The main principle behind the cloud is simple—deliver IT functionality as a service. In order to do so, three components are required as shown in Figure 3-2:

- The service consumer, who:

  - Does not need to own the software or hardware assets that are behind the cloud service

  - Does not need to know how to integrate or maintain the cloud service

  - Orders only the needed service, in the right quantities and specifications, and pays only for its use

---

 Note

> The payment for services consumed is true in public cloud environments, whereas private clouds do not always implement metering, show-back, and charge-back.

- The service provider, who:

  - Owns and manages all IT assets that power the cloud service

  - Assumes all functions, costs, and risks of the components that make up the service

- The cloud service, published via a cloud service catalog, which is a contract that binds the consumer to the provider

ITaaS: A paradigm shift

For many IT organizations and customers, delivering Information Technology as a Service (ITaaS) is a paradigm shift. It means changing the business approach from a technology focus to a service focus:

- The technology-centric model includes:

  - Asset ownership

  - Fixed costs

  - Operating on the build-to-order model, where building the configuration or solution may take weeks to months

  - Dedicated and static environment (there is no easy way to scale up or down)

  - Control and risk adversity, setting standards

  - Best technology components

- The service-centric model includes:

  - Pay-per-use consumption model and cost transparency

  - Variable costs based on what is being consumed by the business at what point in time

- Operating on the configure-to-order model, where the configuration of the desired service takes minutes to hours

- Shared and scalable infrastructure and software/application baseline

- Flexibility and scalability

- Risk management

- Consumer and user experience

What is cloud computing?

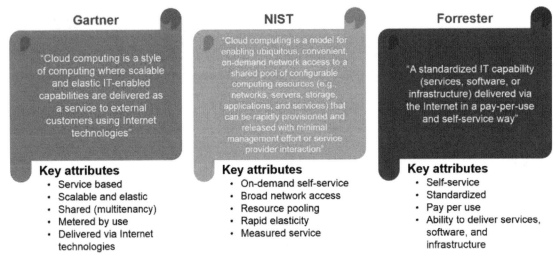

Figure 3-3 Cloud computing definitions

There are several industry definitions of cloud computing. Definitions from Gartner, National Institute of Standards and Technology (NIST), and Forrester are shown in Figure 3-3.

Gartner

Gartner's definition of cloud computing can be explained using Gmail as an example. According to Gartner, the key attributes of cloud computing include the following:

- **Service based**—When users want to send or receive email, they do not have to think about the infrastructure that underlies the Gmail system. They do not need to install an email client on their computer or concern themselves with setting up the service. They simply log in to their Gmail account.

- **Scalable and elastic**—The service can grow to accommodate new users and the storage space per user can be expanded endlessly.

- **Shared**—The Gmail servers and storage systems are shared between multiple different users. This characteristic is referred to as "multitenancy."

- **Metered by use**—Gmail is free (paid for by advertising), but if users run out of storage space, they could buy additional storage.

- **Delivered via Internet technologies**—Gmail is hosted on servers that are owned by Google, and the service is delivered over the Internet.

 Note

To access the cloud definition from Gartner, scan this QR code or enter the URL into your browser.

http://www.gartner.com/it-glossary/cloud-computing/

NIST

In many respects, NIST's definition of cloud computing is similar to Gartner's definition. However, the NIST definition includes applications and services.

The essential characteristics of cloud, according to NIST, are the following:

- **On-demand self-service**—Computing capabilities such as server time and network storage can be provisioned unilaterally and automatically. Consumers do not need support from service providers.

- **Broad network access**—Customers can access capabilities, which are available over the network, using standard mechanisms that are compatible with heterogeneous thin or thick client platforms. For example, customers can use smartphones, tablets, workstations, and so on.

- **Resource pooling**—Computing resources from the provider, including storage, processing, memory, and network bandwidth, are pooled so that multiple consumers can access them. With this multitenant model, various physical and virtual resources are dynamically assigned and reassigned according to consumer demand. Typically, customers do not know where the provided resources are located. However, they may be able to identify the location at a higher level of abstraction such as country, state, or data center.

- **Rapid elasticity**—To keep pace with demand, resources can be elastically provisioned and released. This can even happen automatically depending on the situation. From the consumer's perspective, provisioning capabilities seem unlimited (even though they are not). These capabilities can be appropriated in any quantity at any time.

- **Measured service**—To maintain optimal performance, cloud systems automatically control usage of resources. The systems use a metering capability at a level of abstraction that depends on the type of service (for example, storage, processing, bandwidth, and active user accounts). Reports on resource use are available for the provider and consumers.

 Note

To access the NIST definition on page 2 of *The NIST Definition of Cloud Computing*, scan this QR code or enter the URL into your browser.

http://csrc.nist.gov/publications/nistpubs/800-145/SP800-145.pdf

---

Forrester

The Forrester definition of cloud computing identifies some important characteristics:

- **Self-service**—This is a fundamental characteristic of cloud computing. It allows users (service consumers) to choose the services that best suit their requirements.

- **Standardized**—Standardization brings many advantages, including operational cost savings and lower costs for provisioning services.

- **Pay per use**—Consumers pay depending on how they use the service. Various payment models exist, and cloud service providers can offer various payment models to match the needs of their customers.

- **Ability to deliver services, software, infrastructure**—Services, software, and infrastructure are important IT components that service providers can deliver. Typically, the term "service" includes any type of service (including software and infrastructure).

 Note

To access the definition from Forrester Research, scan this QR code or enter the URL into your browser.

https://www.forrester.com/Cloud-Computing

Cloud service models

Service models

| IaaS | PaaS | SaaS | XaaS |

Figure 3-4 Cloud service models

Figure 3-4 shows common service models. The service model determines the type of resources that are provided to the consumer. For example, do consumers receive an application SaaS, or do they use a platform to develop and run their own applications (PaaS)? Or do consumers purchase access to a platform with an operating system but no installed applications (IaaS)?

Details of these common service models include:

- **IaaS**—HPE customers are moving to Helion to increase agility, reduce costs, and support innovation in an open architecture. Thus, Helion is commonly used to accelerate application testing and development by providing private cloud IaaS solutions.

The NIST definition of IaaS is:

The capability provided to the consumer is to provision processing, storage, networks, and other fundamental computing resources where the consumer is able to deploy and run arbitrary software, which can include operating systems and applications. The consumer does not manage or control the underlying cloud infrastructure but has control over operating systems, storage, deployed applications, and possibly limited control of select networking components such as host firewalls.

The cloud infrastructure service model is simple. IaaS delivers a computer infrastructure, typically a virtualized environment, as a service. Rather than purchasing servers, software, data center space, networking equipment, and power and cooling systems, users instead buy those resources as a service. The service is typically billed on a utility computing basis, and the cost is calculated based on the level of activity. Usually, users can buy services by the hour, day, week, or month.

Amazon Elastic Compute Cloud (Amazon EC2) is one example of an IaaS provider. Users can install and run their own applications on a computing environment with an operating system. Amazon EC2 is a central part of Amazon Web Services (AWS), the Amazon.com cloud computing platform.

- **PaaS**—The NIST definition of PaaS is:

  The capability provided to the consumer is to deploy onto the cloud infrastructure consumer-created applications using programming languages and tools supported by the provider, such as Java, Python, or .NET. The consumer does not manage or control the underlying cloud infrastructure, network, servers, operating systems, or storage, but the consumer has control over the deployed applications and possibly application hosting environment configurations.

  Cloud platform services deliver a computing platform, solution stack, or both. PaaS enables companies to deploy applications acquired or created by users with provided programming languages and tools, without the cost or complexity of buying and managing underlying hardware and software layers. The consumer does not control the underlying infrastructure (networking, servers, operating systems, or storage) but does have control over the deployed applications. A virtual machine that supports Linux, Apache, MySQL, and PHP/Perl/Python (commonly known as the *LAMP stack*) is one example of PaaS. Google App Engine is another example of PaaS. It offers a full development stack for people who want to develop and host applications on Google's infrastructure. Google App Engine supports applications written in several programming languages.

- **SaaS**—The NIST definition of SaaS is:

  The capability provided to the consumer is to use the provider's applications running on a cloud infrastructure. The applications are accessible from various client devices through either a thin client interface such as a web browser or a program interface. The consumer does not manage or control the underlying cloud infrastructure, network, servers, operating systems, storage, or even individual application capabilities, with the possible exception of limited user-specific application configuration settings.

  Cloud application services provide access to software applications over the Internet. Users do not need to develop an application or install and run a purchased application on their own computers. A company that uses SaaS does not need to manage the underlying infrastructure of applications—or the cost of providing the infrastructure in the first place. A good example of SaaS is Salesforce. Salesforce customers can access the Salesforce sales, service, marketing, and many other applications using an Internet connection.

The main difference between SaaS and PaaS is that SaaS service consumers do not develop the applications; instead, they simply access applications that have been developed by someone else. With PaaS, however, consumers develop and deploy their own applications using software development kits and programming languages.

Public, private, and hybrid deployment models

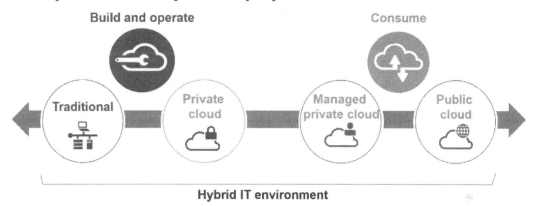

Figure 3-5 Deployment models

Deployment models, as shown in Figure 3-5, characterize the cloud implementation based on who manages the resources that deliver the service.

- **Private cloud**—In a private cloud, services are provided to specific users such as members of departments within a company. Services are delivered over the network, but data remains private because it stays behind the company firewall rather than being transmitted over the Internet. The solution might still feature multitenancy and shared underlying resources, but the multiple tenants are departments within the company. Although a private cloud requires an investment in infrastructure and incurs ongoing maintenance costs, it can be made highly available and flexible. Retaining data behind corporate firewalls can help to address common enterprise concerns regarding security, governance, availability, and control.

- **Public cloud**—In a public cloud (also known as an *external cloud*), all cloud assets that underlie the services are owned and operated by a provider, and users access the services over the Internet. Public clouds feature multitenancy, where resources are shared by all users. Users pay for the services on a utility computing (pay-per-use) basis. Users do not have to invest in the acquisition and maintenance of the computer infrastructure. The users could be employees of different companies, located in different countries, legal jurisdictions, or time zones. In fact, anyone with a credit card and access to the Internet is capable of consuming public cloud services.

  Unlike a private cloud scenario, in which data resides securely behind the company firewall, the public cloud requires users' data to travel over the public Internet and reside in a location unknown to the users. This model clearly raises concerns for many enterprises regarding security, governance, availability, and control. Sensitive business data could be stored in a different country with a different legal system, potentially with barriers to remediation in the event of an outage or loss of data.

IaaS, PaaS, and SaaS can be public cloud solutions. The entire infrastructure is owned by the provider—Amazon (in the IaaS example), Google (in the PaaS example), or Salesforce.com (in the SaaS example). The services are delivered over the Internet, and users are charged for the services on a pay-per-use basis.

- **Hybrid cloud**—Providers such as Amazon, Microsoft, and Google make a massive amount of resources available to users. However, resources in a private cloud are limited. For example, a company with a private cloud solution might be limited to resources provided by that solution. Although the solution, when correctly sized and configured, can provide a large amount of resources, no private cloud can match the resources of the Amazon, Microsoft, or Google data centers. To overcome this potential resource limitation, a hybrid cloud environment makes use of services sourced from internal and external providers.

One way of thinking about this approach is convenience versus control. The public cloud provides convenience, and the private cloud gives control. If you need to define a very specific environment for your customer's applications, and you need to control that environment tightly, private cloud is appropriate. If convenience or ease of access is important, public clouds are more appropriate. A hybrid cloud provides flexibility to use both private clouds and public clouds. A combination of public and private cloud can provide the right destination for the right applications at the right cost.

Consuming resources and services

When it comes to consuming resources and services from service providers, several options are available:

- **Managed cloud infrastructure services**—These cloud solutions are maintained and operated by a service provider. They are often augmented by consulting and professional services to increase value gained from cloud, expedite on boarding, and ensure optimization of the solutions for ongoing management.

  - **Managed virtual private cloud**—This secure, multitenant private cloud allows multiple organizations to share the cloud infrastructure. It is managed (and typically owned) by a third party.

  - **Managed private cloud**—This secure, dedicated private cloud runs on infrastructure that is dedicated to a single client organization (tenant), but managed by a third party. The underlying infrastructure can be delivered from a customer's facility or from a facility owned by the service provider.

- **Public cloud**—This cloud option encompasses services from providers who offer them primarily for public use. Security considerations are typically different for services (applications, storage, and other resources) that are made available for a public audience over the Internet. Generally, public cloud service providers such as Amazon, Microsoft, and Google own and operate the infrastructure at their data centers.

The road to hybrid begins with private cloud

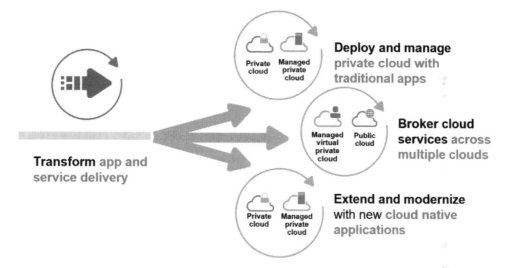

Figure 3-6 The road to hybrid

Hybrid is sometimes viewed as a future state where public cloud meets private cloud. On the contrary, enterprises using cloud computing services are most likely already in a hybrid cloud configuration. Whether they use SaaS or deploy applications to a cloud platform, those services most likely are connecting back to at least one resource in their data center. For example, enterprises using SaaS-based sales force automation software usually connect this application to their on-premises enterprise resource planning (ERP), finance, or eCommerce systems.

Cloud is not a single-stage transformation, as indicated by Figure 3-6. It is a paradigm shift with an expansive scope that requires planning and a phased approach. HPE believes that the road to hybrid cloud begins with a private cloud infrastructure, which includes servers, storage, networking hardware, and the associated cloud computing, cloud management, and cloud security software.

With a hybrid approach based on an end-to-end, integrated private cloud solution that works with heterogeneous IT environments, enterprises can improve time-to-solution, leverage current IT investments, and reduce the costs associated with delivery and management of existing workloads without disrupting critical, ongoing IT operations.

Hybrid approach to cloud

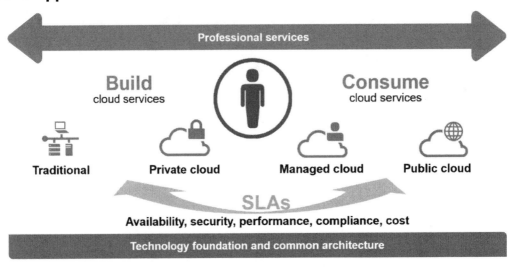

Figure 3-7 Hybrid approach to cloud

A recent HPE study revealed that:

- About 75% of IT executives plan to pursue a hybrid delivery model.

- For 72% of executives, it is important that the cloud implementation permits portability of workloads between cloud models.

- About 65% of executives are concerned with vendor lock-in.

By embracing a hybrid cloud, companies can address IT service delivery challenges in the most efficient and cost effective way. Businesses can continue to leverage traditional IT resources and take advantage of the capabilities offered across private, managed, and public cloud implementations, as depicted in Figure 3-7. For example, with traditional IT, customers might spend indefinitely without achieving their goals. With cloud implementations, service-level agreements (SLAs) guarantee that companies receive a solution that meets regulatory compliance requirements and provides set levels of availability, security, and performance.

To move toward a hybrid cloud, customers can segment their portfolio of applications according to their respective SLA requirements. Each type of application should be aligned to the appropriate deployment models, thus optimizing the use of internal and external resources.

In this hybrid world, the CIO and IT department expand their traditional role as the builders of services. They become both builders and brokers, creating a seamless experience for users, independent of service sources.

Customers can accelerate adoption of hybrid infrastructure by choosing to consume resources from service providers. For most enterprise customers, this approach complements their hybrid IT strategy, where services are composed using traditional IT, private cloud, and public cloud. Services are combined based on customers' data security and compliance, staffing, time to market, funding, and other factors. For example, sensitive data might be kept in-house, using traditional IT or private cloud, and nonsensitive data might be placed in a public cloud.

Accelerating the journey to the hybrid cloud video—Learner activity

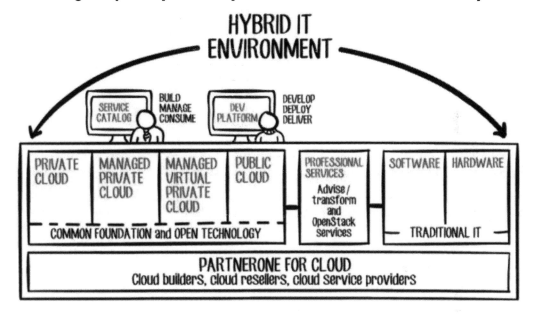

Figure 3-8 Journey to hybrid cloud

Watch this eight-minute video explaining how to accelerate the journey to the hybrid cloud with Helion by scanning Figure 3-8 into the My Learning HPE app on your mobile device. After watching the video, answer these questions.

1. What three goals have been identified by IT leaders for their business transformation?

    _____

    _____

    _____

2. What types of cloud environments does Helion provide?

    _____

    _____

    _____

    _____

3. What are the three principles of the Helion portfolio?

_____

_____

_____

Accelerating the journey to the hybrid cloud video activity — Solution

1. What three goals have been identified by IT leaders for their business transformation?

   - **Choose the right destination for the right application**

   - **Become an internal service provider**

   - **Support enterprise developers with tools and environment**

2. Explain the competitive threats to the traditional IT organizations.

   - **Private cloud, managed private cloud, and managed virtual private cloud**

3. When the traditional IT organization cannot quickly and efficiently support new requests from the LOBs, where do these LOBs turn?

   - **Openness, security, agility**

HPE Helion portfolio

Transformation to an open, hybrid cloud

 Cloud software   Integrated solutions

Figure 3-9 HPE Helion

The Helion portfolio of products, services, and solutions helps enterprises and service providers transform to an open, hybrid cloud. Helion offers cloud software, integrated solutions, and managed services as illustrated in Figure 3-9.

- Cloud software helps customers manage hybrid clouds, develop and deploy cloud-native applications, and support AWS-compatible private clouds:

  - HPE Helion CloudSystem (flexible, capable, and open private cloud software)

  - HPE Cloud Service Automation (cloud management platform)

- HPE Helion OpenStack (open-source IaaS)

- HPE Helion Development Platform (open source PaaS)

- HPE Helion Eucalyptus (AWS-compatible private cloud software)

- Integrated solutions include all the hardware and software needed to deploy a cloud solution, based on HPE servers, storage and networking hardware, and HPE cloud management software:

  - HPE Helion CloudSystem Solution (fully integrated hybrid cloud solution)

  - HPE Helion Rack (OpenStack-based private cloud solution)

  - HPE Helion Content Depot (OpenStack Swift-based object storage solution)

- HPE Helion Managed Cloud Services provide the infrastructure foundation for critical application workloads requiring high levels of security, performance, availability, continuity, responsiveness, and quality.

HPE Helion CloudSystem

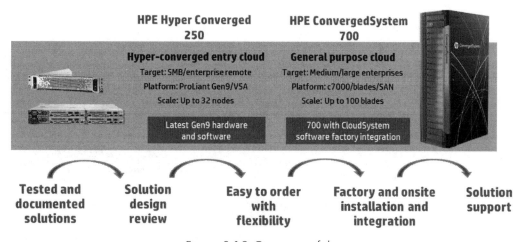

Figure 3-10 Current portfolio

Helion CloudSystem is a fully integrated, end-to-end, private cloud solution built for traditional and cloud-native workloads. It delivers automation, orchestration, and control across multiple clouds. It includes cloud management software, servers, storage, and networking, combined with installation services, making it quick and easy to deploy a private cloud. Helion CloudSystem is powered by OpenStack technology and developed with an emphasis on automation and ease-of-use.

For maximum flexibility, Helion CloudSystem is available as a software package or delivered as part of an integrated HPE ConvergedSystem solution. You can deploy it on existing infrastructure to increase ROI or as an integrated system to minimize time to value.

Depending on the customer's cloud workload needs, they can choose from these ConvergedSystem solutions, as shown in Figure 3-10:

- Customers can deploy Helion CloudSystem on HPE Hyper Converged 250, a hyper-converged platform with compute and local storage

  - Target market: SMB and enterprise with remote and regional cloud needs

  - Platform: HPE ProLiant Gen9 servers and HPE StoreVirtual VSA

  - Scale: up to 32 nodes

- Customers can deploy Helion CloudSystem on HPE ConvergedSystem 700 for cloud, an intelligent compute and SAN with the latest release of HPE OneView and ProLiant Gen9

  - Target market: medium and large enterprises

  - Platform: HPE BladeSystem c7000 enclosures and ProLiant half-height server blades, SAN (EVA P6000, StoreServ, or other HPE SAN or an existing SAN)

  - Scale: up to 100 server blades

HPE Helion CloudSystem Foundation and Enterprise

Helion CloudSystem Enterprise
Advanced infrastructure and application services

- Hybrid cloud management with Service Marketplace and designed based on Cloud Service Automation and Operations Orchestration
- Enterprise-class lifecycle management for application services including compliance[2]

HPE Cloud Service Automation and HPE Operations Orchestration— delivered as virtual appliances

| Helion CloudSystem Foundation | Matrix Operating Environment |

Helion CloudSystem Foundation
Core infrastructure services and PaaS

- Push-button activation of compute and storage nodes
- Out-of-the-box networking services
- Built on Helion OpenStack[1]
- Delivered via virtual appliances
- Helion Development Platform

HPE infrastructure management—HPE OneView[2]

Built on ConvergedSystem or on HPE and third-party infrastructure

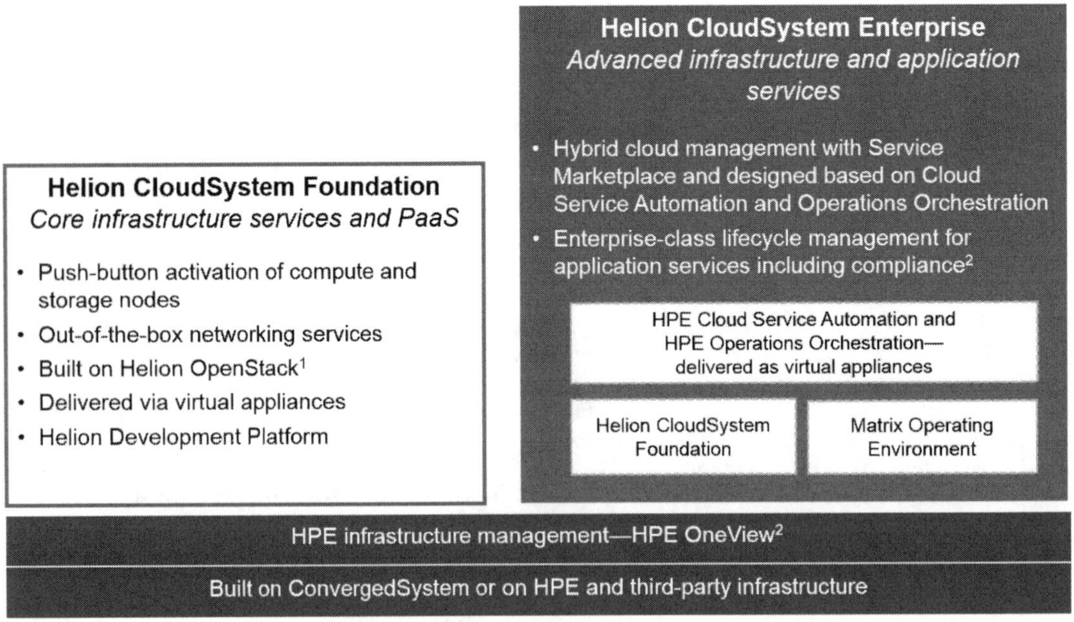

Figure 3-11 HPE Helion CloudSystem Foundation and Enterprise

As shown in Figure 3-11, Helion CloudSystem 9 includes similar packaging as the previous version (8.x). This means customers can choose between two versions of Helion cloud software: Helion CloudSystem Foundation or Helion CloudSystem Enterprise.

- **Helion CloudSystem Foundation**—This preintegrated software appliance provides core infrastructure services and a PaaS. It includes the following:

  - Push-button activation of compute and storage nodes

  - Out-of-the box networking services

  - HPE Helion Development Platform

  Helion CloudSystem Foundation is built on Helion OpenStack, and it includes Nova, Cinder, Neutron, Keystone, Glance, Horizon, Swift, and Heat projects from OpenStack.

- **Helion CloudSystem Enterprise**—This solution provides advanced infrastructure and application services such as the following:

  - Hybrid cloud management within the service marketplace and based on HPE Cloud Service Automation and Operations Orchestration software

  - Enterprise-class lifecycle management for application services including compliance (optional)

Helion CloudSystem Enterprise includes Cloud Service Automation and Operations Orchestration, both delivered on a virtual appliance called the CloudSystem Enterprise appliance.

The main difference with the Foundation package is the inclusion of the Helion Development Platform, which enables customers to develop cloud-native applications. When customers upgrade from Foundation to Enterprise, the package includes both the Helion CloudSystem Foundation and HPE CloudSystem Matrix solutions as providers. That gives customers the choice of managing either provider with Cloud Service Automation.

As an infrastructure platform, both packages can run on ConvergedSystems with HPE OneView as the infrastructure management solution. They can also be built on other HPE or third-party hardware infrastructure.

Helion CloudSystem Foundation is built on Helion OpenStack, which includes:

- **Nova**—An OpenStack compute component that is a cloud computing fabric controller (the main part of an IaaS system).

- **Cinder**—An OpenStack block storage component that provides persistent block-level storage devices for use with OpenStack compute instances. The block storage system manages the creation, attaching, and detaching of block devices to servers.

- **Neutron**—An OpenStack networking component used as a system for managing networks and IP addresses. It ensures the network will not be the bottleneck or limiting factor in cloud deployments and provides users with a self-service way of creating their own networks and routers.

- **Keystone**—An OpenStack identity service that provides a central directory of users mapped to the OpenStack services they access. It acts as a common authentication system across the cloud operating system and can integrate with directory services such as LDAP.

- **Glance**—An OpenStack image service that provides discovery, registration, and delivery services for disk and server images. Stored images can be used as templates or to store and catalog an unlimited number of backups.

- **Horizon**—An OpenStack dashboard component that provides administrators and users a graphical interface to access, provision, and automate cloud-based resources. It allows third-party products and services (billing, monitoring, and additional tools) to be used.

- **Swift**—An OpenStack object storage component that is a scalable redundant storage system. Objects and files are written to multiple disks throughout servers in the data center, with the OpenStack software responsible for data replication and integrity.

- **Heat**—An OpenStack orchestration service that facilitates the provisioning of multiple composite cloud applications using templates.

Managing and using CloudSystem Foundation

Administration:
Helion CloudSystem Operations
Console

Consumption:
Helion OpenStack user portal

Figure 3-12 CloudSystem Foundation

CloudSystem Foundation provides two graphical user interfaces for management, administration, and consumption as shown in Figure 3-12:

- The **Helion CloudSystem Operations Console** is the administrative UI that provides access to the primary resource management areas, including compute, networking, and storage resources, along with appliance administration.

- The **Helion OpenStack user portal** is the OpenStack Horizon-based UI that allows users to configure and deploy virtual machine instances.

Managing and using CloudSystem Enterprise

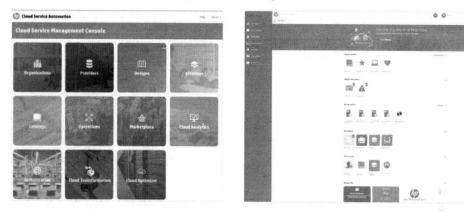

Administration:
Cloud Service Automation

Consumption: Marketplace Portal

Figure 3-13 CloudSystem Enterprise

CloudSystem Enterprise provides two graphical user interfaces for management, administration, and consumption as shown in Figure 3-13:

- The **Cloud Service Automation (CSA)** interface is the administrative UI that provides access to tasks such as configuring organizations, administering users, creating designs, and managing catalogs.

- The **Marketplace Portal** is used by consumers (service users) to access the published service catalog and subscribe to services. The Marketplace Portal and service catalog are provided by CSA.

HPE Helion OpenStack

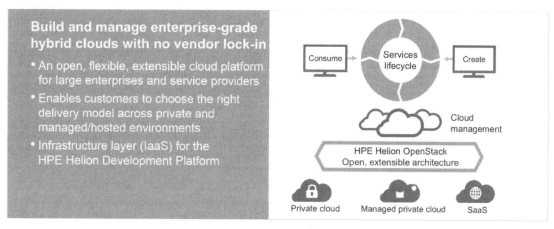

Figure 3-14 HPE Helion OpenStack

Many customers have been interested in OpenStack technology, but they lack the time and resources needed to fully implement the architecture. This includes working through the large number of configuration options and various dependencies that typically come with a unique configuration process. Many early adopters of the OpenStack technology have been excited by the benefits, but they face serious challenges in adapting it for their business needs.

Helion OpenStack was developed to help customers address the complexities of installing an open-source project in a large business environment. As shown in Figure 3-14, the Helion OpenStack enterprise-grade distribution provides customers with an IaaS platform that is easy to install and maintain. The platform meets expected business requirements for security, high availability, reliability, and scalability within their cloud.

With Helion OpenStack, customers can rely on expert guidance when deploying and managing an open-source hybrid cloud solution. This expertise will help simplify the path to successful cloud computing, without vendor lock-in.

Helion OpenStack is an open, extensible cloud platform that enables large businesses and service providers to more easily build, manage, and use hybrid clouds. It offers the flexibility to choose the right delivery model across private and traditional IT environments. And lastly, Helion OpenStack delivers leading open-source cloud computing technology in a resilient, cost-effective, and maintainable solution.

HPE value-add to OpenStack

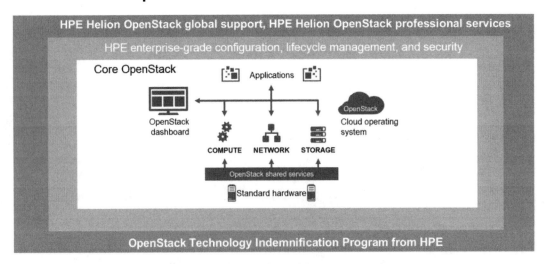

Figure 3-15 HPE value-add to OpenStack

As shown in Figure 3-15, through Helion OpenStack, HPE offers an enterprise-grade solution with:

- Reduced risk and enhanced security at the product level
- OpenStack Technology Indemnification Program from HPE

- Helion OpenStack worldwide 24/7 support included in each server license

- Increased agility and ease of use with Helion OpenStack lifecycle management and configuration

OpenStack software runs on top of standard hardware, helping customers create an open-source cloud in a software-defined data center. With the OpenStack technology, you can create pools of compute, network, and storage. Then, you can create the cloud environment from a GUI (OpenStack Horizon dashboard) or from applications.

HPE adds to the OpenStack core services product-level enhancements by providing enterprise-level capabilities that are not always available in the base OpenStack technology components. For example, the HPE configuration enhances compute and storage functionality by adding unique configuration intellectual property to the basic Nova and Cinder compute configurations found in the OpenStack Trunk. HPE also streamlines the integration of VMware ESXi with OpenStack technology. These configurations enhance user capabilities, maintain portability, and avoid vendor lock-in.

Typical use cases

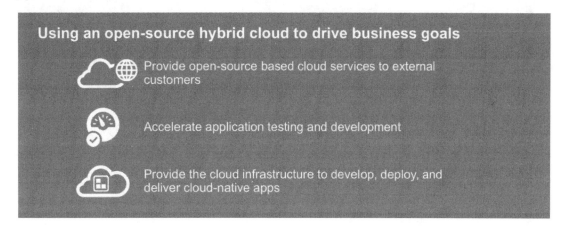

Figure 3-16 Use cases

Customers want to increase agility of their IT solutions, reduce costs, and support innovation using an open architecture. Figure 3-16 shows common use cases for Helion OpenStack. One primary use case is to provide a complete set of development services that allow developers to rapidly deploy cloud-native applications or applications that are designed to run in the cloud. Companies can use Helion OpenStack cloud technology to speed up software development cycles, reducing the time spent on direct interaction with IT services, environment preparation, deployment, and testing.

Another common use case for Helion OpenStack is to accelerate application testing and development, most commonly by providing private cloud IaaS solutions.

Finally, forward-looking IT organizations are transforming themselves to provide cloud-based services to external customers more quickly. In the past, IT organizations focused on building custom cloud services. Now, they can use Helion OpenStack cloud technologies to offer classic outsourcing, cloud-based private services, and application hosting from a single provider.

Use cases that are a good fit for Helion OpenStack

Green zone	Red zone
– Customer is interested in open-source technology and wants to **build an OpenStack foundation in their environment**	– Customer wants to **run classic apps in a traditional IT scale-up environment** (for example, run Oracle on Linux, SQL Server on Windows); lead with Helion CloudSystem
– Customer wants to **optimize for cloud-native environments** and does not need a bridge to traditional IT	– Customer is **running Hyper-V as primary hypervisor**; lead with Helion CloudSystem
– Customer sees the **need for a development environment** to attract and retain their developers	– Customer has **multiple data center deployment needs**; lead with Helion CloudSystem
– Customer wants to **avoid vendor lock-in and reduce overall TCO** by moving from virtualization to a hybrid cloud environment	– Customer use cases are **heavily dependent on third-party components requiring custom integration**; lead with HPE Services
– Customer **talks about DevOps**	

Figure 3-17 Good fit and bad fit use cases

Figure 3-17 provides a list of customer use cases or needs that are correctly matched with the Helion OpenStack software platform (the green zone), and which situations are not aligned (the red zone). If a customer need falls under the red zone, you should recommend a different product or solution, most likely Helion CloudSystem or HPE Services.

 Note

Green and red zones may change over time because products and solutions mature and expand their capabilities.

Helion OpenStack and CloudSystem positioning

Table 3-2 Helion OpenStack and CloudSystem positioning

Cloud infrastructure management	Helion OpenStack	CloudSystem Foundation 9.0	CloudSystem Enterprise 9.0
Foundational IaaS to deliver rapid infrastructure provisioning	✓	✓	✓
Integrated IaaS and PaaS	✓	✓	✓
Hybrid cloud management capabilities		✓	✓
Heterogeneous management with combined IaaS, PaaS, and SaaS services			✓
Product features	**Helion OpenStack**	**CloudSystem Foundation 9.0**	**CloudSystem Enterprise 9.0**
Turnkey solution with out-of-the-box intuitive UI/self-service customer portal and easy-to-use designer tools			✓
Advanced cloud features such as infrastructure and application service for multitier architecture			✓
Physical "bare-metal" provisioning offering "close to the trunk" solution	✓		
Highly configurable solution with included hypervisor	✓		

Table 3-2 summarizes the cloud infrastructure management and product features of Helion OpenStack, Helion CloudSystem Foundation 9.0, and Helion CloudSystem Enterprise 9.0. You can make purchase recommendations based on the needs of customers in the green zone for these products.

 Note

The hybrid cloud management capabilities include cloud bursting to Microsoft Azure, Amazon Web Services, HPE CloudAgile bursting partners, and other CloudSystem clouds.

Optimized HPE Helion OpenStack cloud solutions

Planning, architecting, and deploying cloud technology can be challenging, especially for highly specialized endeavors that require sophisticated solutions. Optimized cloud solutions from HPE provide complete offerings that include everything customers need to be successful when facing a demanding use case. Built, tested, and validated by HPE cloud experts, the optimized cloud solutions are based on industry-leading ProLiant servers and cloud software products.

- **HPE Helion Self-Service HPC**—HPE Helion Self-Service HPC is an OpenStack-based high-performance computing (HPC) solution that provides scalable performance to meet the intensive demands of design, simulation, and analysis workloads. This flexible, easy-to-use solution delivers HPC as a service, giving customers access to powerful and secure computing capabilities whenever and wherever they need them. It is integrated with Ansys and built on HPE Apollo or HPE Cluster Platforms. To extend the availability of valuable HPC assets, Helion Self-Service HPC incorporates a customizable, self-service portal that streamlines the management of complex HPC resources. Customers who require quick time to value may benefit the most from HPC. Businesses in these industries are ideal HPC targets:

  - Manufacturing

  - Oil and gas or energy

  - Entertainment and media

  - Healthcare and pharmaceutical

  - Financial services

- **HPE Helion Content Depot**—HPE Helion Content Depot is an integrated, turnkey object storage solution that delivers massive scalability and simplifies the storage and management of ever-increasing data volumes on an open platform. Helion Content Depot is a highly available, peta byte-scale platform that securely stores vast amounts of unstructured data and is optimized for durability, availability, and concurrency. It is built on ProLiant servers, HPE Networking, Helion OpenStack, and Swift. It is priced to compete with similar solutions from Amazon, Google, and Microsoft.

- **HPE Helion Rack**—HPE Helion Rack is an open private cloud solution that supports rapid infrastructure provisioning and can run multiple secure and high-performance cloud-native workloads. This pretuned and pretested solution is built on commercial-grade Helion OpenStack and incorporates the flexibility and performance of industry-leading ProLiant servers. It streamlines the development and provisioning of cloud-native applications.

 Note

To access more information including videos for each solution, scan this QR code or enter the URL into your browser.

http://www8.hp.com/us/en/cloud/helion-portfolio.html

Assessment activity

Knowing the terminology

For this activity, match each definition with its corresponding term.

1. Draw lines to match the definitions to the correct terms.

Definition	**Corresponding term**
The process of collecting server, storage, network, and application availability and performance data	Provisioning
The process of creating a service from a service template	Service template
A set of actions that execute customer-specific IT tasks such as approvals, manual OS deployment, manual storage provisioning, and notifications	Workflow
A design blueprint that specifies the requirements for an infrastructure service in terms of server groups, networks, and storage; contains customization; may include applications	Monitoring

2. Draw lines to match the definitions to the correct terms.

Definition	**Corresponding term**
When a service is no longer needed, its resources are returned to the pool of available resources; if cross-charge or billing is needed, a final invoice is generated	Orchestration
The process of planning, provisioning, customizing, and configuring applications; ongoing patch management; and retirement of used resources	Service catalog
The execution of a predefined workflow used in the process of provisioning and managing infrastructure	De-provisioning
A repository of service templates	Application lifecycle management

3. Draw lines to match the definitions to the correct terms.

Definition	Corresponding term
The process of making services available via a network	Hosting
The ability to dynamically expand or contract a computing resource based on demand	Brokering
Providing access to services from different resource pools, such as from traditional IT or private and public clouds, often to minimize downtime	Elasticity
The process of making a service available based on multiple providers	Bridging

4. Draw lines to match the definitions to the correct terms.

Definition	Corresponding term
A cloud computing environment in which an organization provides and manages some resources and has others provided externally	Private cloud
The cloud infrastructure is made available to the general public or a large industry and is owned and operated by the organization selling cloud services	Multi-tenancy
Compute nodes that host users' application workloads, as opposed to servers that run management software	Public cloud
Software architecture where a single instance of the software runs on a server for multiple clients; each client receives a customized virtual application instance	Hybrid cloud
The cloud infrastructure is operated for a single organization; can be owned or operated by that organization or by a third party (on- or off-premise)	Tenant array

5. Draw lines to match the definitions to the correct terms.

Definition	Corresponding term

Cloud service model providing processing, storage, network, or other fundamental computing resources to the consumer	IaaS
Cloud service model providing development platform for consumer to develop applications using programming languages or tools from the provider	SaaS
Cloud service model providing access to applications hosted on the cloud infrastructure	PaaS

Knowing the terminology activity answers

1. Corresponding term: Definition

   a. **Provisioning: The process of creating a service from a service template**

   b. **Service Template: A design blueprint that specifies the requirements for an infrastructure service in terms of server groups, networks, and storage; contains customization; and may include applications**

   c. **Workflow: A set of actions that execute customer-specific IT tasks such as approvals, manual OS deployment, manual storage provisioning, and notifications**

   d. **Monitoring: The process of collecting server, storage, network, and application availability and performance data**

2. Corresponding term: Definition

   a. **Orchestration: The execution of a predefined workflow used in the process of provisioning and managing infrastructure**

   b. **Service catalog: A repository of service templates**

   c. **De-provisioning: When a service is no longer needed, its resources are returned to the pool of available resources; if cross-charge or billing is needed, a final invoice is generated**

   d. **Application lifecycle management: The process of planning, provisioning, customizing, and configuring applications; ongoing patch management; and retirement of used resources**

3. Corresponding term: Definition

    a. **Hosting: The process of making services available via a network**

    b. **Brokering: Providing access to services from different resource pools, such as from traditional IT or private and public clouds, often to minimize downtime**

    c. **Elasticity: The ability to dynamically expand or contract a computing resource based on demand**

    d. **Bridging: The process of making a service available based on multiple providers**

4. Corresponding term: Definition

    a. **Private cloud: The cloud infrastructure is operated for a single organization; can be owned or operated by that organization, or by a third party (on- or off-premise)**

    b. **Multitenancy: Compute nodes that host users' application workloads, as opposed to servers that run management software**

    c. **Public cloud: The cloud infrastructure is made available to the general public or a large industry and is owned and operated by the organization selling cloud services**

    d. **Hybrid cloud: A cloud computing environment in which an organization provides and manages some resources and has others provided externally**

    e. **Tenant array: Software architecture where a single instance of the software runs on a server for multiple clients; each client receives a customized virtual application instance**

5. Corresponding term: Definition

    a. **IaaS: Cloud service model providing processing, storage, network, or other fundamental computing resources to the consumer**

    b. **SaaS: Cloud service model providing access to applications hosted on the cloud infrastructure**

    c. **PaaS: Cloud service model providing development platform for consumer to develop applications using programming languages or tools from the provider**

Helion Cloud Services enable the new style of business and IT

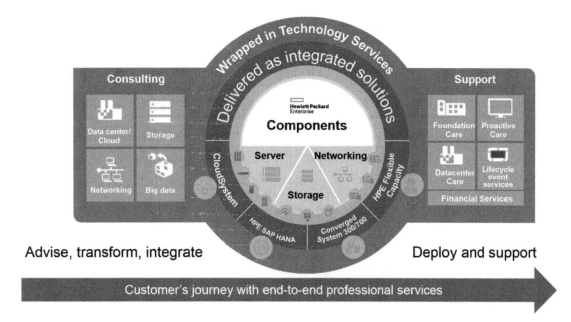

Figure 3-18 The new style of business and IT

As shown in Figure 3-18, HPE provides a comprehensive portfolio of professional services to advise, transform, and manage clients' journey to a hybrid infrastructure and on-demand IT. HPE takes an end-to-end lifecycle approach, addressing customers' needs in the areas of people, processes, and technology. The HPE Services portfolio includes solutions and products for the cloud, security, mobility, and big data analytics.

HPE Services fall into three major categories:

- **HPE Services**—HPE provides infrastructure technology outsourcing, applications, and industry services, including business process outsourcing to more than 1700 business and government clients in 90 countries.

- **HPE Consulting Services**—Technology consulting services provide IT design, planning, implementation, integration, and maintenance services and support for organizations and government agencies. These services are tightly aligned and optimized for the HPE enterprise product portfolio. HPE consulting services include primarily data center and technology consulting services, including IT infrastructure, cloud, mobility, security and risk management, and big data.

- **HPE Support Services**—This simple and standardized portfolio of support services spans all technologies and is offered uniformly in all geographies. It is provided by HPE Technology Services. It includes HPE Care Pack Services, HPE Datacenter Care, HPE Foundation Care, and HPE Proactive Care Services, which cover hardware support, lifecycle event, installation and deployment, software, and mission-critical services.

Customers can become a broker of apps and services

Deploy private cloud

Create a single catalog of
all services and apps

Extend and modernize
for cloud-native apps

Build: HPE Helion CloudSystem Consume: HPE Helion Managed Private Cloud and HPE Helion Managed Virtual Private Cloud	HPE Cloud Service Automation HPE Propel	HPE Helion OpenStack HPE Applications Transformation to Cloud services

Figure 3-19 Broker of apps and services

Many IT organizations are looking to become an internal service provider that can broker applications and services regardless of where they are hosted. HPE is able to help these organizations with the solutions as shown in Figure 3-19. Some customers are building their own private cloud using Helion CloudSystem. Others want to consume a private cloud that is built and managed by an outside vendor. For this, they can use Helion Managed Private Cloud or Helion Managed Virtual Private Cloud.

To create a single catalog of all IT services and applications, customers can use Cloud Service Automation, which provides cloud management software that automates and simplifies deployment and management of hybrid IT services. Another option is HPE Propel, a self-service portal for delivering all IT services to users.

To extend and modernize cloud-native applications with web-scale, hyper-converged infrastructure, there is Helion OpenStack and Helion Application Transformation services.

 Note

To get more information about these products, solutions, and services, scan the appropriate QR code below or enter the URL into your browser.

Helion CloudSystem: http://www.hp.com/go/cloudsystem

Helion Managed Private Cloud and Helion Managed Virtual Private Cloud: http://www8.hp.com/us/en/business-solutions/solution.html?compURI=1762950#.VmVJEnnruUn

Cloud Service Automation: http://www.hp.com/go/csa

HPE Propel: http://www.hp.com/go/propel

Helion OpenStack: http://www8.hp.com/us/en/cloud/hphelion-open-stack.html

Helion Application Transformation services: http://www8.hp.com/us/en/cloud/services-application-transformation.html

Empower developers to create business solutions

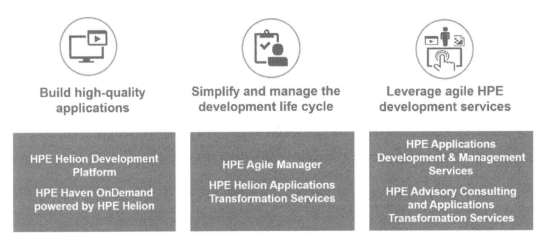

Figure 3-20 Empower developers

In the final phase of the journey to a hybrid infrastructure and on-demand IT, developers are empowered to create business solutions. Why is this important? This is because application developers can create or contribute to business value faster than ever before.

To make developers more productive, customers need an open platform and management software and services for application development and delivery. As shown in Figure 3-20, HPE offers a variety of solutions to support developers:

- **Build high-quality applications**—HPE Helion Development Platform and HPE Haven OnDemand powered by Helion

- **Simplify and manage the development lifecycle**—HPE Agile Manager and Helion Application Transformation Services

- **Assist with application development**—HPE Applications Development and Management Services and HPE Advisory Consulting and Application Transformation Services

 Note

To get more information about these products, solutions, and services, scan the QR code or enter the URL into your browser.

http://www.hp.com/go/cloudsystem

HPE infrastructure consulting and support

To maximize the value of HPE ConvergedSystem, two categories of services are available from HPE:

- **ConvergedSystem Consulting**—Services from consulting to implementation provide capabilities to plan, build, and operate converged infrastructures for virtualization, cloud, big data, and mobility needs.

- **Support services**—Services include standardized support (HPE Foundation Care), a hands-on approach (Proactive Care), and individualized and comprehensive support (HPE Datacenter Care).

  - **Foundation Care**—Provides support for HPE servers, storage, and networking products, as well as third-party software for the leading x86 operating systems.

  - **Proactive Care**—Combines smart technology support with a hands-on approach, providing consultation, recommendations, and reporting to prevent issues and quickly resolve problems.

  - **Datacenter Care**—Provides a single point of contact (with dedicated account team) and consulting expertise that delivers the right people, processes, and technology to meet specific business goals.

HPE Proactive Care support

A complete system includes not only hardware and software, but also services and support to ensure that the deployment is successful and can be effectively managed through its lifecycle. HPE Proactive Care is included with HPE Verified Reference Architectures. Table 3-3 shows the features and benefits of HPE Proactive Care support. This level of service adds specific proactive elements to further drive efficiency in the delivery of support services and improved response times. These services use HPE Insight Remote Support as a foundation to provide secure 24×7 monitoring, diagnostics, and notifications. Proactive Care includes both reactive and proactive components.

Table 3-3 HPE Proactive Care support features and benefits

	Feature	**Benefit**
Reactive support	Single point of contact to HPE experts for end-to-end case ownership and management	Reduces time to repair and identifies key roles Maximizes uptime and helps with service-level agreements
Proactive support	Semiannual proactive scan and recommendations Quarterly incident and trend reporting	Identifies KPIs Provides insight into potential issues Tracks progress against KPIs
Enabled through market-leading automation	24×7 "live" problem monitoring, diagnostics, and notifications	Helps identify problems before they occur. Automates issue reporting for rapid intervention

Reactive features include hardware and software support through HPE Advanced Solution Center Technical Solution Specialists to deliver a premium call experience:

- Single point of contact for end-to-end case ownership and call management

- Enhanced response times

Proactive features are delivered through HPE Advanced Solution Center Technical Account Managers. This provides support for ConvergedSystem updates and creates an opportunity for a live discussion including ConvergedSystem recommendations:

- Annual support plan

- Firmware release and software patching analysis and recommendations (two times per year)

- Proactive scan and recommendations (two times per year)

- Incident reporting that provides quarterly history and trend analysis for continuous improvements

Services for a managed private cloud

HPE offers managed services for customers who want to shift from a CAPEX model to an OPEX model without managing infrastructure. HPE managed services include virtual private clouds, private clouds, and applications:

- **HPE Helion Managed Private Cloud**—A complete technology platform for private cloud built on HPE CloudSystem. With its preintegrated hardware and software, you can create virtual pools of servers, storage, and networking resources on demand. HPE private cloud services simplify applications and business processes deployment by providing IT capacity and standardization. It has four predefined virtual server configurations for lighter workloads not requiring dedicated servers, and three predefined dedicated physical server configurations for business-critical production workloads. These configurations offer fast cloud services implementation and, as a managed service, benefit from automatic and seamless technology, and security updates and improvements.

- **HPE Helion Managed Virtual Private Cloud**—An agile and cost-effective way to scale infrastructure to meet business requirements in a managed environment leveraging complete, open, and integrated systems. HPE Helion Managed Virtual Private Cloud provides infrastructure services without the high cost of owning and managing data center equipment. This multitenant cloud platform is designed and built for the enterprise so business applications such as SAP, Microsoft Dynamics, and Oracle can be deployed in a secure cloud environment.

- **HPE Helion Managed Cloud Broker**—A managed service that allows businesses to provision, access, consolidate, and securely control services across multiple cloud workloads and providers. The HPE Helion Managed Cloud Broker service provides IT administrators with control and instant visibility over an organization's IT assets, from traditional IT to private and public clouds, and allows the orchestration of all of these assets to improve responsiveness, financial management, and user satisfaction. Deploy it on a traditional infrastructure, or on dedicated HPE Helion Managed Private or multitenant Virtual Private Cloud Services.

Note

These managed solutions are currently not available to HPE partners. The Helion Managed Private and Virtual Private Cloud Services will be available to partners in the future.

HPE Financial Services

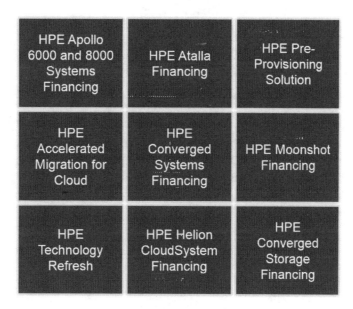

Figure 3-21 HPE Financial Services

The transition to hybrid infrastructure and on-demand IT does not happen overnight. Customers developing their investment strategy should ask:

- Should the company pay for ownership or usage?

- How can the company avoid being locked in to technologies that might become quickly outdated?

- How can the business maximize the value of its current IT equipment?

The HPE Financial Services shown in Figure 3-21 can provide a customized investment strategy for each customer's journey. These services focus on the following:

- Moving away from a costly ownership model, freeing up capital for innovation (controlling funding requirements such as shifting from CAPEX to OPEX).

- Implementing a refresh cycle that allows for scaling up or down, based on business needs.

- Retiring legacy systems, ensuring data security, and maximizing end-of-life value.

Note

For additional information about HPE Financial Services, scan this QR code or enter the URL into your browser.

http://www8.hp.com/us/en/hp-financial-services/index.html

HPE Partner Ready Certification and Learning

HPE Partner Ready Certification and Learning provides training and certification for the most sought-after IT disciplines, including convergence, cloud computing, software-defined networking, and security. The Partner Ready Certification and Learning program provides learning opportunities directly through HPE Education Services and indirectly through third-party partners. Training is offered in many ways—self-paced, instructor-led virtual, instructor-led on-premise, and remote lab classes. HPE Partner Ready Certification and Learning offers the following:

- Content covering a full range of skill levels from foundational to master

- Personalized learning plans and resources

- Certifications that command some of the highest pay premiums in the industry

- A focus on end-to-end integration, open standards, and emerging technologies

- Maximum credit for certifications you already hold

- A supportive global community of IT professionals

- A wide range of curricula from HPE

 Note

For more information about HPE Partner Ready Certification and Learning, scan this QR code or enter the URL into your browser.

http://certification-learning.hpe.com/tr/index.html

HPE Education Services

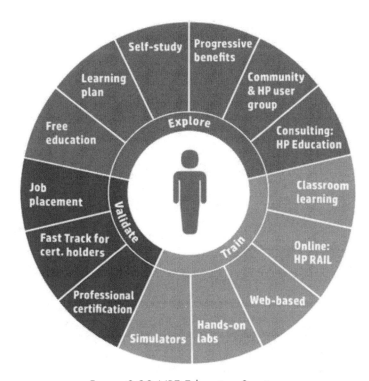

Figure 3-22 HPE Education Services

As shown in Figure 3-22, HPE Education Services provide customers and channel partners:

- **Certification training**—HPE Education Services provides in-depth technical and skills-based training plus preparation for HPE certification exams. Cloud and data center based training courses are offered in the following categories:

  - Cloud computing

  - Vendor-neutral OpenStack

  - Helion OpenStack

  - Helion CloudSystem

  - Cloud Service Automation

  - HPE Eucalyptus

- **Consulting**—HPE Education Consulting Services increase user awareness of new tools and processes. These services include:

  - HPE Workforce IT Transformation Solutions

  - HPE Workforce Analysis Service

  - HPE Workforce IT Transformation Services

Finding the right service

Figure 3-23 HPE Services homepage

Finding and matching the right service with customer needs can be a daunting task. The HPE Services homepage, shown in Figure 3-23, lists services by topic, such as Enterprise Services, Consulting Services, and Support Services. You can also browse service offerings for cloud, security, mobility, and big data analytics.

Note

To access the HPE Services homepage, scan this QR code or enter the URL into your browser.

https://www.hpe.com/us/en/services.html

Mapping Helion products to customer needs

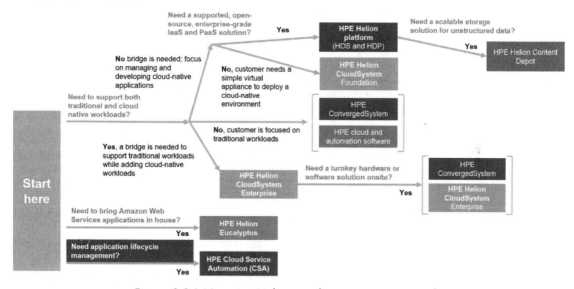

Figure 3-24 Mapping Helion products to customer needs

Figure 3-24 maps Helion products to customer needs. It is always critical to understand the customer requirements before you can recommend the correct product or solution. Begin at the left edge of the graphic to see how different customer situations align with relevant products and solutions.

Review: Helion portfolio

Cloud software

- HPE Helion CloudSystem
- HPE Cloud Service Automation
- HPE Helion OpenStack
- HPE Helion Development Platform
- HPE Helion Eucalyptus

Integrated solutions

- HPE Helion CloudSystem Solution
- HPE Helion Rack
- HPE Helion Content Depot

Managed services

- HPE Helion Managed Virtual Private Cloud
- HPE Helion Managed Private Cloud
- HPE Helion Managed Cloud Applications

Transformation to open, hybrid cloud

HPE Helion Professional Services provide traditional cloud and OpenStack technology consulting

Figure 3-25 Helion portfolio

As shown in Figure 3-25, HPE offers a broad portfolio of products, solutions, and services to help enterprises and service providers transform to an open, hybrid cloud:

- HPE offers integrated solutions that include all the hardware and software customers need to deploy a cloud solution. These solutions are based on HPE industry-leading servers, storage, and networking hardware, along with HPE cloud management software.

- HPE offers cloud software for managing hybrid clouds, developing and deploying cloud-native applications, and supporting private clouds that are compatible with Amazon Web Services.

- HPE offers managed services for customers who want to shift from a CAPEX model to an OPEX model and do not want to manage the infrastructure. Managed services from HPE include virtual private clouds, private clouds, and applications such as SAP, Microsoft 365, and Oracle.

- HPE Helion Professional Services offer cloud computing expertise for planning and deploying cloud projects.

Resources

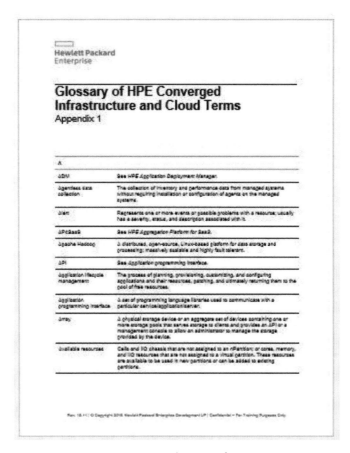

Figure 3-26 Glossary of terms

In the appendix, you can find a glossary of HPE Converged Infrastructure and cloud terms shown in Figure 3-26. Use this appendix to study the converged infrastructure and cloud terminology.

Learning check

1. What are cloud-native applications?

    _____

    _____

    _____

    _____

2. The private cloud provides convenience, and the public cloud gives control.

    ☐ True

    ☐ False

3. What is the main difference between Helion CloudSystem Foundation and Helion CloudSystem Enterprise?

    _____

    _____

    _____

    _____

4. Which categories of services does HPE offer to help customers maximize the value of HPE ConvergedSystem?

    _____

    _____

    _____

    _____

Learning check answers

1. What are cloud-native applications?

    - **Cloud-native applications are the following:**
        - **Unique to each customer because of the different ways applications will support the business**
        - **Written to use open-source application program interfaces, tools, and languages**

- **Built for hybrid application deployment**
- **Stateless, meaning they maintain their state internally, as opposed to relying on another layer, such as a database**

- **The intent of cloud-native applications is to allow users to change the underlying infrastructure without having to rewrite the application**

2. The private cloud provides convenience, and the public cloud gives control

    a. True

    b. False

3. What is the main difference between HPE Helion CloudSystem Foundation and HPE Helion CloudSystem Enterprise?

    - **The main difference with the Foundation package is the inclusion of the Helion Development Platform, which enables customers to develop cloud-native applications. In addition, CloudSystem Enterprise includes CSA software by default and CloudSystem Foundation does not**

4. Which categories of services does HPE offer to help customers maximize the value of HPE ConvergedSystem?

    - **The two categories are:**

        - **ConvergedSystem Consulting—Services from consulting to implementation provide capabilities to plan, build, and operate converged infrastructures for virtualization, cloud, big data, and mobility needs**

        - **Support services—Services include standardized support (HPE Foundation Care), a hands-on approach (HPE Proactive Care), and individualized and comprehensive support (HPE Datacenter Care)**

Summary

- The main principle behind the cloud is simple—deliver IT functionality as a service. This is a fundamental shift for IT, meaning the business approach is shifting from a technology focus to a service focus.

- Deployment models characterize the cloud implementation based on who manages the resources that deliver the service. Cloud deployment types include:

    - **Private cloud**—In a private cloud, services are provided to specific users such as members of departments within a company.

- **Public cloud**—In a public cloud, all cloud assets that underlie the services are owned and operated by a provider, and users access the services over the Internet.

- **Hybrid cloud**—Providers such as HPE, Amazon, and Google make a massive amount of resources available to users.

- The Helion portfolio of products, services, and solutions helps enterprises and service providers transform to an open, hybrid cloud. Helion offers cloud software, integrated solutions, and managed services.

- Helion Cloud Services offer support for customers who want to:

  - Become an internal service provider that can broker applications and services

  - Consume a private cloud that is built and managed by an outside vendor

  - Create a single catalog of all IT services and applications

  - Extend and modernize cloud-native applications

  - Build high-quality applications

  - Shift from a CAPEX model to an OPEX model without managing infrastructure

  - Develop a customized investment strategy

Hands-on exercises

You now have the opportunity to get some hands-on experience with some tools that would be used to size a customer solution. There are three exercises to complete:

1. Using the HPE Power Advisor Tool

2. Using the HPE Converged Infrastructure Solution Sizer Suite

3. Using the HPE Sizer for Server Virtualization

Exercise 1—Using the HPE Power Advisor Tool

Objectives

After completing this exercise, you should be able to:

- Install the HPE Power Advisor tool and select the input voltage

- Select a rack and an enclosure and then configure the enclosure

- Select and install server blades in the enclosure and then configure the server blades

- Create, review, and save a power report

Requirements

To complete this exercise, you need:

- Internet access

- A supported Internet browser

 **Important**

Microsoft Internet Explorer 11 is supported only with Microsoft Windows 10.

Introduction

Power Advisor is a tool provided by HPE (http://www.hp.com/go/hppoweradvisor) to assist in the estimation of power consumption and in the proper selection of components, such as power supplies at the system, rack, and multirack levels. Additional features of the tool include a condensed bill of materials (BOM), a cost of ownership calculator, and a power report.

 Note

Screenshots in this document are for your reference and might differ slightly from the ones you will see on your screen during these exercises.

Part 1—Installing the Power Advisor tool and selecting the input voltage

This exercise demonstrates how to install the Power Advisor tool and select the input voltage.

 **Important**

> The system requirements for Microsoft Silverlight (required for the Power Advisor) are
> at:http://www.microsoft.com/getsilverlight/get-started/install/default.aspx

1. On your PC, open an Internet browser and navigate to: **http://www.hp.com/go/HPPowerAdvisor**

2. Click **Online HP Power Advisor**. Note that you can also download Power Advisor and install it on your local machine.

 Note

> If you are using Google Chrome, see the special instructions at the bottom of the website.
> If a Microsoft Silverlight dialog box appears, click **Run** to install it.

3. On the HP Power Advisor License Agreement screen, click **I Agree**.

4. On the Profile Information dialog box (if one appears), enter your name, email address, and country, and click **OK** (Figure 3-27).

Figure 3-27 Profile Information

5. The main Power Advisor screen appears. Select the data center input power voltage (for this example, select **220VAC**) and click **Go** (Figure 3-28).

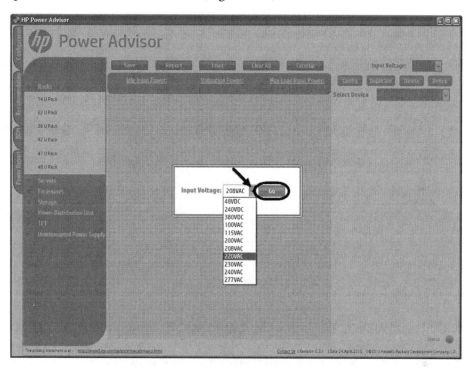

Figure 3-28 Power Advisor—Input Voltage

Part 2—Selecting a rack and an enclosure and then configuring the enclosure

This exercise demonstrates how to select a rack and an enclosure, place the enclosure in the rack, and then configure the enclosure. The enclosure configuration includes infrastructure components such as fans, power phase, and redundancy, OA modules, interconnects, and power supplies.

1. To select a rack, on the left-hand side of the screen click **Racks**. Select a rack (for this example, select **42 U Rack**), enter the information needed, and click **OK** (Figure 3-29).

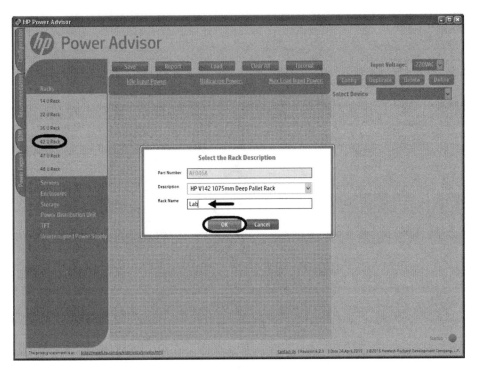

Figure 3-29 Power Advisor—Racks

 Note

Notice the options at the left-hand side of the screen, which enable you to select enclosures.

2. The rack appears in the working area, with its height represented in rack units (U) (Figure 3-30).

 Note

The U measurement for racks refers to "rack units," where 1U = 1.75 in or 44.45 mm.

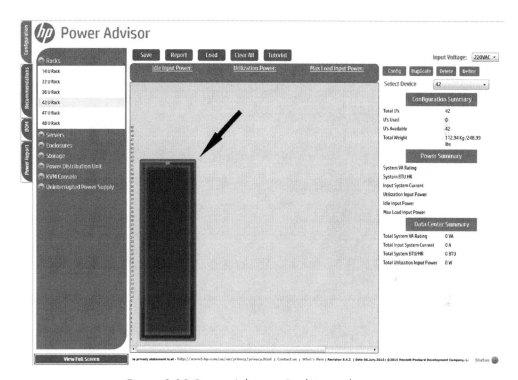

Figure 3-30 Power Advisor—Rack in working area

3. To select an enclosure, on the left-hand side of the screen, click **Enclosures**. Select the **BladeSystem c7000** enclosure to add to the rack (Figure 3-31).

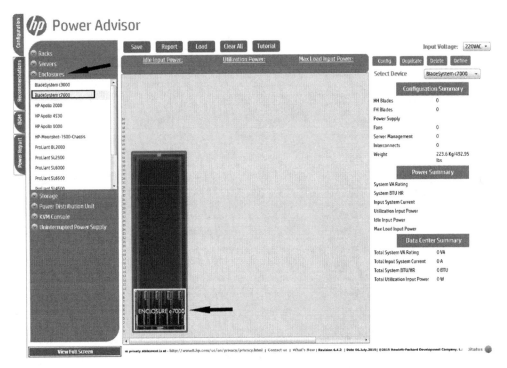

Figure 3-31 Power Advisor—Enclosure

4. To configure the enclosure, click the enclosure to highlight it (a colored border appears). On the top right-hand side of the screen, click the **Config** button (Figure 3-32).

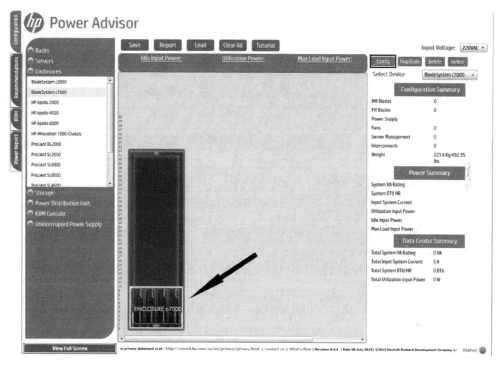

Figure 3-32 Configure the enclosure

5. The BladeSystem c7000 screen appears. Click **Config** (Figure 3-33).

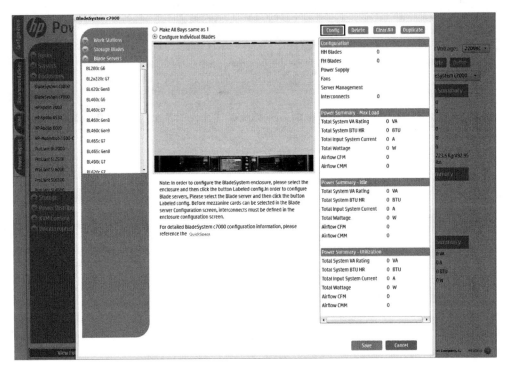

Figure 3-33 Configure BladeSystem c7000

6. Configure the following enclosure components:

    - General Configuration section

        - Select the number of cooling fans (notice the Maximum Fans limit): **10**

        - Select the power phase (**Three**) and power redundancy settings (**N+N**).

        - Select **Redundant Onboard Administrators**.

    - Interconnects section

        - Select the interconnects used in the enclosure as shown in Figure 3-34.

    - Power Supply section

        - Select the type of power supply (**HP 2650W Platinum Hot Plug Power Supply Kit**).

        - Click the **Add** button to add the required number and type of power supplies. For N+N power redundancy, the number must be even. Add **6** of these power supplies.

After the configuration is finished, click **Save** (Figure 3-34).

Figure 3-34 Save the configuration

Part 3—Selecting and installing the server blades in an enclosure and then configuring the server blades

This exercise demonstrates how to select and install the server blades in an enclosure and then how to configure the server blades with internal components such as processors, memory, internal storage, and mezzanine cards.

1. From the BladeServers panel on the left-hand side of the screen, click the server blade you want to use in the enclosure (**BL460c Gen9**).

 Note

> The Configuration section is empty. The server blade does not yet have processors, memory, internal storage, mezzanine cards, or other components.

2. By default, you can configure the enclosure with individual server blades. For this exercise, select **Make All Bays same as 1** to replicate the selected server blade in all bays (Figure 3-35).

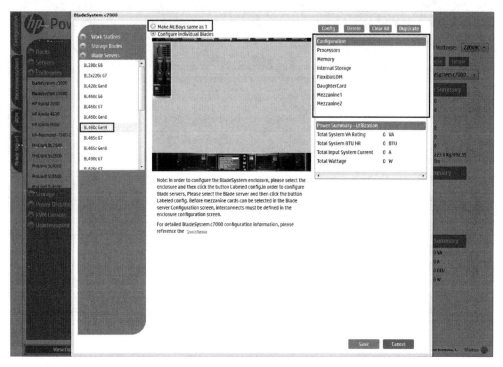

Figure 3-35 BladeSystem c7000—Replicate the selected server blade

3. The following example has all device bays populated with "empty" HPE ProLiant BL460c Gen9 server blades. Click **Save** (Figure 3-36).

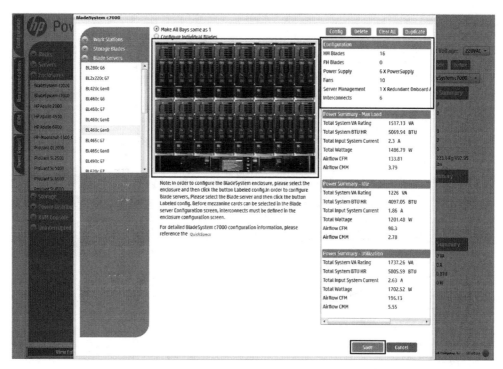

Figure 3-36 BladeSystem c7000 — Save

4. Configure the server blades with internal components. Click **Config** to return to the Configuration screen (Figure 3-37).

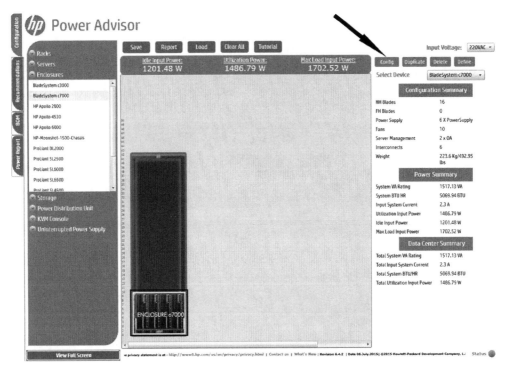

Figure 3-37 Power Advisor—Click Config

5. Highlight the first server blade (in bay 1). Under the Configuration heading, notice the empty values for Processors, Memory, Internal Storage, and so forth. Click **Config** (Figure 3-38).

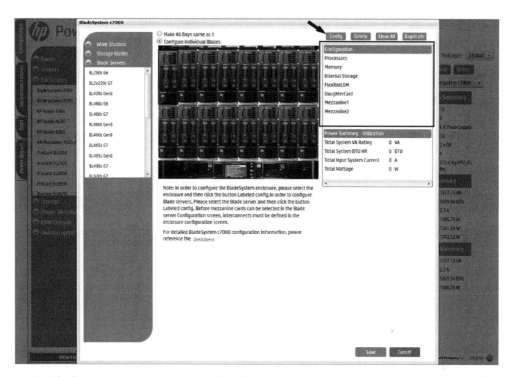

Figure 3-38 BladeSystem c7000—Empty values for Processors, Memory, Internal Storage, and so forth

6. On the left-hand side of the screen, select the appropriate processors, memory, storage, FlexibleLOM, daughter card, and mezzanine card settings and click **Add** after selecting each component. On the right-hand side of the screen, review the selected options and click **Save**. Notice that the allowed quantities depend on the type of server blade (Figure 3-39).

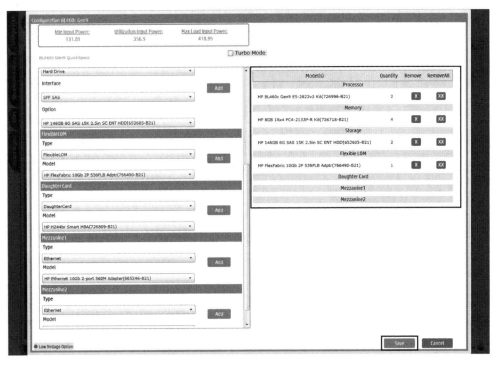

Figure 3-39 Configure BL460c Gen9

7. To populate the remaining server blades within this enclosure with the same internal components, highlight the first server blade and click **Make All Bays same as 1**. Click **Save** again (Figure 3-40).

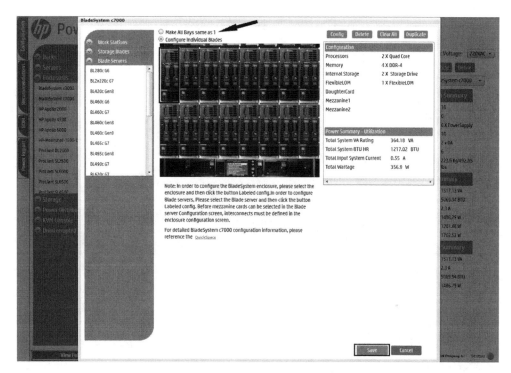

Figure 3-40 Populate the remaining server blades

8. At the top of the screen, click the **Save** button. Saved configurations can be loaded later if you do not exit the tool (Figure 3-41).

! Important

You can save the configuration to your desktop. However, after you exit, your configuration will be lost and you will have to start over.

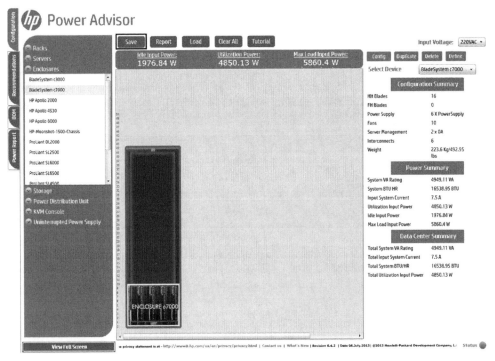

Figure 3-41 Power Advisor—Save the configuration

Part 4—Creating, reviewing, and saving a power report

This exercise demonstrates how to create, review, and save a power report; and generate and export a BOM.

1. On the left-hand side of the page, click the **Power Report** tab. In the appropriate fields, enter the approximate costs per kWh and Server Lifecycle in years. Select the **Total Cost of Ownership** check box to see the complete costs, and then click the **Generate Report** button (Figure 3-42).

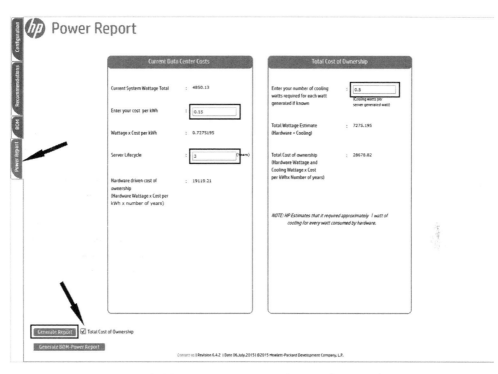

Figure 3-42 Power Report—Total Cost of Ownership

2. Export the generated report as an HTML and save the file to your desktop (Figure 3-43).

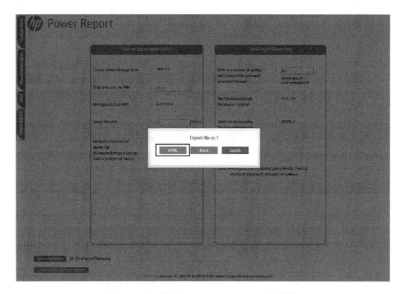

Figure 3-43 Export report as HTML

3. Generate a BOM by clicking the **BOM** tab and then click **Generate BOM** (Figure 3-44).

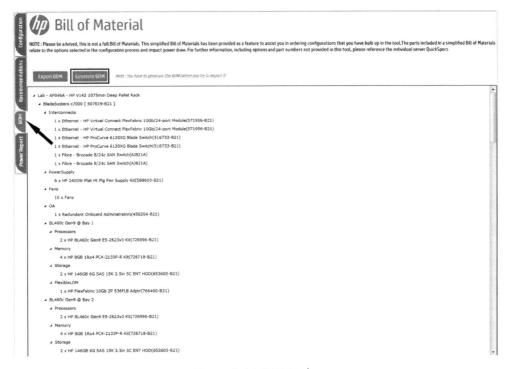

Figure 3-44 BOM tab

4. After the report is generated, click **Export BOM** and save the file in HTML format (for example) to your desktop (Figure 3-45).

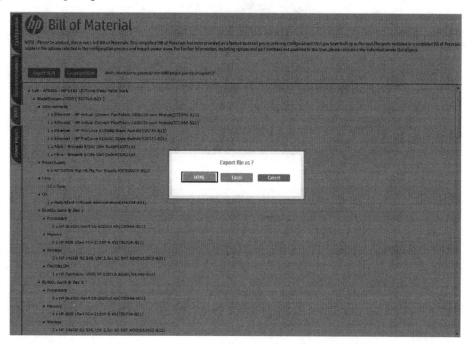

Figure 3-45 BOM tab—Export file as HTML

5. Close the Power Advisor by clicking the **X** in the upper right-hand corner. Examine the saved reports on the desktop by double-clicking them to open them in a web browser.

 Note

The reports might open automatically in some browsers (Figure 3-46).

Lab - AF046A - HP V142 1075mm Deep Pallet Rack - 1		
507019-B21 - BladeSystem c7000 - 1		
Model Name	**PartNumber**	**Quantity**
Fans		
Fans		10
OA		
Redundant Onboard Adminstrators	456204-B21	1
Interconnect		
Ethernet - HP Virtual Connect FlexFabric 10Gb/24-port Module	571956-B21	1
Ethernet - HP Virtual Connect FlexFabric 10Gb/24-port Module	571956-B21	1
Ethernet - HP ProCurve 6120XG Blade Switch	516733-B21	1
Ethernet - HP ProCurve 6120XG Blade Switch	516733-B21	1
Fibre - Brocade 8/24c SAN Switch	AJ821A	1
Fibre - Brocade 8/24c SAN Switch	AJ821A	1
PowerSupply		
HP 2400W Plat Ht Plg Pwr Supply Kit	588603-B21	6
727021-B21 - BL460c Gen9 - 1		
Model Name	**PartNumber**	**Quantity**
Processors		
HP BL460c Gen9 E5-2623v3 Kit	726996-B21	2
Memory		
HP 8GB 1Rx4 PC4-2133P-R Kit	726718-B21	4
Hard Drive		
HP 146GB 6G SAS 15K 2.5in SC ENT HDD	652605-B21	2
FlexibleLOM		
HP FlexFabric 10Gb 2P 536FLB Adptr	766490-B21	1
727021-B21 - BL460c Gen9 - 2		
Model Name	**PartNumber**	**Quantity**
Processors		
HP BL460c Gen9 E5-2623v3 Kit	726996-B21	2
Memory		
HP 8GB 1Rx4 PC4-2133P-R Kit	726718-B21	4
Hard Drive		
HP 146GB 6G SAS 15K 2.5in SC ENT HDD	652605-B21	2
FlexibleLOM		
HP FlexFabric 10Gb 2P 536FLB Adptr	766490-B21	1
727021-B21 - BL460c Gen9 - 3		
Model Name	**PartNumber**	**Quantity**
Processors		
HP BL460c Gen9 E5-2623v3 Kit	726996-B21	2
Memory		
HP 8GB 1Rx4 PC4-2133P-R Kit	726718-B21	4
Hard Drive		

Figure 3-46 Bill of materials

Current Data Center Costs

Utilization Input Power total	4850.13
Your cost per kW/hr	0.15
Wattage x cost per kW/h	0.7275195
Server Lifecycle	3
Hardware driven cost of ownership (Hardware Wattage x Cost per kW/h x number of years)	19119.21
# of cooling watts required for each watt generated	0.5

Total Cost of Ownership

Total Wattage Estimate (Hardware + Cooling)	7275.195
Total Cost of ownership (Hardware Wattage and Cooling Wattage x cost per kWh x number of years)	28678.82

Data Center Summary

Line Voltage	220 VAC
BTU HR	16538.95 BTU
System Current	7.5 A
Total Utilization Input Power	4850.13 W
VA Rating	4949.11 VA
Total Idle Input Power	0 W
Total Max Load Input Power	0 W

Lab - AF046A - HP V142 1075mm Deep Pallet Rack

Rack Level Summary

Line Voltage	220 VAC
VA Rating	4949.11 VA
BTU HR	16538.95 BTU
System Current	7.5 A
Utilization Input Power	4850.13 W
Idle Input Power	1976.84 W
Max Load Input Power	5860.4 W
System weight (Kg)	336.54 Kg
System weight (lbs)	741.94 lbs

Name	Location	Height	Weight(Kg)	Weight(lbs)	VA Rating (VA)	BTU HR (BTU)	System Current (A)	Utilization Input Power (W)	Idle Input Power (W)	Max Load Input Power (W)
BladeSystem c7000	1	10	223.6 Kg	492.95 lbs	4949.11	16538.95	7.5	4850.13	1976.84	5860.4

Figure 3-47 Power Report

Resources

For more information, refer to the following resources (Figure 3-47):

- **Websites**—HPE Power Advisor homepage: **http://www.hp.com/go/HPPowerAdvisor**

- **Documents**—Power Advisor installation and usage guide available at the HPE Power Advisor homepage

Exercise 2—Using the HPE Converged Infrastructure Solution Sizer Suite
Objectives

After completing this exercise, you should be able to access, download, install, and use the HPE solution sizers that are available through the HPE Converged Infrastructure Solution Sizer Suite (CISSS).

Requirements

To complete this exercise, you need:

- A supported Internet browser

- Internet access

Introduction

The CISSS contains all of the HPE solution sizers, conveniently packaged and available through a single-file download. The suite also consolidates the bill of materials (BOM) generated by multiple sizers.

 Note

> Screenshots in this document are for your reference and might differ slightly from the ones you see on your screen during these exercises.

Part 1—Accessing, downloading, and installing the sizer

This exercise demonstrates how to access, download, and install the CISSS.

1. On your PC, open an Internet browser and navigate to: **http://sizers.houston.hp.com/sb/installs/Converged_Infrastructure_Solution_Sizer_Suite.zip**

 Note

> If this URL does not work, try http://h71019.www7.hp.com/ActiveAnswers/secure/71110-0-0-225-121.html or use Google to search for ActiveAnswers.

2. Download the zip file with the sizer inside of it. Alternatively, you might need to click **HP Converged Infrastructure Solution Sizer Suite**.

3. Depending on how you downloaded and saved the zip file, it might open automatically. If the file does not open, open it manually, and then double-click the **Converged_Infrastructure_ Solution_Sizer_Suite.exe** file.

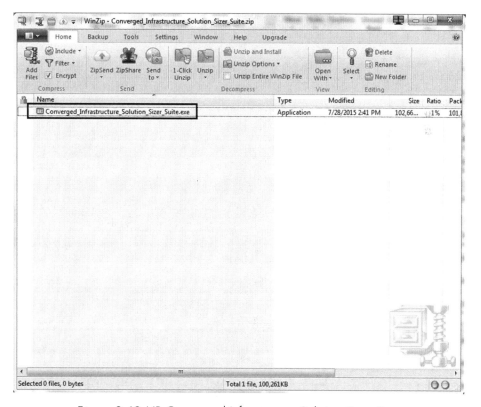

Figure 3-48 HP Converged Infrastructure Solution Sizer Suite

4. The Install Shield Wizard begins installation. Answer questions by selecting the default response to complete the installation (Figure 3-48).

5. The installation creates an icon on your desktop.

Part 2—Using the CISSS

This exercise demonstrates how to launch CISSS and use Sizer Manager. It also provides the available sizing tools for application solutions.

1. Launch the sizer by clicking the desktop icon or the **Start → All Programs** menu.

2. At the Profile Information screen, complete the requested fields and click **OK**.

Figure 3-49 Profile information

3. The sizer launches (Figure 3-49).

4. The CISSS Home page rotates through the welcoming screens and displays the list of supported application solution sizers.

Using the Sizer Manager

The Sizer Manager lists the available tools and allows them to be installed.

5. From the Go menu, click **Sizer Manager** (Figure 3-50).

Figure 3-50 The Go menu

6. Review the information on the screen. How many and which sizers are currently installed?

    _____

    _____

    _____

    _____

7. Select and install one of the uninstalled sizers (Figure 3-51).

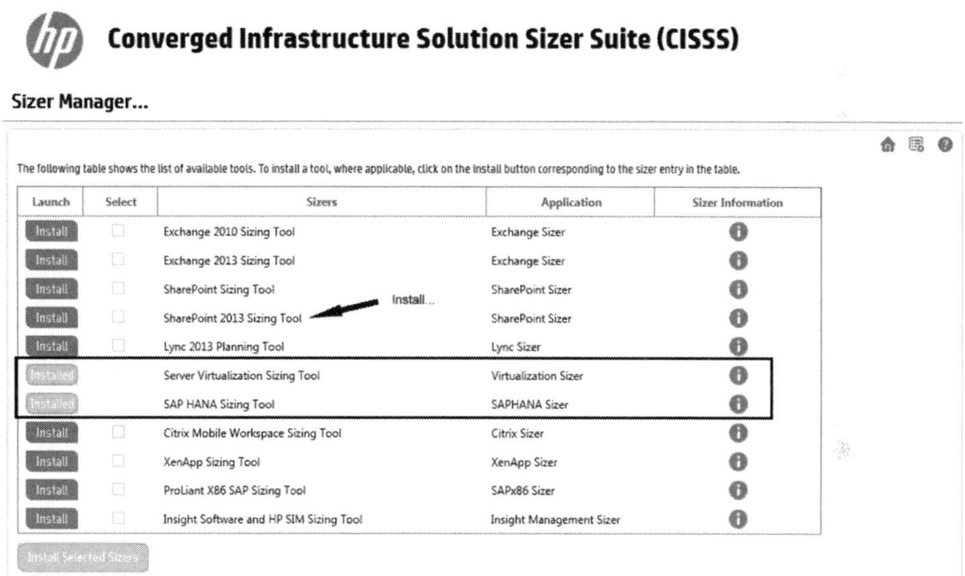

Figure 3-51 CISSS—Sizer Manager

8. After the installation completes, verify the installation using the Sizer Manager (Figure 3-52).

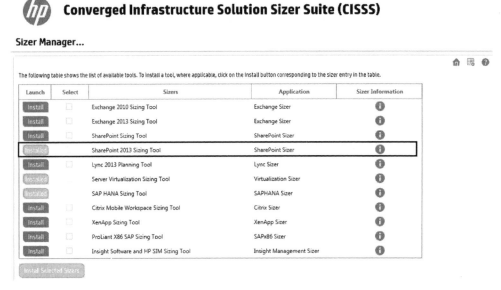

Figure 3-52 CISSS Sizer Manager — List of available tools

Sizing an individual application solution

From the list of available sizing tools, you can select to launch an application sizer. You can also place your favorite sizers in the Favorites box for easy access.

9. From the Go menu, click **Size an application solution** (Figure 3-53).

Figure 3-53 The Go menu—Size an application solution

10. You should have at least two application sizers available: Server Virtualization Sizing Tool and SAP HANA Sizing Tool. Select the **SAP HANA Sizing Tool** by clicking the corresponding **Launch** button.

 Note

You can add either or both sizers to the Favorite Sizers list by clicking the corresponding star in the Fav. Sizer column. Your action is followed by a dialog box confirming addition of the sizer in the Favorite Sizers list (Figure 3-54).

Figure 3-54 Size an application solution

11. If there are new updates to be installed, the suite prompts you (Figure 3-55). It is recommended that you install all updates. Click **Yes**. Also, confirm that you do want to save the session/workload.

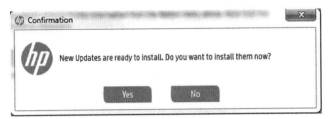

Figure 3-55 Confirmation for installing

12. At the SAP HANA Sizer Home screen, explore the options.

What happens if you click **Contact us**?

_____

What happens if you click **Additional Information**?

_____

_____

_____

What happens if you click **Reference Architecture**?

_____

_____

_____

How many and which Reference Architectures are listed?

_____

_____

_____

_____

What happens when you click **BOM for a Reference Architecture**?

_____

_____

_____

When you open the corresponding BOM, what actions can you take?

_____

_____

_____

13. At the SAP HANA Sizer Home screen, click **Build Solution**.

14. At the Build your own solutions—1 of 3 screen, enter the customer and requestor information and click **Next** (Figure 3-56).

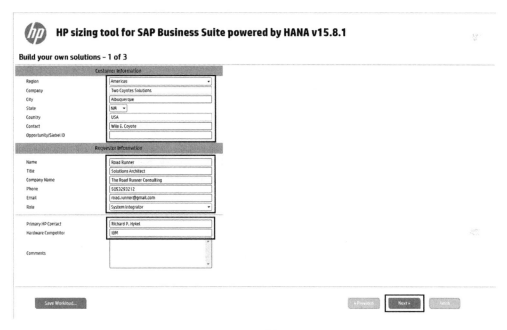

Figure 3-56 HP sizing tool for SAP Business Site

15. At the Build your own solutions—2 of 3 screen, enter arbitrary but somewhat realistic values for the uncompressed database size, number of years, and growth percentage. Click **Next** (Figure 3-57).

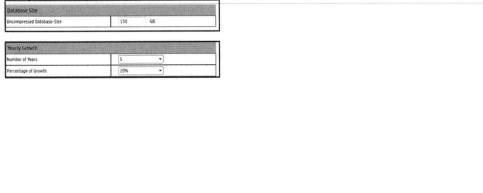

Figure 3-57 HP sizing tool—Build your own solutions

 Note

At any point, you can click **Save Workload** to save your work in an XML file.

---

16. At the Build your own solutions—3 of 3 screen, select one of the options presented: Test/Quality Assurance System, Development System, or SandBox System. Specify the percentage of the database size compared to production, and whether the system is dedicated or combined (shared). Click **Finish**.

 Note

In the following example, Test/QA and SandBox are combined/stacked on the same system, whereas Development is on a dedicated system (Figure 3-58).

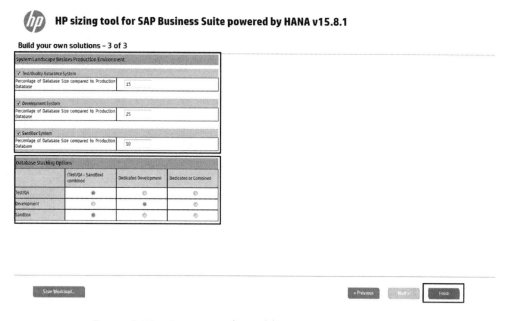

Figure 3-58 HP sizing tool—Build your own solutions 3 of 3

17. The Solution Alternatives screen appears. It shows the solution cost (standard worldwide list price), solution details, and a number of options such as Configuration Details, Display BOM, Customize, Additional Information, and Graphical Representation. Explore each option and save the results on the desktop or in the C:\Temp\CISSS folder (Figure 3-59).

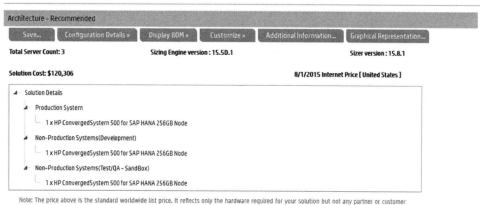

Figure 3-59 Solution Alternatives

What solution was recommended?

_____

How many solutions/systems were recommended?

_____

What was the cost of the entire solution?

_____

Which file type did the sizer use to save your results?

_____

18. Return to the CISSS Home page by clicking the home (🏠) icon.

Sizing an HPE ConvergedSystem solution

Using a list of available sizing tools that recommend HPE ConvergedSystem, you can start sizing an HPE ConvergedSystem.

19. At the CISSS Home page from the Go menu, click **Size an HP ConvergedSystem solution** (Figure 3-60).

Figure 3-60 CISSS Home page—Size an HP ConvergedSystem solution

20. From the Size an HP ConvergedSystem solution screen, launch the **Server Virtualization Sizing Tool** (Figure 3-61).

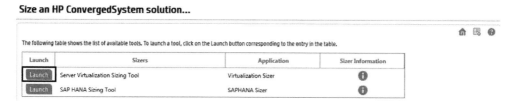

Figure 3-61 Size an HP ConvergedSystem

21. At the Virtualization Sizer Home screen, explore the available options. Select **Hyper-V** and click **Build ConvergedSystem Solution** (Figure 3-62).

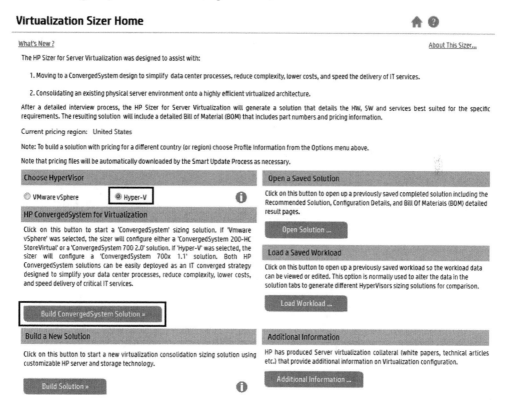

Figure 3-62 Virtualization Sizer Home

22. At the HP ConvergedSystem 700x v1.1 for Microsoft screen, enter reasonable input values and click **Next**. You can use the sample values from the following screenshot, if you wish. Click the icon to read an explanation of each option (Figure 3-63).

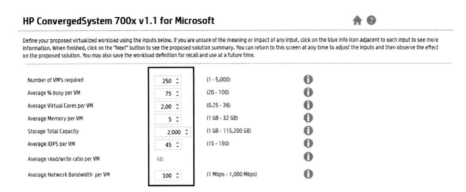

Figure 3-63 HP ConvergedSystem 700x v1.1

23. At the HP ConvergedSystem Configuration screen, click the **Production** tab to configure the production servers. Select the following information and click **Finish**.

- **HP ConvergedSystem 700x 3PAR StoreServ 7400 4N Storage 48TB Embedded—Single Rack for HyperV** for the ConvergedSystem model

- Number of servers per enclosure:**12**

- Memory per server: RAM Size **256** GB and DIMM Size **16** GB

- Local drives: **HP 900GB 6G SAS 10K 2.5in SC ENT HDD**

Note

Experiment with different options and notice how the results vary. Use the **Click here to reset all customizations** option (Figure 3-64).

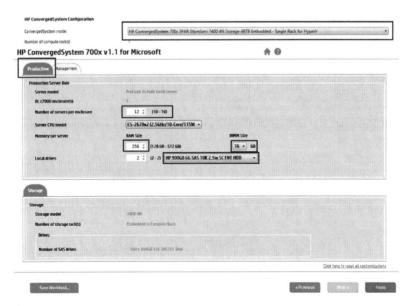

Figure 3-64 HP ConvergedSystem Configuration

24. Click the **Management** tab to configure the management servers and write down the suggested options (Figure 3-65).

Server model:_____

Number of servers:_____

Memory per server:_____

Number and size of local drives:_____

Total number of SAS drives required by this solution:_____

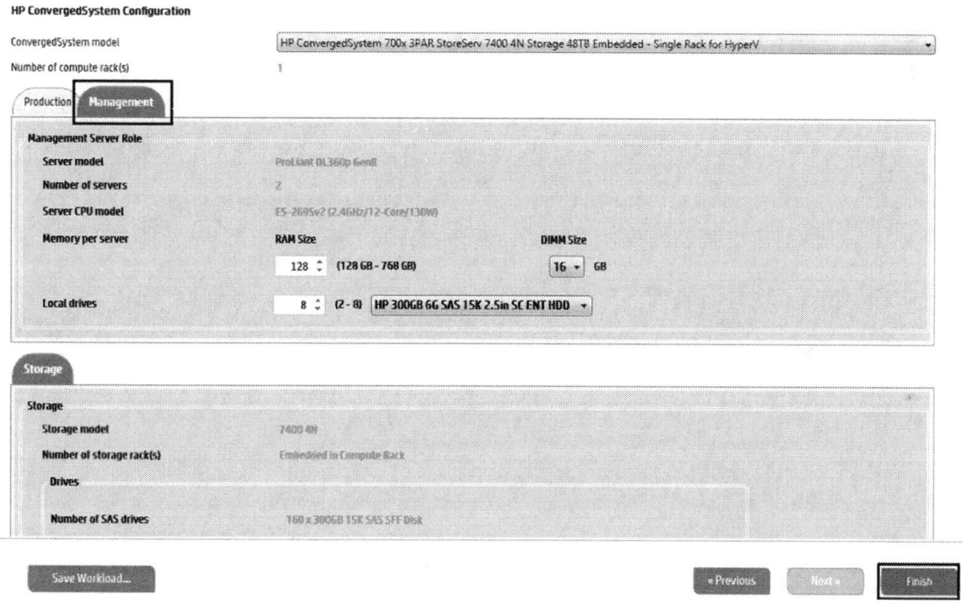

Figure 3-65 HP ConvergedSystem—Management tab

25. Accept the defaults and click **Finish**.

26. The recommended solution appears.

How many compute racks are recommended?_____

How many c7000 enclosures are recommended?_____

How many server blades are recommended?_____

27. Explore the available options, such as Configuration Details, Additional Information, and Graphical Representation.

28. Save the configuration on the desktop or in the C:\Temp\CISSS folder (Figure 3-66).

Figure 3-66 Save the configuration

29. Return to the CISSS Home screen.

Combining application solutions

After two or more solutions have been saved, the BOM for these solutions can be combined into a single list. The resulting combined BOMs can be saved as a Microsoft Excel or Word file.

30. At the CISSS Home page from the Go menu, click **Combine application solutions** (Figure 3-67).

Figure 3-67 CISSS Home page—Combine application solutions

31. The Combine application solutions screen appears.

    How many and which solutions/applications are displayed on your screen?

    _____

    _____

32. Select **SAP HANA Sizer** and **Virtualization Sizer** (or two other sizers, if your options differ) and click **Combine** (Figure 3-68).

Combine application solutions...

The following table shows the list of available tools to combine the solutions. Select the application(s) corresponding to the entry in the table, and click on the 'Combine' button.

Select	Solution	Application	Sizer Information
☐	SAP HANA	SAPHANA Sizer	ⓘ
☐	Session	Virtualization Sizer	ⓘ

Combine

Figure 3-68 Combine application solutions—**SAP HANA Sizer** and **Virtualization Sizer**

33. The Consolidated BOM screen appears. Explore the different options on this screen, and when finished, return to the CISSS Home screen (Figure 3-69).

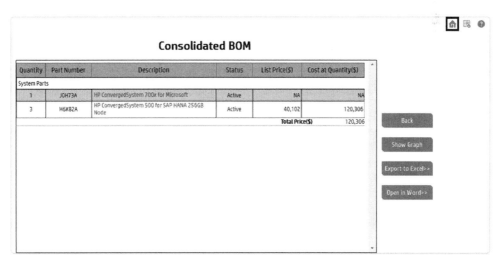

Consolidated BOM

Quantity	Part Number	Description	Status	List Price($)	Cost at Quantity($)
System Parts					
1	J0H73A	HP ConvergedSystem 700x for Microsoft	Active	NA	NA
3	H6X82A	HP ConvergedSystem 500 for SAP HANA 256GB Node	Active	40,102	120,306
			Total Price($)		120,306

Back

Show Graph

Export to Excel>>

Open in Word>>

Figure 3-69 Consolidated BOM screen

Sizing that starts with a Reference Architecture

HPE Reference Architectures that correspond with the installed sizing tools are displayed under this option. Reference Architectures may be downloaded as PDF files, and the BOM for the Reference Architecture may be opened as an Excel spreadsheet.

Note

Some Reference Architectures may be used as a starting point to size a custom solution. If available, next to the Reference Architecture title select **Customize this solution**. The appropriate sizer will be launched and the solution can be customized within the sizer.

34. At the CISSS Home page from the Go menu, click **Size starting with a Reference Architecture** (Figure 3-70).

Figure 3-70 Reference Architecture

35. A screen showing the available reference architectures appears (Figure 3-71).

How many and what types of reference architectures (and BOMs) are available to you?

_____

_____

_____

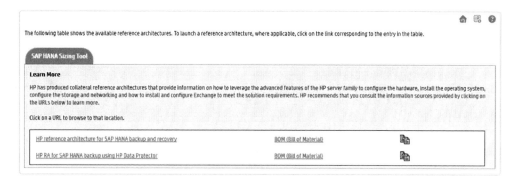

Figure 3-71 Available Reference architectures

36. Click **BOM (Bill of Materials)** for one of the available reference architectures to display the associated BOM (Figure 3-72).

SAP HANA Sizing Tool

Consolidated - Bill Of Material (BOM)

Quantity	Part Number	Description	Status	List Price($)	Cost at Quantity($)
2	QW938A	HP SN3000B 24/24 FC Switch	Active	14300.0	28600
2	JG505A	HP 5920 Series Switch	Active	1.0	2
1	BB857A	No Short Desc Available	Obsolete	125000.0	125000
1	BW904A	HP 642 1075mm Shock Intelligent Series Rack	Active	1899.0	1899
1	AK379A	HP StoreEver MSL2024	Active	4000.0	4000
1	709943-001	HP DL380p Gen8 E5-2690v2 Perf US Svr	Active	11179.0	11179
				Total Price $	170,680

This is an approximate bill of materials. Please contact your HP Representative for a complete configuration.

Open Spreadsheet>>   Send Mail »   Close

Note:
* Microsoft(R) Excel must be installed on your machine in order to view the spreadsheet.
* Once you open the spreadsheet, you will be able to save it to your hard disk.
* If Microsoft(R) Excel is not installed on your machine, you will be provided an option to save the spreadsheet for later use. Please rename the file extension to '.xls' if you want the file to be opened in Microsoft(R) Excel at a later point in time.

Figure 3-72 HP SAP HANA Sizing Tool

37. Click the name for one of the available reference architectures to display the associated technical white paper. Review the information in the document (Figure 3-73).

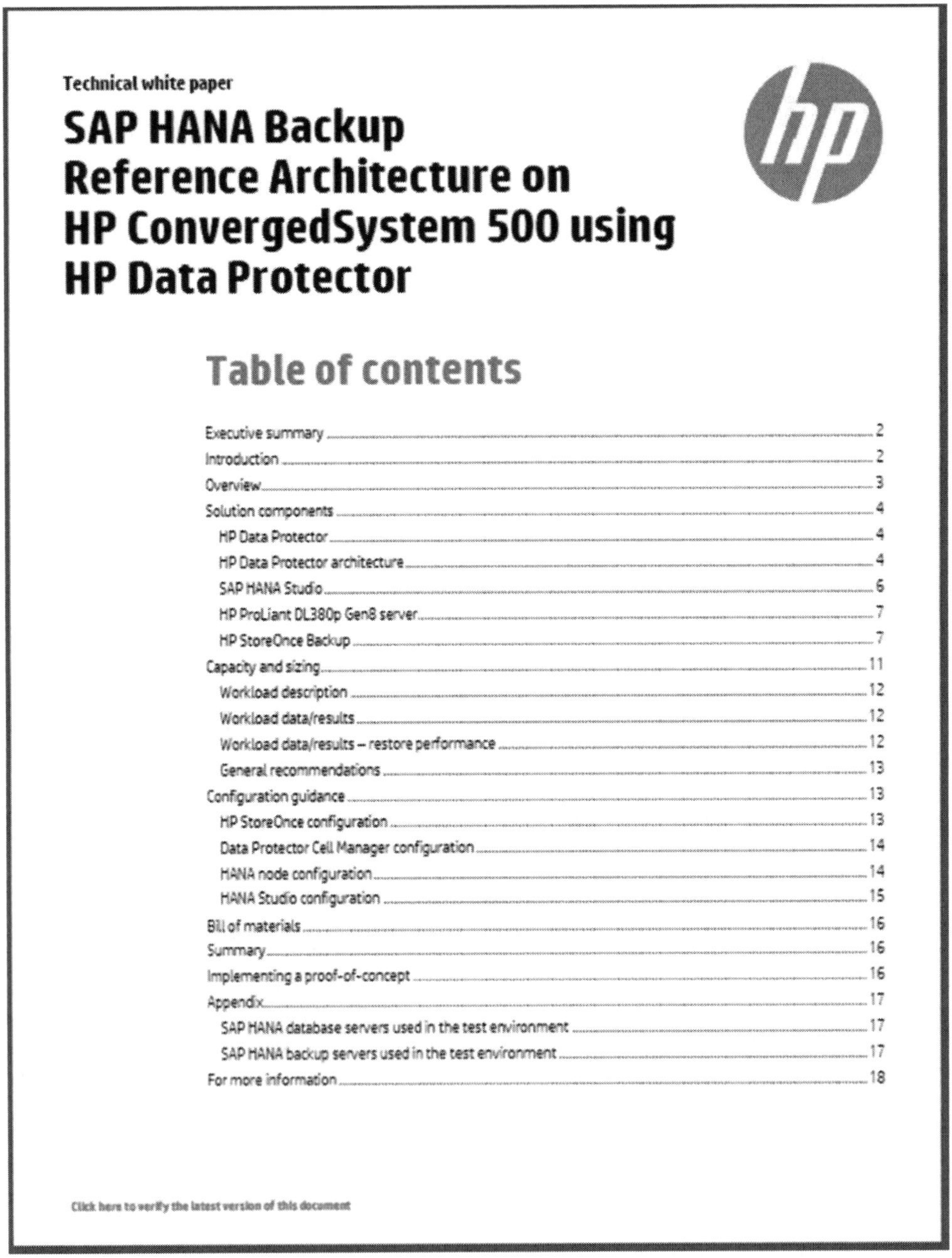

Technical white paper

SAP HANA Backup Reference Architecture on HP ConvergedSystem 500 using HP Data Protector

Table of contents

Click here to verify the latest version of this document

Figure 3-73 HP SAP HANA—Backup

38. Return to the CISSS Home screen.

Sizing Power solution

HPE Power Advisor can be launched using this option. The HPE Power Advisor provides power information for HPE server and storage solutions. After launching HPE Power Advisor, users configure a solution using a drag-and-drop interface. This tool runs independent of CISSS after it is launched.

Loading a solution

Solutions can be saved by using the sizers launched from the Size an application solution and Size an HPE ConvergedSystem solution options. These features allow the details from a saved solution to be loaded.

39. At the CISSS Home page from the Go menu, click **Load Solution** (Figure 3-74).

Figure 3-74 CISSS Home page—Load Solution

40. A Load a Sizer Solution from File screen appears. If necessary, navigate to the location where you previously saved your solutions, select one of the saved solutions, and click **Open** (Figure 3-75).

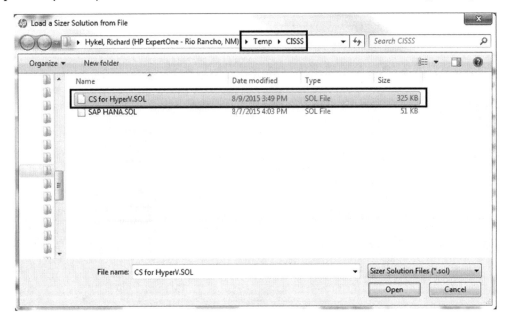

Figure 3-75 Load a Sizer Solution from File screen

41. The solution appears.

42. Return to the CISSS Home screen.

Using Favorite Sizers and Recently Saved Solutions options

For an added convenience, sizers can be added to the Favorite Sizers list by using the Size an application solution option. Solutions can be saved when using the Size an application and Size an HP ConvergedSystem features, and these saved solutions appear in the Recently Saved Solutions list, from which they can be quickly selected.

• Which favorite sizers are displayed on your home screen?

_____

_____

_____

_____

_____

- Which recently saved solutions are displayed on your home screen?

_____

_____

_____

_____

_____

Exercise 3 — Using the HPE Sizer for Server Virtualization

Objectives

After completing this exercise, you should be able to access, download, install, and use the HPE Sizer for Server Virtualization tool.

Requirements

To complete this exercise, you need:

- A supported Internet browser

- Internet access

Introduction

HPE Sizer for Server Virtualization is an automated, downloadable tool that provides quick and helpful sizing guidance for "best-fit" HPE server and storage configurations running in VMware vSphere 5.0 or Microsoft Hyper-V R2 environments. The tool is intended to assist with the planning of virtual server deployment projects on HPE server and storage technologies. It enables you to quickly compare different solution configurations and produces a customizable server and storage solution complete with a detailed bill of materials (BOM) that includes part numbers and prices.

The Sizer for Server Virtualization tool allows users to create new solutions, open existing solutions, or use other types of performance data collecting tools, such as the Microsoft Assessment and Planning tool (MAP), to build rich virtualized configurations based on HPE server and storage technologies. The tool allows rapid comparison of virtualized solutions using various HPE server and storage choices and offers a choice of vSphere or Hyper-V virtualization technologies in a common tool.

 Note

Screenshots in this document are for your reference and might differ slightly from the ones you see on your screen during these exercises.

Part 1—Accessing, downloading, and installing the sizer

This exercise demonstrates how to access, download, and install the Sizer for Server Virtualization tool (Figure 3-76).

 Note

If you completed the Using the HPE Converged Infrastructure Solution Sizer Suite exercise, you have already installed the HPE Sizer for Server Virtualization tool. From the Go menu, click **Size an application solution**. Select and launch the **Server Virtualization Sizing Tool**. If this is the case, omit this exercise and proceed with the next exercise.

Figure 3-76 Server Virtualization Sizing Tool

1. If you did not already install CISSS in a previous exercise. on your PC, open an Internet browser and navigate to: **http://sizers.houston.hp.com/sb/installs/Converged_Infrastructure_Solution_Sizer_Suite.zip**

 Note

If this URL does not work, use Google to search for **ActiveAnswers** or try **http://h71019.www7.hp.com/ActiveAnswers/secure/71110-0-0-225-121.html**. Alternatively, you can navigate to **http://sizers.houston.hp.com/sb/installs/HP_Unified_Sizer_for_Server_Virtualization.zip** and save and open the zip file and install the Sizer for Server Virtualization.

---

2. Download the zip file with the sizer inside of it. Alternatively, you might need to click **HP Converged Infrastructure Solution Sizer Suite**.

3. If the zip file does not open automatically, open it manually and double-click the **Converged_Infrastructure_Solution_Sizer_Suite.exe** file.

4. The Install Shield Wizard begins installation. Answer questions by selecting the default response to complete the installation.

Part 2—Using the Sizer for Server Virtualization tool

1. To use the Sizer for Server Virtualization tool, launch the sizer using the desktop icon or the **Start → All Programs** menu.

2. At the Profile Information screen, complete the requested fields and click **OK** (Figure 3-77).

Figure 3-77 Profile Information screen

3. The sizer launches. On the Virtualization Sizer Home screen, review the displayed information. The Open Solution and Load Workload options enable you to retrieve a saved solution or workload, respectively. The Build ConvergedSystem Solution option uses either the ConvergedSystem 200-HC StoreVirtual or ConvergedSystem 700 2.0 for VMware sizing or the ConvergedSystem 700x 1.1 for Hyper-V sizing. Click **Build Solution** to build a new virtualization consolidation sizing solution that does not use HPE Converged Systems (Figure 3-78).

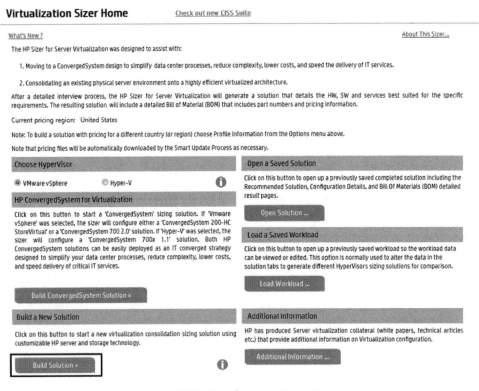

Figure 3-78 Virtualization Sizer Home

4. On the Servers to Consolidate screen, review the Servers to Consolidate Interview section. Click **Link to example spreadsheet here** to see sample consolidation data (Figure 3-79).

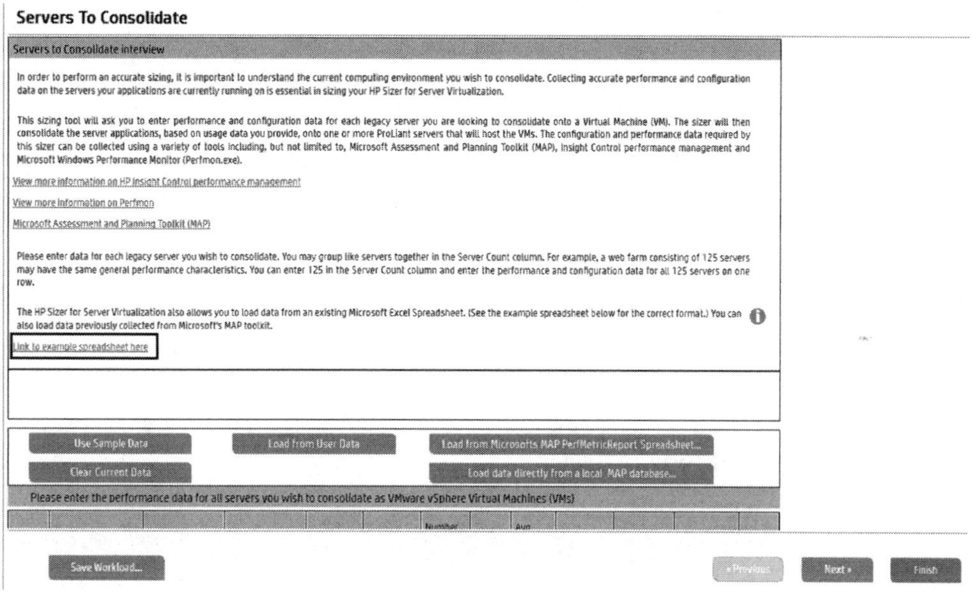

Figure 3-79 Servers to Consolidate

What data points are included in this spreadsheet for each legacy server?

_____

_____

_____

_____

_____

_____

_____

_____

_____

_____

5. Close the spreadsheet.

6. Click **Use Sample data** (Figure 3-80).

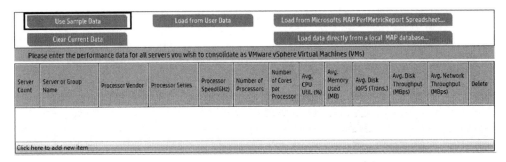

Figure 3-80 Use Sample data

7. The sizer imported the sample data you just viewed. You can delete any of the entries, add new entries, and edit any of the values (Figure 3-81).

Servers To Consolidate

Server Count	Server or Group Name	Processor Vendor	Processor Series	Processor Speed(GHz)	Number of Processors	Number of Cores per Processor	Avg. CPU Util. (%)	Avg. Memory Used (MB)	Avg. Disk IOPS (Trans.)	Avg. Disk Throughput (MBps)	Avg. Network Throughput (MBps)	Delete
1	Legacy-server-1	AuthenticAMD	Opteron 800	1.8	2	1	87.8	512	900	25	80	X
1	Legacy-server-2	GenuineIntel	Xeon	1.9	2	2	56.7	256	750	33	120	X
1	Legacy-server-3	GenuineIntel	Xeon 3400	1.2	2	1	90.2	128	150	12	55	X
1	Legacy-server-4	AuthenticAMD	Opteron 800	1.6	4	2	97.3	1024	1400	88	110	X
1	Legacy-server-5	AuthenticAMD	Opteron 1200	2.1	2	2	59.8	512	527	29	64	X
1	Legacy-server-6	GenuineIntel	Xeon 6500	2.2	4	2	22.3	1024	225	67	102	X
1	Legacy-server-7	GenuineIntel	Pentium 4	2	2	1	98.9	31	90	15	22	X
1	Legacy-server-8	AuthenticAMD	Opteron 1200	1.4	2	1	96.3	128	212	19	41	X
1	Legacy-server-9	GenuineIntel	Xeon 7000	2.1	4	2	69	512	546	45	91	X
1	Legacy-server-10	AuthenticAMD	Opteron 100	1.1	2	1	44.5	32	122	9	12	X
1	Legacy-server-11	GenuineIntel	Xeon 8400	2.2	2	4	64.1	256	789	37	112	X
1	Legacy-server-12	GenuineIntel	Generic Intel	2.1	4	4	25.3	512	654	28	88	X
1	Legacy-server-13	AuthenticAMD	Opteron 8200	1.9	4	2	22	512	768	23	65	X
1	Legacy-server-14	GenuineIntel	Pentium P5	1	1	1	99	32	85	11	16	X
1	Legacy-server-15	AuthenticAMD	Generic AMD	1	1	1	90	32	44	11	14	X
1	Legacy-server-16	AuthenticAMD	Opteron 1300	1	1	1	95	32	44	11	14	X
1	Legacy-server-17	GenuineIntel	Xeon 7200	2.1	4	2	46.7	1024	555	33	87	X
1	Legacy-server-18	GenuineIntel	Xeon 8400	2.1	4	4	36	1024	602	44	77	X
1	Legacy-server-19	AuthenticAMD	Opteron 8400	2	4	4	11.2	1024	52	44	19	X
1	Legacy-server-20	AuthenticAMD	Opteron 8200	1.9	4	2	19.9	1024	45	33	21	X

Click here to add new item

Figure 3-81 Servers to consolidate

8. Make arbitrary changes and click **Save Workload**. Save the .WL file on the desktop or in the C:\ Temp folder. Click **Next**.

9. On the Storage Performance and Capacity Options screen, review the values the sizer calculated based on your input. Change some values and click **Next**.

- Read/Write Ratio: **75%**

- Do you want to set up LUNs for a "Boot from SAN" configuration for your host servers?: **checked**

- Space reserved for the Boot from SAN LUN: **25** GB (Figure 3-82)

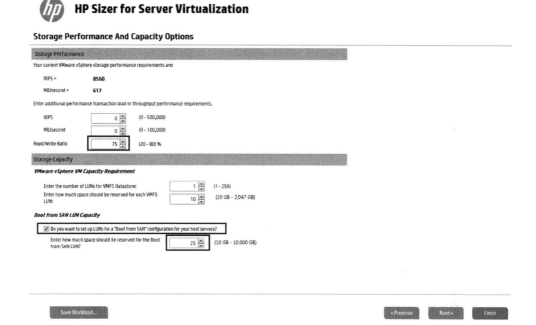

Figure 3-82 HP Sizer for Server Virtualization—Storage Performance and Capacity Options

10. On the Target Server Resource Utilization screen, review the proposed target server utilization levels. Arbitrarily adjust a few options and click **Next**.

What happens if you exceed the HPE recommended values (Figure 3-83)?

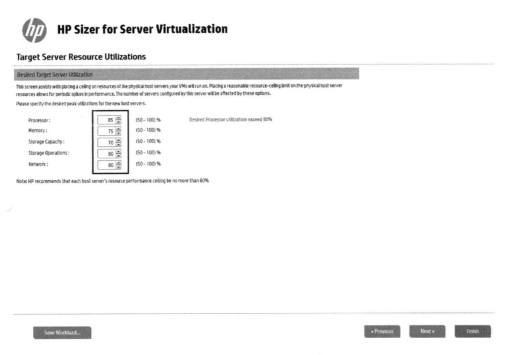

Figure 3-83 Target Server Resource Utilization screen

11. On the Role Server Options screen, review the presented options. Select the various drop-down menus and review the alternative options available. Leave all values as **Sizer recommendation** and click **Next** (Figure 3-84).

Figure 3-84 Role Server Options screen

12. On the Role Storage Options screen, ensure that the **Sizer Recommendation** option is selected for the storage architecture.

What other storage architecture options are available?

_____

_____

_____

13. Accept all of the other defaults and click **Finish** (Figure 3-85).

Figure 3-85 Role Server Options—Storage Architecture

14. The sizer calculates and displays the recommended solution.

    How many servers are recommended?_____

    What is the recommended RAM?_____

    What storage architecture is the sizer recommending?_____

    _____

    How many disk drives are included in the storage recommendation?_____

15. On the Solution Alternatives screen, experiment with the different options available such as the
    Configuration Details, **Display BOM**, **Customize**, **Additional Information**, and **Graphical
    Representation**.

    Does the sizer recommend a rack as part of the solution?_____

    If so, what kind of rack is recommended?_____

16. To illustrate how you can adjust different parameters, in the left-hand navigation pane, select the
    Role Storage Options menu option to change the storage architecture to either iSCSI-based or
    Fibre Channel-based SAN. Observe how the sizer displays the available options. Experiment
    with different options and values and click **Finish** to generate the recommendations (Figure 3-86).

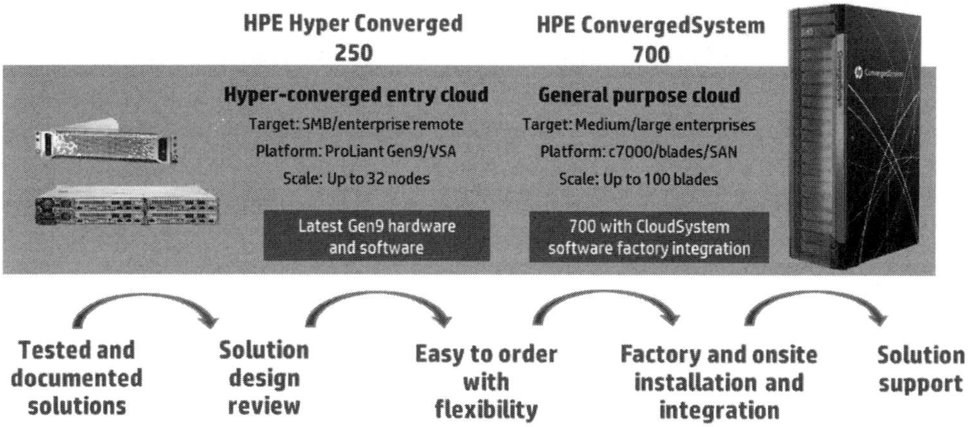

Figure 3-86 Solution Alternatives—Role Storage Options

17. Finally, save your solution on your desktop or another convenient location.

4 CloudSystem and OpenStack Fundamentals

WHAT'S IN THIS CHAPTER FOR YOU?

After completing this chapter, you should be able to:

✓ Discuss why customers look for open-source and cloud solutions and explain how HPE CloudSystem and OpenStack software can provide benefits

✓ Provide an overview of OpenStack and its core technologies

✓ Describe the cloud storage types of OpenStack

✓ Discuss how Hewlett Packard Enterprise (HPE) contributes to OpenStack projects

HPE Helion CloudSystem

Before proceeding with this section, assess your existing knowledge by attempting to answer the following questions.

Assessment questions

1. How many versions of CloudSystem can you name?

_____

_____

2. Can you name any of the target use cases for CloudSystem?

_____

_____

Assessment question answers

1. How many versions of CloudSystem can you name?

    a. **HPE Helion CloudSystem Foundation**

    b. **HPE Helion CloudSystem Enterprise**

2. Can you name any of the target use cases for CloudSystem?

    a. **Rapid infrastructure provisioning**

    b. **Deploy traditional applications to cloud**

    c. **Broker and manage multiple clouds**

    d. **Modernize software development lifecycle**

HPE Helion CloudSystem

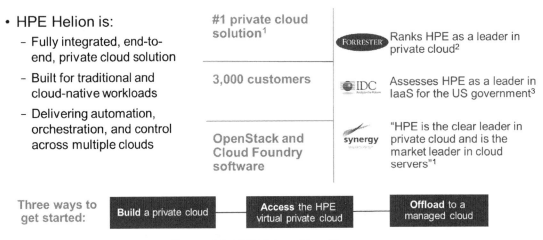

Figure 4-1 HPE Helion CloudSystem

Helion CloudSystem is the most complete, integrated, and open cloud solution on the market. The end-to-end service lifecycle management solution is powered by OpenStack technology and developed with an emphasis on automation and ease of use. Created for enterprises and service providers, Helion CloudSystem gives you the ability to effortlessly provision applications and their corresponding infrastructure resources.

Whether customers are transitioning existing environments to cloud or just starting their journey, Helion CloudSystem offers a rapid, reliable, and cost-effective path with built-in extensibility for future growth and development as shown in Figure 4-1.

 Notes

1. Synergy Research Group, March 2015

2. The Forrester Wave: Private Cloud Solutions in China, Q1 2015

3. IDC MarketScape: US Government Private Cloud IaaS 2014 Vendor Assessment

Helion CloudSystem management platform

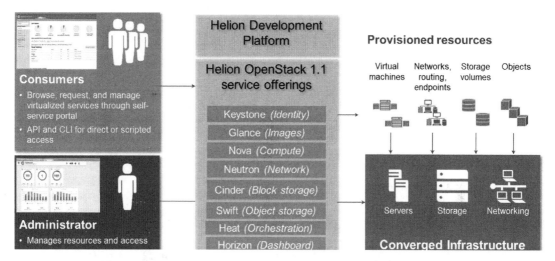

Figure 4-2 Heterogeneous management platform

As shown in Figure 4-2, Helion CloudSystem provides a common management platform across multiple environments that is very easy to use. Users can design, provision, and manage services from multiple providers. These capabilities enable customers to achieve high productivity and business agility.

Cloud Service Automation (CSA) is integral to these capabilities. CSA is a component of Helion CloudSystem Enterprise that provides heterogeneous, extensible, and enterprise-grade cloud service lifecycle management to design and orchestrate sophisticated cloud services. CSA embraces existing automation assets with an enhanced orchestration engine, which provides IT with an informed, transparent delivery of secure, compliant services for the hybrid cloud.

Support for multiple hypervisor, storage, compute, networking, and public cloud vendors means users are in control. If customers choose to use hybrid service delivery, they can run on-premises services in the private cloud and simultaneously manage off-premises services or bursting to external public cloud providers.

HPE Helion CloudSystem portfolio

Helion CloudSystem Enterprise
Advanced infrastructure and application services

- Hybrid cloud management with Service Marketplace and designed based on Cloud Service Automation and Operations Orchestration
- Enterprise-class lifecycle management for application services including compliance[3]

HPE Cloud Service Automation and HPE Operations Orchestration— delivered as virtual appliances

| Helion CloudSystem Foundation | Matrix Operating Environment |

Helion CloudSystem Foundation
Core infrastructure services and PaaS

- Push-button activation of compute and storage nodes
- Out-of-the-box networking services
- Built on Helion OpenStack[1]
- Delivered via virtual appliances
- Helion Development Platform

HPE infrastructure management—HPE OneView[2]

Built on ConvergedSystem or on HPE and third-party infrastructure

Figure 4-3 CloudSystem portfolio

Two CloudSystem packages are available, as shown in Figure 4-3:

- Helion CloudSystem Foundation for core infrastructure services

- Helion CloudSystem Enterprise for advanced infrastructure as well as application and platform services

If a customer is just starting their journey to a hybrid infrastructure and on-demand IT, they might need core infrastructure services and Platform-as-a-Service (PaaS). In this case, Helion CloudSystem Foundation is the right solution.

For a customer who needs a more comprehensive cloud solution, Helion CloudSystem Enterprise is the optimal choice.

 Notes

1. Includes Nova, Cinder, Neutron, Keystone, Glance, Horizon, Swift, and Heat

2. Optional

3. Optional advanced service automation through HPE Helion CloudSystem Enterprise PAA (Platform, Applications and Analytics) add-on software, including automated patch and compliance management with HPE Server Automation (SA) and sophisticated database and middleware management with HPE Database and Middleware Automation software.

Helion CloudSystem Foundation

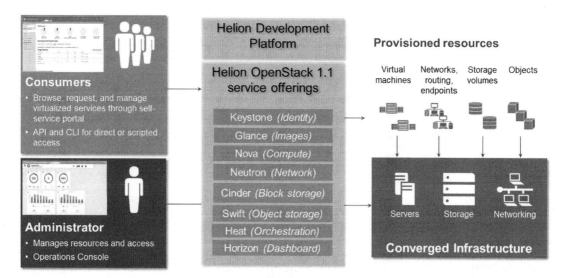

Figure 4-4 Helion CloudSystem Foundation

Helion CloudSystem Foundation is based on the Helion OpenStack distribution of the OpenStack cloud software. It integrates hardware and software to deliver core IaaS provisioning and lifecycle management of compute, network, and storage resources as shown in Figure 4-4.

Helion CloudSystem Foundation is the ideal entry point for businesses interested in straightforward, rapid deployment of cloud Infrastructure-as-a-Service (IaaS) solutions. The preintegrated software appliance delivers core cloud infrastructure services in an enterprise-grade environment for rapid return on investment (ROI).

Helion CloudSystem Foundation delivers these benefits:

- Easy step toward private cloud

  - Simple and consistent user interfaces for increased productivity

  - Quick delivery of simple infrastructure services within minutes

  - Fast and easy installation via software appliance delivery model

- Openness

  - Built on enterprise-grade OpenStack technology, enabling easy integration and customization

  - Open application program interfaces (APIs) provided for both administrative and cloud service functions, enabling highly automated cloud delivery

- Proven technologies

  - HPE Helion, proven in thousands of deployments worldwide, used to accelerate deployment time, ensure reliability, and extend easily

  - Ability to seamlessly upgrade to Helion CloudSystem Enterprise

The CloudSystem environment can be managed from its Operations Console and CLIs. You can develop, deploy, and scale cloud applications using the OpenStack user portal and the OpenStack APIs and CLIs. You can also deploy the Helion Development Platform on top of Helion CloudSystem Foundation to use its PaaS features in cloud applications.

Helion CloudSystem Enterprise

The Helion CloudSystem Enterprise package builds on the capabilities of HPE Helion CloudSystem Foundation. CloudSystem Enterprise introduces an Enterprise appliance that includes HPE Cloud Service Automation and Operations Orchestration software.

Helion CloudSystem Enterprise is a hybrid cloud platform designed to deliver advanced infrastructure services, in addition to platform and application services. This end-to-end, enterprise-level solution delivers advanced infrastructure services and optional platform and application services for hybrid cloud environments. It is available as a software or as an integrated combination of storage, servers, networking, and software. Customers can build and consume private and hybrid cloud services best suited to their business and technology objectives. Helion CloudSystem Enterprise was built with integration in mind so that customers can incorporate cloud services with existing IT assets.

Helion CloudSystem Enterprise delivers these benefits:

- Superior flexibility and variety

  - Is built on enterprise-grade OpenStack technology

  - Blends management of traditional and cloud-native workloads, all from one cloud management platform

  - Leverages multiple resource pools to deliver cloud services in the most optimized fashion

  - Allows customers to build the cloud environment that best serves the business with bursting capabilities to the preferred public cloud provider

  - Expedites service provisioning and deployment with predefined services and maintains ownership and control over resources and security or compliance obligations

- Easy step toward private and hybrid cloud

  - Accelerates development, testing, and deployment of cloud services with an optimized, ready-to-run infrastructure solution based on the HPE ConvergedSystem model

- Minimizes delays and limitations with a built-in, defined approach that unifies and stream-lines previously disparate processes

- Allows customers to choose from an extensive selection of cloud entry points and services for ideal positioning and deployment options

- Provides capability to provision services with pre-engineered solutions and a holistic design

- Increases productivity with intuitive, consumer-inspired user interfaces

- Boosts performance with proactive, expert support services

- Advanced, enterprise-level capabilities

  - Rapid provisioning of advanced infrastructure and application services

  - Enterprise-class lifecycle management for hybrid cloud services

  - Sophisticated marketplace portal to easily browse, request, and manage cloud services

  - Drag-and-drop designer tools for rapid definition of multitier cloud services

  - Broad support for multiple hypervisors, operating systems, and business applications

  - Built on Cloud Service Automation and Operations Orchestration software

HPE Helion CloudSystem 9.0 features

Helion CloudSystem 9.0 builds on previous versions of CloudSystem and expands cloud capabilities for bringing traditional environments to cloud. It supports emerging technologies, including:

- Enterprise-ready, highly available architecture

- Flexible and portable service designs

- Expanded support for Microsoft Hyper-V, in addition to Red Hat KVM and VMware ESX

- Expanded support of physical provisioning via HPE OneView

- Decreased time to delivery via prebuilt solutions

- Intelligent targeting of workloads to infrastructure

- Installation of the latest cloud management software within hours

Opportunities provided with version 9.0 include the following:

- Rapid development and deployment of applications through the Helion Development Platform

- Management of Amazon Web Services (AWS) workloads repatriated to private cloud through support for Helion Eucalyptus

- Expanded support for more storage-centric use cases

Helion CloudSystem 9.0 is available as a stand-alone software supporting a multiple-vendor hardware environment or as a fully integrated blade-based or a hyper-converged infrastructure with Converged System. Helion CloudSystem 9.0:

● Integrates Helion OpenStack and the Helion Development Platform to provide customers an enterprise-grade, open-source platform for cloud-native application development and infrastructure

● Simultaneously supports multiple cloud environments, including AWS, Microsoft Azure, OpenStack technology, and VMware, with the ability to fully control the workloads where they reside

● Supports unstructured data through the OpenStack object storage (Swift) project

● Includes version 4.5 of CSA software (CloudSystem Enterprise only), providing the management capabilities to control hybrid cloud environments and a built-in path to support distributed compute, efficient object storage, and rapid cloud-native application development

HPE Financial Services IT investment and consumption offerings are available to help enterprises acquire Helion CloudSystem 9.0.

Target use cases

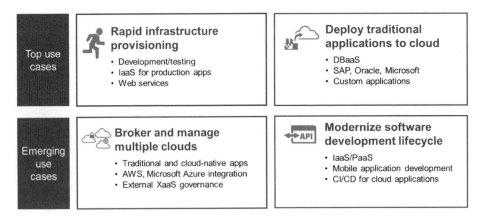

Figure 4-5 Helion CloudSystem use cases

As shown in Figure 4-5, the top use cases for Helion CloudSystem are rapid infrastructure provisioning and deployment of traditional, on-premises applications to cloud:

● **Rapid infrastructure provisioning**—Helion CloudSystem 9.0 expands support for provisioning across a range of popular hypervisors, containers, and physical servers. Third-party servers are also more broadly supported with this release. With these expanded capabilities, Helion

CloudSystem 9.0 provides powerful new ways to turn today's heterogeneous data center infrastructure into a nimble, cloud-based IaaS.

- **Deploy traditional applications to cloud**—Companies can optimize service price and performance and leverage the unique value of various cloud infrastructure offerings from HPE with Helion CloudSystem 9.0. It includes the ability to match a service component to a particular resource pool. Now, for example, the database tier of a traditional application can be directed to land on a blade with SAN storage, the app server can be directed to a ProLiant DL and P4000 configuration, and the web servers can be connected to a scalable ProLiant SL server pool.

Emerging use cases include bridging and brokering multiple accounts by IT service providers and development and deployment of cloud-native applications.

- **Broker and manage multiple clouds**—With Helion CloudSystem 9.0, services can be deployed across multiple different clouds, and portable service designs are bound to particular providers at deployment time to reduce design sprawl. Helion Eucalyptus enables unparalleled hybrid cloud management capabilities, including seamless movement of workloads, tools, and processes between on-premises resources and Amazon Web Services.

- **Modernize software development lifecycle**—Helion CloudSystem 9.0 gives developers new technologies and tools they need to address rapidly changing application development requirements. The Helion Development Platform is now included with Helion CloudSystem 9.0. This PaaS technology is ideally suited to cloud-native application development. Additionally, developers can leverage OpenStack Swift object storage in Helion CloudSystem 9.0 to build applications that help businesses capture, store, and manage today's unprecedented (and ever-growing) volumes of unstructured data.

Helion CloudSystem options

Figure 4-6 Helion CloudSystem options

For maximum flexibility, Helion CloudSystem is available as a software package or delivered as part of an integrated ConvergedSystem solution as shown in Figure 4-6. The software option to create a private cloud on existing infrastructure helps customers increase ROI by utilizing existing hardware. The integrated system option helps customers minimize time to value with pre-built and tested solutions.

For a brand-new cloud environment, HPE recommends the fully integrated Helion CloudSystem solution, delivered on ready-to-use, workload-optimized ConvergedSystem infrastructure. Depending on the customer's cloud workload needs, they can choose from:

- **HPE Helion CloudSystem on ConvergedSystem 700x for cloud (updated intelligent compute and SAN supporting the latest release of HPE OneView and ProLiant Gen9)**— Generally used in larger enterprise scale deployments

- **HPE Helion CloudSystem on ConvergedSystem 250-HC StoreVirtual (hyper-converged platform with compute and local storage)**—Targeting opportunities where smaller footprint, predictable, linear scaling, and lower storage cost and complexity are valued

Both integrated solutions are available with the customer's choice of Helion cloud software—either Helion CloudSystem Foundation or Helion CloudSystem Enterprise.

Learning check

1. Match the components with the correct CloudSystem 9.0 offering:

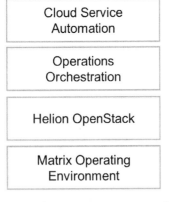

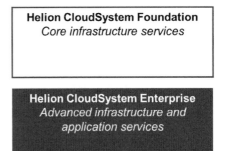

2. What are the primary use cases for Helion CloudSystem Foundation 9.0? (Select two.)

   a. Rapid infrastructure provisioning

   b. Modernizing software development lifecycle

   c. Brokering and managing multiple clouds

   d. Deploying traditional applications to cloud

Learning check answers

1. Match the components with the correct CloudSystem 9.0 offering:

   a. CSA: Helion CloudSystem Enterprise

   b. Operations Orchestration: Helion CloudSystem Enterprise

   c. Helion OpenStack: Helion CloudSystem Foundation and Enterprise

   d. Matrix Operating Environment: Helion CloudSystem Enterprise

2. What are the primary use cases for Helion CloudSystem Foundation 9.0? (Select two.)

   a. Rapid infrastructure provisioning

   b. Modernizing software development lifecycle

   c. Brokering and managing multiple clouds

   d. Deploying traditional applications to cloud

OpenStack software

Before proceeding with this section, assess your existing knowledge about OpenStack by attempting to answer the following questions.

Assessment activity

1. How many OpenStack projects or components can you name?

_____

_____

_____

_____

_____

_____

2. Can you name any of the HPE innovations and contributions to OpenStack?

_____

_____

_____

_____

_____

_____

Assessment activity answers

1. How many OpenStack projects or components can you name?

   a. **Neutron**

   b. **Keystone**

   c. **Ceilometer**

   d. **Glance**

   e. **Nova**

   f. **Trove**

   g. **Swift**

   h. **Cinder**

   i. **Horizon**

   j. **Heat**

   k. **Ironic**

2. Can you name any of the HPE innovations and contributions to OpenStack?

   a. **Security**

   b. **Integrated high availability and scalability support**

   c. **Added lifecycle management capabilities**

   d. **Cloud orchestration and automation**

   e. **Hybrid cloud support**

   f. **Specific independent hardware vendor (IHV) and independent software vendor (ISV) plug-ins**

g. **Indemnification**

h. **Added certification**

Hybrid is the new reality

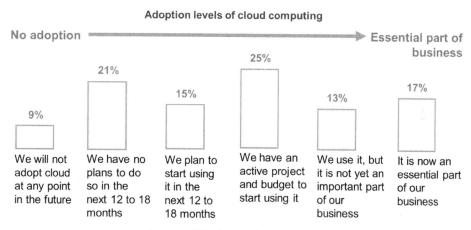

Figure 4-7 Cloud adoption levels

Due to shifts in new workloads and applications, enterprises are dramatically expanding their use of cloud technologies. Figure 4-7 shows findings from a recent Frost & Sullivan global study, which reported that 70% of enterprises will adopt some level of hybrid cloud during the next 18 months. An HPE study revealed that 75% of IT executives plan to pursue a hybrid delivery model. Gartner also supports these findings, predicting that nearly half of large enterprises will have hybrid cloud deployments by the end of 2017.

Hybrid IT is a combination of public, hosted/managed, and private cloud environments, along with traditional IT. Better business outcomes can be achieved using a well-reasoned distribution of resources and solutions across these computing models.

What are customers looking for today?

Figure 4-8 Customer goals

As shown in Figure 4-8, on this journey to a hybrid infrastructure and on-demand IT, customers want to:

- **Avoid vendor lock-in**—Customers do not want to be locked into a vendor's technology and pace of innovation. Instead, they want to be able to control costs and follow their own roadmap. These abilities can be limited when they cede control to a vendor.

- **Have workload portability**—Customers want to be able to move a workload or an application to the required deployment model, including public, private, or managed cloud. Furthermore, they only want to code that workload or application once.

- **Lower total cost of ownership (TCO)**—Customers want to invest resources in actions that drive their business forward.

An application-centric cloud

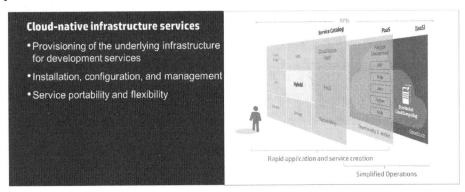

Figure 4-9 Cloud-native infrastructure

IT must enable cloud-native application development. The term "cloud-native" means creating applications that are specifically designed to run in a cloud environment. To support cloud-native application development, IT solutions need to meet a few major capabilities as shown in Figure 4-9. Most IT organizations provide virtual machines (VMs) to developers. IT manages the underlying infrastructure, leaving it up to developers to install, configure, and manage middleware, databases, and development tools. In the future, IT needs to take on a broader role, which would allow developers to focus on coding instead of installing, configuring, and managing developer services.

Open-source software: The new infrastructure

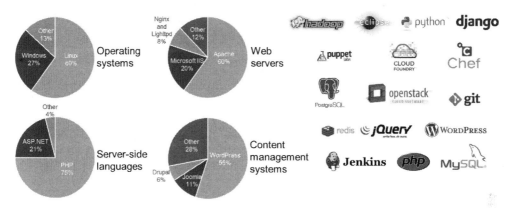

Figure 4-10 Open-source software

Customer goals (avoid vendor lock in, have workload portability, and lower costs) are turning them toward open-source software. Open-source software is rapidly becoming the new infrastructure to resolve customers' current challenges.

According to market share statistics from the Jordan Open Source Association and shown in Figure 4-10, leading open-source assets include:

- Linux is the leading operating system with 60% market share.

- Apache is the leading web server with 45% market share.

- Similarly, PHP and WordPress are the domain leaders for server-side languages and content management systems.

Why open source and OpenStack software?

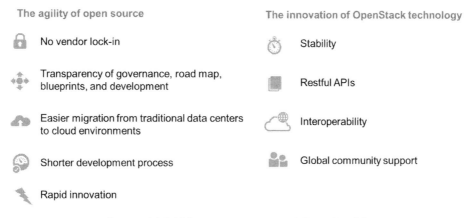

Figure 4-11 Why open source and OpenStack?

HPE needed a product to use as the foundation for its own cloud operating system and cloud delivery platform. The company required a product that was developed using an open-source model. Although proprietary public clouds have been successful, none of the existing options are available for private cloud deployments, making them inappropriate for hybrid cloud computing.

Proprietary public clouds are "black boxes," meaning that customers do not know what software and hardware are being used. To be able to burst into another proprietary cloud when the demand for computing capacity spikes, customers need to redesign their infrastructure, which is expensive and time-consuming.

Figure 4-11 describes some of the key benefits to open source and OpenStack.

Open-source platforms are not tied to a single vendor's technology or roadmap and offer:

- No vendor lock-in with access to a rich ecosystem offering a choice of value-added solutions

- Transparency of governance, roadmap, blueprints, and development

- Easier migration from traditional data centers to cloud environments

- Shorter development process by removing the need to negotiate new license agreements or commit engineering time and resources to implement patches and changes

- Rapid innovation through contributions from subject matters experts across multiple use cases

The key highlights of the OpenStack software include the following characteristics:

- **Stability**—With OpenStack software, there are roadmaps with long-term vision and commitment. There is a clearly defined process for code submissions, reviews, removal, and maintenance. So far OpenStack technology has maintained a steady stream of releases approximately every six months. Considering the size and the complexity of the source code, this is an amazing development pace.

- **Representational state transfer(REST) APIs**—This set of simple, well-defined, and programmable RESTful APIs make it easier to build on an innovative, modular architecture. All of the code for the OpenStack technology is freely available under the Apache 2.0 license, the most permissive open-source license. Anyone can run it, build on it, or submit changes back to the project.

- **Interoperability**—The interoperability of OpenStack prevents vendor lock-in and makes it easier to work between a variety of OpenStack clouds.

- **Global community support**—The OpenStack project has a global community of more than 10,000 supporters, developers, and users in nearly 100 countries, actively participating in code development, online discussions, and deployment. More than 200 companies are involved in the OpenStack ecosystem, including HPE, Rackspace, IBM, NTT, Dell, Canonical, Red Hat, and VMware.

Learning check

1. What are customers looking for when transitioning to a hybrid infrastructure?

_____

_____

_____

_____

2. What are the key highlights of OpenStack software?

_____

_____

_____

_____

Learning check answers

1. What are customers looking for when transitioning to a hybrid infrastructure?
   - **To follow their own road map**
   - **Freedom from vendor lock in**
   - **Lower TCO**
   - **Workload portability**
   - **Improved data security**

2. What are the key highlights of OpenStack software?
   - **OpenStack software offers stability, including road maps with long-term vision and commitment**
   - **RESTful APIs allow users to build on an innovative, modular architecture**
   - **Interoperability prevents vendor lock-in and makes it easier to work between a variety of OpenStack clouds**
   - **Global community support includes more than 10,000 supporters, developers, and users in nearly 100 countries, actively participating in code development, online discussions, and deployment**

Introducing OpenStack

OpenStack is an open-source cloud computing software platform that controls large pools of compute, storage, and networking resources throughout a data center. The software is designed to produce and maintain a massively scalable public and private IaaS cloud on standard hardware. It can be managed through a dashboard that gives administrators control and empowers users to provision resources through a web interface.

OpenStack began as a project jointly launched by NASA and Rackspace. The initial code was provided from the NASA Nebula platform and the Rackspace cloud files platform. The free, open-source (Apache license) software is governed by the OpenStack Foundation, which promotes the development, distribution, and adoption of the OpenStack cloud operating system. A growing number of industry leaders, including HPE, are committed to developing, testing, and documenting the software required to support OpenStack.

 Note

For more information on OpenStack, scan this QR code or enter the URL into your browser.

http://www.openstack.org/

OpenStack technology

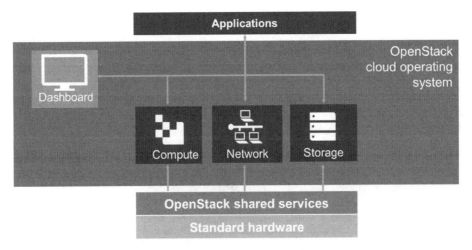

Figure 4-12 OpenStack technology

OpenStack provides cloud design flexibility, with no proprietary hardware or software requirements and the ability to integrate with legacy systems and third-party technologies. As shown in Figure 4-12, it is designed to manage and automate pools of compute, network, and storage resources and can work with widely available virtualization technologies, as well as bare-metal and high-performance computing (HPC) configurations. The major building blocks are:

- **Dashboard**—The OpenStack dashboard gives administrators and users a GUI to access, provision, and automate cloud-based resources. The extensible design makes it easy to plug in and expose third-party products and services, such as billing, monitoring, and additional management tools. The dashboard can be branded for service providers and other commercial vendors who want to use it. The dashboard is just one way to interact with OpenStack resources. Developers can automate access or build tools to manage their resources using the native OpenStack API or the Amazon Elastic Compute Cloud (EC2) compatibility API.

- **Compute**—Administrators often deploy OpenStack compute using one of multiple supported hypervisors in a virtualized environment. Linux kernel-based virtual machine (KVM) and Citrix XenServer are popular choices for hypervisor technology and recommended for many use cases. Linux container technology such as LXC is also supported when users want to minimize virtualization overhead and achieve greater efficiency and performance. In addition to different hypervisors, OpenStack supports ARM and alternative hardware architectures.

- **Network**—Many devices in today's data center networks are divided into virtual machines and virtual networks. The number of IP addresses, routing configurations, and security rules can grow to millions. Traditional network management techniques cannot provide a truly scalable, automated approach to managing these next-generation networks. At the same time, users expect more control and flexibility with quicker provisioning. OpenStack Networking is a pluggable, scalable, API-driven system for managing networks and IP addresses that ensures the network will not be the bottleneck in a cloud deployment. It gives users true self-service capabilities, even over network configurations. With this functionality, administrators and users can increase the value of existing data center assets.

- **Storage**—OpenStack supports object and block storage with deployment options depending on use. Object storage provides a fully distributed, API-accessible storage platform that can be integrated directly into applications or used for backup, archiving, and data retention. It is ideal for cost-effective, scale-out storage. Block storage allows block devices to be exposed and connected to compute instances for expanded storage, better performance, and integration with storage platforms.

OpenStack benefits

The OpenStack cloud computing platform meets the needs of public and private clouds regardless of size, by being simple to implement and massively scalable.

OpenStack provides:

- Community-driven support with contributions from more than 2000 developers and more than 180 participating organizations

- Modular and plug-in architecture compatible with various hypervisors, block storage systems, network implementations, and hardware

- On-demand virtual machines (VMs) with provisioning and snapshot capabilities

- Multitenancy support for setting user quotas

OpenStack releases

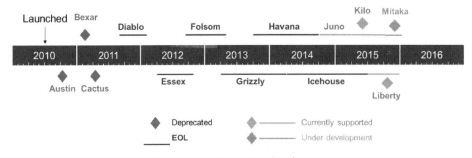

Figure 4-13 OpenStack releases

OpenStack began as the Nebula project, which was the initial Nova code developed by NASA in 2009. Rackspace contributed the Swift (object storage) code to this project as part of a joint effort in 2010. More than 25 contributors created the OpenStack Foundation in mid-2010. The first release (Austin)occurred three months after OpenStack was created.

Releases are scheduled about every six months and named alphabetically, as shown in Figure 4-13.

New features in each release

OpenStack releases are numbered using a YYYY.N time-based scheme. For example, the numbering scheme for the first release of 2016 would be 2016.1.

During the development cycle, the release is identified using a codename. Those codenames are ordered alphabetically: Austin was the first release; Bexar was the second; Cactus was the third, and so on. The OpenStack Foundation membership chooses these code names by voting. Normally, code-names are related to the cities or counties near where the corresponding OpenStack Design Summit took place.

Note

For more information about OpenStack release naming conventions, scan this QR code or enter the URL into your browser.

https://wiki.openstack.org/wiki/Release_Naming

For more information about which new features are in each OpenStack release, visit the Release webpage and click the corresponding release hyperlink. To access this webpage, scan this QR code or enter the URL into your browser.

http://docs.openstack.org/releases/

OpenStack high-level architecture

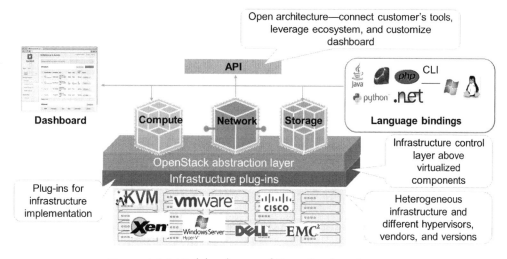

Figure 4-14 High-level view of OpenStack architecture

The compute, network, and storage figures in Figure 4-14 represent the major components of OpenStack. Figure 4-14 also shows how storage, networking, and hypervisors from various vendors can be incorporated into OpenStack through the OpenStack abstraction layer.

Several infrastructure plug-ins for the abstraction layer already exist for some common services. The OpenStack APIs enable you to easily develop plug-ins for the growing number of cloud components that can be integrated with OpenStack.

OpenStack services—Activity

These graphics contain definitions of OpenStack services along with their names. Use the OpenStack Wiki to find information about these services. Draw lines between each definition and the correct OpenStack service name.

 Note

To access information about these OpenStack services, scan this QR code or enter the URL into your browser. Then, scroll down to the bottom of the page to see the main and supporting services (Figure 4-15).

https://www.openstack.org/software/

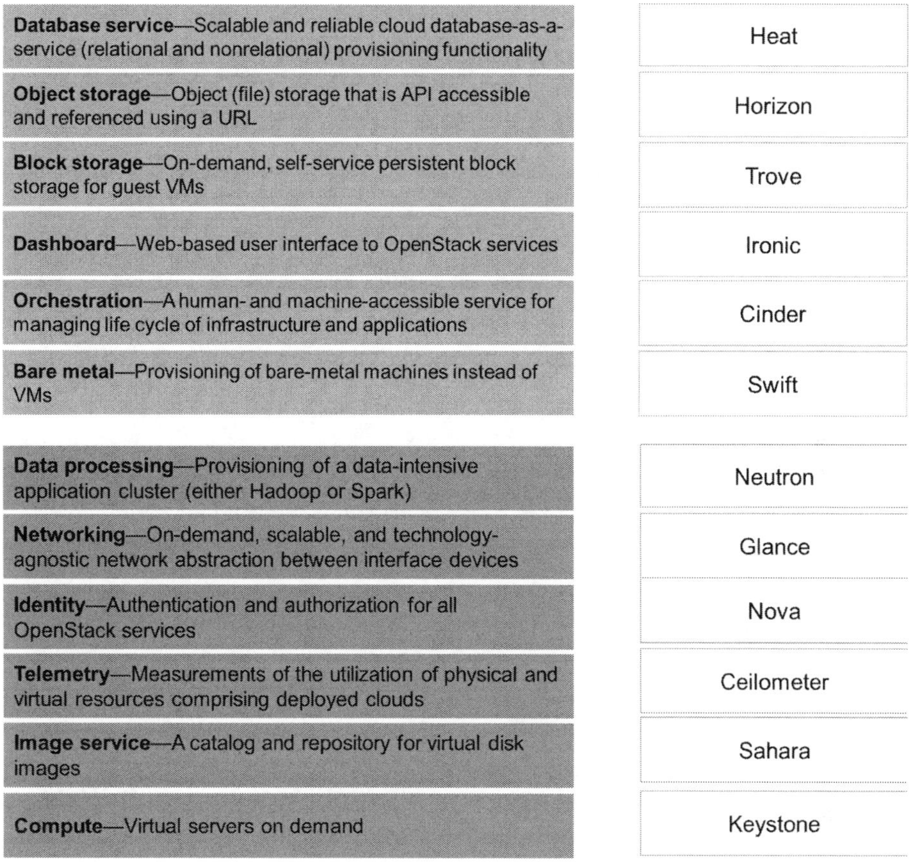

Database service—Scalable and reliable cloud database-as-a-service (relational and nonrelational) provisioning functionality	Heat
Object storage—Object (file) storage that is API accessible and referenced using a URL	Horizon
Block storage—On-demand, self-service persistent block storage for guest VMs	Trove
Dashboard—Web-based user interface to OpenStack services	Ironic
Orchestration—A human- and machine-accessible service for managing life cycle of infrastructure and applications	Cinder
Bare metal—Provisioning of bare-metal machines instead of VMs	Swift
Data processing—Provisioning of a data-intensive application cluster (either Hadoop or Spark)	Neutron
Networking—On-demand, scalable, and technology-agnostic network abstraction between interface devices	Glance
Identity—Authentication and authorization for all OpenStack services	Nova
Telemetry—Measurements of the utilization of physical and virtual resources comprising deployed clouds	Ceilometer
Image service—A catalog and repository for virtual disk images	Sahara
Compute—Virtual servers on demand	Keystone

Figure 4-15 The main and supporting services

OpenStack services activity—answers

1. OpenStack services

   - **Database service: Trove**

   - **Object storage: Swift**

   - **Block storage: Cinder**

   - **Dashboard: Horizon**

   - **Orchestration: Heat**

   - **Bare metal: Ironic**

2. OpenStack services

- **Data processing: Sahara**

- **Networking: Neutron**

- **Identity: Keystone**

- **Telemetry: Ceilometer**

- **Image service: Glance**

- **Compute: Nova**

OpenStack conceptual architecture

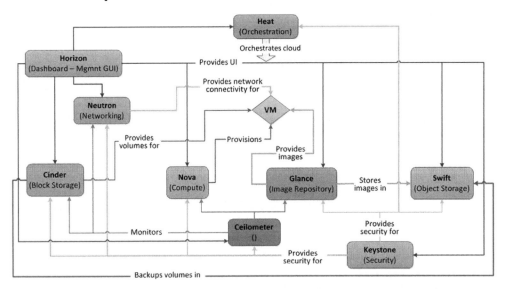

Figure 4-16 OpenStack conceptual architecture

OpenStack provides an IaaS solution through a set of interrelated services. Each service offers an API that facilitates this integration. Depending on the business need, users can install some or all services. Figure 4-16 shows how the core components of OpenStack are interconnected.

- **Swift** provides object storage. It allows users to store or retrieve files (although users cannot mount directories as they could with a file server). Several companies provide commercial storage services based on Swift, including KT, Rackspace, and Internap. Swift is also used internally at many large companies to store data.

- **Glance** provides a catalog and repository for virtual disk images. These disk images are mostly used in OpenStack Compute to provide a boot resource for VMs. Glance has a client-server architecture and provides a REST API through which requests to the server are performed.

Note

REST is a software architecture style for building scalable web services. RESTful systems typically communicate over HTTP or HTTPS with the same verbs (get, post, put, delete, and so on). Web browsers use these verbs to send and retrieve data to and from remote servers.

- **Nova** provides virtual servers on demand. Rack space and HPE provide commercial compute services built on Nova, and it is used internally at companies such as MercadoLibre and NASA (where it originated).

- **Horizon** provides a modular, web-based user interface for performing most cloud operations. This allows users to launch an instance, assign IP addresses, and set access controls.

- **Keystone** provides authentication and authorization for all OpenStack services. It also provides a catalog of services within a particular OpenStack cloud.

- **Neutron** provides network connectivity as a service between interface devices managed by other OpenStack services (typically Nova). The service allows users to create their own routers and networks and attach interfaces to them. Neutron has a pluggable architecture to support many popular networking vendors and technologies.

- **Cinder** provides persistent block storage to VMs.

- **Ceilometer** provides utilization metering for OpenStack deployments.

- **Heat** is a template-based orchestration solution for provisioning the resources required to deploy a cloud.

- **Sahara** provides capabilities to provision and scale Hadoop clusters in OpenStack by specifying parameters such as Hadoop version, cluster topology, and hardware details for the cluster nodes.

- **Ironic** is an integrated OpenStack service that provisions bare-metal machines instead of virtual machines. It is based on the Nova bare-metal driver.

- **Trove** provides scalable and reliable cloud database-as-a-service functionality for both relational and nonrelational database engines.

Basic OpenStack physical configuration

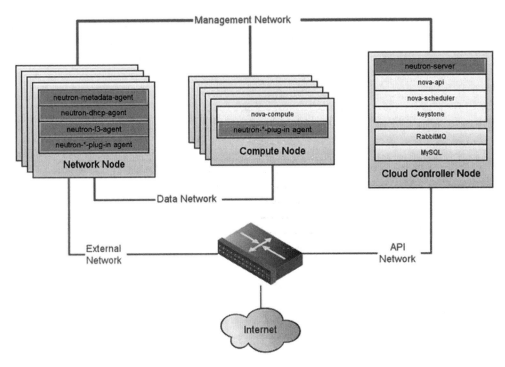

Figure 4-17 Basic OpenStack physical components

Figure 4-17 shows an example of a three-node configuration including the basic components of an OpenStack installation. All basic components can exist on the same device or horizontally distributed, with multiple network, compute, and controller nodes.

OpenStack is designed to be massively horizontally scalable, which allows all services to be distributed widely. However, to better conceptualize the primary components, the diagram groups OpenStack into three primary functional components:

- **Network node**—A node that generally provides the virtual bridging, DHCP server, virtual routing services, and plug-in agents for the associated network hardware

- **Compute node**—A node that provides the hypervisor, communicates with the cloud controller node, and uses the Neutron plug-in agent to communicate with the network node

- **Cloud controller node**—A functional unit that provides the OpenStack Nova, Keystone, Glance, Cinder, Swift, and dashboard services along with hosting of the databases and message queue services

There are four types of functional networks:

- **Management network**—Provides communications between OpenStack services. This network is used by the administrator to access the nodes and for inter-service communication. Every node is on this network.

- **API network**—Supports the standard create, read, update, and delete (CRUD) operations.

- **Data network**—Provides the data stream between the network and compute nodes. This network is used for inter-VM communication. Depending on which network virtualization mechanism is being used, the packets are VLAN tagged packets or encapsulated packets (VxLAN or GRE).

- **External network**—Provides external user access to OpenStack cloud services. This network allows the VMs to access the Internet and the user to access the VMs from the Internet (providing the appropriate access controls are configured correctly).

These networks are normally implemented as physically separated networks and can exist on a single server or can be distributed across multiple servers throughout the cloud.

Identity service (Keystone)

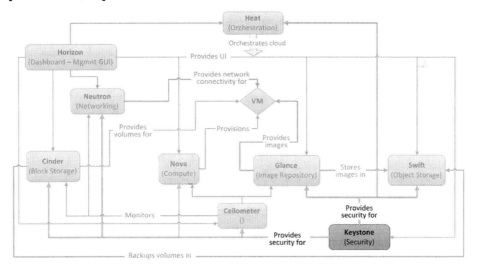

Figure 4-18 Identity service (Keystone)

The simplest way to authenticate a user is to ask for credentials, such as a username, password, or key, and check them against a database. This method may not be practical in situations with a large number of separate services, which is the case with OpenStack because each service uses its own authentication method. A solution to this disparate authentication process is a central authentication and authorization system, which is provided by Keystone, as shown in Figure 4-18.

Keystone is an OpenStack service that provides an identity token catalog and policy services. It is designed specifically for tenants in the OpenStack family of services that implement the OpenStack identity API. This diagram illustrates how Keystone provides those services for most of the OpenStack services through the API network.

You can query user information and determine user permissions for the specific OpenStack services. Keystone supports the creation of collections of users (Groups) and collections of tenants, users, and roles (Domains).

OpenStack dashboard (Horizon)

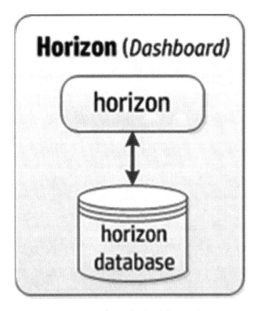

Figure 4-19 OpenStack dashboard (Horizon)

As illustrated by Figure 4-19, the OpenStack dashboard (Horizon) provides administrators and users with a graphical user interface (GUI; dashboard) to access, provision, and automate cloud-based resources. The extensible design makes it easy to plug in and expose third-party products and services such as billing, monitoring, and additional management tools.

The dashboard is an extensible web application that enables cloud administrators and users to control their compute, storage, and networking resources:

- For cloud administrators, the dashboard provides an overall view of the size and state of the cloud. You can create users and projects, assign users to projects, and set limits on the resources for those projects.

- For cloud users, the dashboard provides a self-service portal to provision their own resources within the limits set by the cloud administrators.

Furthermore, the OpenStack dashboard:

- Is stateless and its high availability can be achieved by using a load balancer

- Supports a subset of OpenStack APIs

- Uses memcached or MySQL to store sessions

Horizon architecture

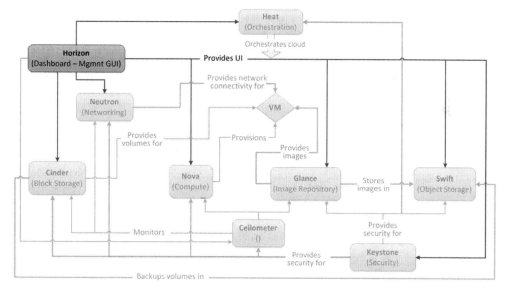

Figure 4-20 Horizon architecture

The OpenStack dashboard provides a convenient interface for requesting services from all major OpenStack projects. As shown in Figure 4-20, it serves as a front-end for the OpenStack project APIs (every dashboard operation invokes API calls to various OpenStack projects).

Compute (Nova)

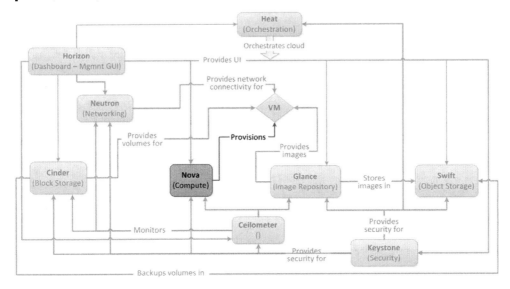

Figure 4-21 Compute (Nova)

Nova is an OpenStack project designed to provide massively scalable, on-demand, self-service access to compute resources. Nova communicates with other OpenStack services as shown in Figure 4-21. Nova is based on these design principles:

- **Component based**—Allows quick changes and additions to the functionality

- **Highly scalable**—Scales to large workloads

- **Fault tolerant**—Isolates processes to avoid cascading failures

- **Recoverable**—Provides easy diagnosis, debugging, and correction of failures

- **Open**—Serves as a reference implementation for a community-driven API

- **API compatible**—Offers compatibility with popular public cloud providers such as Amazon EC2

Cloud storage types

Before proceeding with this section, attempt to answer these questions to assess your existing knowledge about OpenStack storage.

Cloud storage types—Activity

In your own words, describe and differentiate between these three storage types. If necessary, you can research these terms online.

- Ephemeral storage

  _____

  _____

- Block storage

  _____

  _____

- Object storage

  _____

  _____

 Note

Chapter 6 of the *OpenStack Operations Guide* discusses storage decisions within OpenStack. It describes ephemeral and persistent storage and further defines object and block storage. It also compares these storage types, including guidelines on choosing the right storage. To view this chapter, scan this QR code or enter the URL into your browser.

http://docs.openstack.org/openstack-ops/content/storage_decision.html

Cloud storage types activity—answers

In your own words, describe and differentiate between these three storage types. If necessary, you can research these terms online.

- Ephemeral (on-instance) storage

  If you deploy only the OpenStack Compute Service (nova) and do not install the block and object storage services, VMs will not have access to any form of persistent storage by default. The VMs' disks and associated data disappear when a virtual machine is terminated

- Block storage

  Cinder is a block storage service for OpenStack. Cinder virtualizes pools of block storage devices and provides users with a self-service API to request and consume those resources without requiring any knowledge of where their storage is actually deployed or on what type of device. Cinder block storage can be considered to be similar in concept to a virtual USB hard drive that can be connected to and disconnected from VMs at will

- Object storage

  Swift is an object storage service for OpenStack. It is a highly available, distributed, object/ binary large object (BLOB) store. Think Dropbox, Amazon S3, and so on

Block storage (Cinder)

Cinder is a block storage service for OpenStack. Cinder virtualizes pools of block storage devices and provides users with a self-service API to request and consume those resources without requiring any knowledge of where their storage is actually deployed or on what type of device.

Block storage can be either iSCSI-based or Fibre Channel-based. The state of the volumes is independent of instances. By default, the primary operating system volume does not come from Cinder, but rather from ephemeral storage; however, you can create a Cinder boot volume.

 Note

> If you deploy only the OpenStack Compute Service (nova) and do not install the block and object storage services, VMs will not have access to any form of persistent storage by default, rather they will boot from ephemeral (on-instance) storage. This means that the VMs' disks and associated data disappear when the virtual machine is terminated.

Cinder supports multiple storage providers within a zone, which might be useful for an Information Lifecycle Management (ILM) strategy and storage tiering.

Cinder can back up volumes to the object storage component (Swift).

Object storage (Swift)

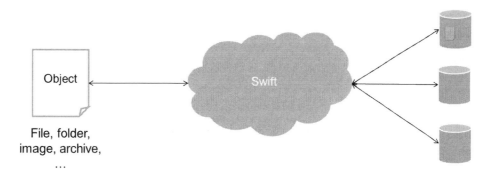

Figure 4-22 Object storage (Swift)

Swift is a highly available, distributed, object/binary large object (BLOB) store and, as illustrated by Figure 4-22, it automatically enables redundancy by providing at least three replicas of users' data. Organizations can use Swift to store high volumes of data efficiently, safely, and inexpensively. Users can create, modify, and retrieve objects and metadata by using the Object Storage API, which is implemented as a set of REST web services.

The core storage system is designed to provide a safe, secure, automatically resized, and network-accessible way to store data. You can store unlimited quantity of files, and each file can be as large as 5 GB. With large object creation (which automatically splits large files into 5 GB chunks), you can upload and store objects of virtually any size.

Swift enables users to store and retrieve files and content through a REST API. Language-specific APIs allow developers to integrate Swift services into their applications. The result is that Swift services are available to any application, not just OpenStack clouds.

Swift features

Swift is a scalable object storage system. It is not a file system in the traditional sense because you cannot mount it as you would with legacy SAN or NAS volumes. Object storage is designed to reliably store billions of objects distributed across standard hardware.

Swift uses a distributed database for redundancy; there is no central database. It provides a RESTful API to compute nodes.

Image service (Glance)

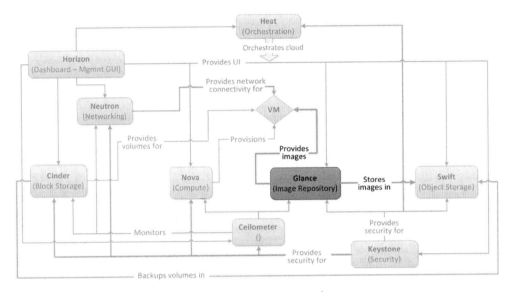

Figure 4-23 Image service (Glance)

The Glance service allows users to upload and discover data assets that are meant to be used with other services such as Nova compute, including images and metadata definitions. Glance image services include discovering, registering, and retrieving VM images and it interacts with other OpenStack services as shown in Figure 4-23.

Glance:

- Provides VMs with an operating system to boot from

- Enables users to choose available images or to create their own from existing servers

- Enables users to store images that can then be used as templates to get new VM instances up and running quickly and consistently

- Supports storing and cataloging of snapshots, which enables VMs to be backed up quickly

Glance has a RESTful API that allows users to query VM image metadata and retrieve the actual images. VM images made available through Glance can be stored in a variety of locations, from simple file systems to object-storage systems such as the OpenStack Swift project.

Glance operations using Horizon

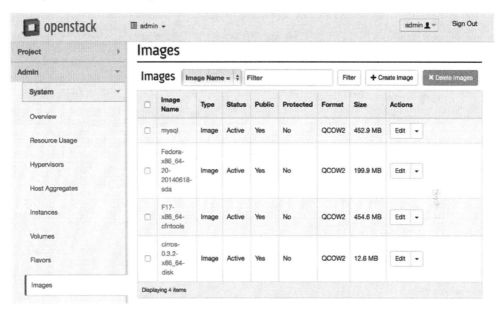

Figure 4-24 Glance operations using Horizon

Most of the Glance operations can be performed using the OpenStack dashboard (Horizon) UI. As shown in Figure 4-24, clicking the Images option in the navigation pane displays a table of available images. From this table, you can:

- Click the **Image Name** link to view details about the selected image.

- View the status of the image, whether it is public, private, or protected, and its disk format.

The Actions column of the table enables you to update some of the image information, including its name, description, architecture, disk format, public availability, and protection.

The Horizon UI also enables you to create new images and delete existing images.

Networking service (Neutron)

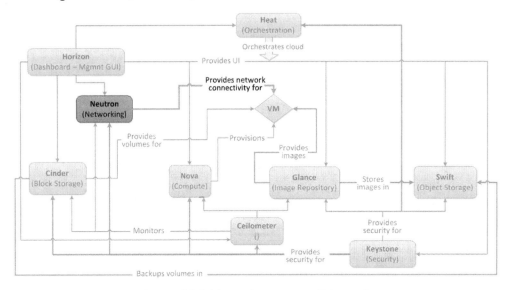

Figure 4-25 Networking service (Neutron)

Neutron provides an OpenStack networking technology that defines the network connectivity through its APIs. Traditional technologies use a static network configuration that must be programmed into each network switch. Neutron uses an SDN infrastructure, which changes how data center networks are defined, provisioned, and consumed. Neutron interacts with other OpenStack services as illustrated by Figure 4-25.

The pluggable architecture enables the configuration and management of physical resources without adding complexity. Neutron allows each tenant to have multiple private networks, and they can choose their own IP addressing scheme.

The key Neutron concepts and terminology include the following:

- **Network**—An isolated Layer 2 (L2) segment, analogous to a VLAN in the physical networking environment

- **Subnet**—A block of IPv4 or IPv6 addresses and the associated configuration state

- **Port**

  - A connection point for attaching a single device, such as a NIC of a virtual server, to a virtual network

  - The associated network configuration, such as the MAC and IP addresses used on that port

  - Compute instances attach to ports

Neutron components

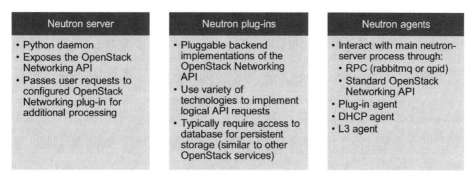

Figure 4-26 Neutron components

Similar to other OpenStack services, Neutron provides cloud administrators with significant flexibility in deciding which services should run on which physical devices. In one scenario, all service daemons can be run on a single physical host for evaluation purposes. In another scenario, each service can have its own physical host and in some cases could be replicated across multiple hosts for redundancy.

As shown in Figure 4-26, the Neutron server uses the neutron-server daemon to expose the Networking API and to pass user requests to the configured Neutron plug-in for additional processing. Typically, the plug-in requires access to a database for persistent storage.

If your deployment uses a controller host to run centralized Compute components, you can deploy the Neutron server on that same host. Because the Neutron service is stand-alone, it could be deployed on its own host as well. Depending on your deployment, the Neutron service can also include these agents:

● **Plug-in agent (neutron-*-agent)** runs on each hypervisor to perform local vswitch configuration. The agent depends on the plug-in that you use, and some plug-ins do not require an agent.

● **DHCP agent (neutron-dhcp-agent)** provides DHCP services to tenant networks. Some plug-ins use this agent.

● **L3 agent (neutron-l3-agent)** provides L3/*network address translation* (NAT) forwarding to provide external network access for VMs on tenant networks. Some plug-ins use this agent. These agents interact with the main Neutron process through RPC (for example, rabbitmq or qpid) or through the standard Networking API.

Neutron relies on the Identity service (Keystone) for authentication and authorization of all API requests.

Telemetry service (Ceilometer)

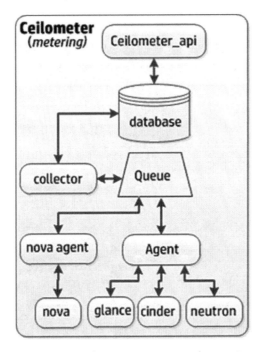

Figure 4-27 Telemetry service (Ceilometer)

Ceilometer collects data from the core OpenStack projects (some are shown in Figure 4-27). It uses agents to poll for data and supports notifications. Data is pushed through a message queue and read by a collector that processes the data and stores it in the database.

Ceilometer consists of these components:

- **A compute (Nova) agent (ceilometer-agent-compute)** runs on each compute node and polls for resource utilization statistics.

- **A central agent (ceilometer-agent-central)** runs on a central management server and polls for resource utilization statistics for resources not tied to instances or compute nodes (such as Glance, Cinder, and Neutron).

- **A collector (ceilometer-collector)** runs on one or more central management servers and monitors the message queues for notifications and for metering data coming from the agent. Notification messages are processed and turned into metering messages; then, they are sent back to the message bus. Metering messages are written to the data store without modification.

- **An alarm notifier (ceilometer-alarm-notifier)** runs on one or more central management servers and enables alarms based on a threshold evaluation of a collection of samples.

- **A datastore** is a database capable of handling concurrent writes (from one or more collector instances) and reads (from the API server).

- **An API server (ceilometer-api)** runs on one or more central management servers to provide access to the data from the datastore.

These services communicate by using the standard OpenStack messaging bus. Only the collector and API server have access to the datastore.

HPE Helion OpenStack

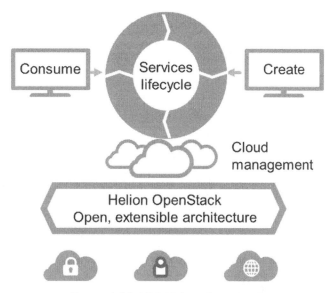

Figure 4-28 HPE Helion OpenStack

Helion OpenStack is an open, extensible cloud platform that enables large enterprises and service providers to easily build, manage, and use hybrid clouds as shown in Figure 4-28. Helion OpenStack enables cost-effective scalability of IT services and business agility. As part of the Helion portfolio, this commercial-grade distribution of OpenStack is backed by a wide range of related cloud products, services, and professional support.

Helion OpenStack:

- Provides customers with the flexibility to choose the right delivery model for their needs across private, hybrid, and traditional IT environments

- Delivers leading open-source cloud computing technology in a resilient, cost-effective, maintainable solution

With Helion OpenStack, you can build and manage enterprise-grade hybrid clouds without vendor lock in. It provides an open, flexible, extensible cloud platform for large enterprises and service providers that enables customers to choose the right delivery model across private and managed/hosted environments.

Customer benefits

Figure 4-29 Customer benefits

As shown in Figure 4-29, with Helion OpenStack, customer benefits include:

- **Enterprise-grade security**—Includes identity management integration with corporate directories, encryption of data in transit, and increased visibility with expanded centralized logging capabilities such as ArcSight integration.

- **Resilient cloud lifecycle management**—Includes integrated high-availability and scalability support (such as StoreServ high-availability enhancements, Ceph integration at the host Linux level, and StoreVirtual VSA integration), multihypervisor workload support (KVM or ESX), active cloud failover for controller services, and backup and recovery.

 Note

Ceph is a fully open source distributed object store, network block store, and file system designed for high reliability, high performance, and extreme scalability from terabytes to exabytes. As such, it provides the capabilities of Cinder and Swift in a single service.

- **Flexible deployment across private, hybrid, and managed/hosted clouds**—Includes targeted node provisioning (meaning you can allocate Swift object storage to storage-optimized servers instead of random servers); certified hardware (ProLiant Gen8 and Gen9 servers, and Moonshot); guest operating system validation (Windows Server 2008 R2 and Windows Server 2012 R2); bare-metal or virtual deployment options for evaluation/proof of concept (POC) environments; and ongoing validation of third-party hardware.

- **Worldwide 24/7 support and professional services**—Includes Foundation Care, HPE Global Cloud Centers of Expertise, local HPE OpenStack experts, environment-level support for the entire data center, and Helion and OpenStack global training and certification offerings.

- **OpenStack Technology Indemnification Program from HPE**—Protects customers using OpenStack code from third-party patent, copyright, and trade-secret infringement claims directed at OpenStack code alone or in combination with Linux code.

Typical use cases

Figure 4-30 Typical use cases

Customers want to increase agility of their IT solutions, reduce costs, and support innovation using an open architecture. Figure 4-30 shows the typical use cases for Helion OpenStack.

One primary use case is for forward-looking IT organizations to provide cloud-based services to external customers more quickly. In the past, IT organizations focused on building custom cloud services. Now, they can use Helion OpenStack cloud technologies to offer classic outsourcing, cloud-based private services, and application hosting from a single provider.

Another common use case for Helion OpenStack is to accelerate application testing and development, most commonly by providing private cloud IaaS solutions.

The final use case is to provide a complete set of development services that allow developers to rapidly deploy cloud-native applications or applications that are designed to run in the cloud. Companies can speed up software development cycles, reducing the time spent on direct interaction with IT services, environment preparation, deployment, and testing.

HPE contributions to OpenStack projects

HPE is one of the leading contributors to OpenStack projects, with this level of participation and recognition:

- Platinum membership

  - Platinum member of the OpenStack Foundation

  - Two board members (one designated platinum corporate sponsorship member and one representative member chosen during an at-large election)

  - Five elected members of the Technical Committee out of 13 total members (no other company has more than two Technical Committee members)

  - Committee membership (Legal Affairs Committee, Incubation Committee, Election Committee, Training Committee, and DefCore Committee)

- Top community contributor

  - First in contributions, code reviews, and number of contributors for Juno

  - Second in lines of code for Juno

  - Nine project team leads (PTLs), which is the first overall rating among contributing organizations, and 38 core reviewers spread out across 18 of the official 22 programs for Juno, which is also the first overall rating

  - HPE PTLs for Keystone, Neutron, Trove, Ironic, Oslo (common libraries), Infra (infrastructure), quality assurance (QA), TripleO (deployment), and Designate (DNS services)

- Dedicated staff

  - Largest contributor by an employer for Juno

  - The only company with dedicated resources to infrastructure

Since joining the OpenStack project, HPE has typically been among the top five contributors, including being the top contributor for the Havana release and second for Icehouse. These contributions represent a massive investment in HPE developer time.

Developers are a significant part of HPE staff assigned to work on OpenStack. However, HPE employees working on OpenStack also include operations personnel, documentation teams, training teams, and QA staff.

HPE also contributes OpenStack cloud accounts required to keep the OpenStack developer projects running. HPE is the only organization other than the OpenStack Foundation itself that provides a dedicated group of employees to the OpenStack developer infrastructure and continuous integration projects.

 Note

Statistics are based on Stackalytics as of October 2014. For more information, scan this QR code or enter the URL into your browser.

http://stackalytics.com

How commitment to OpenStack technology helps HPE customers

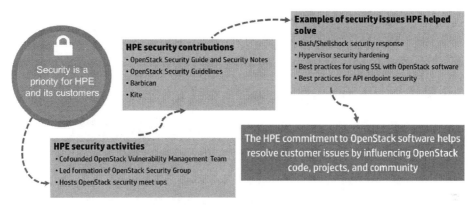

Figure 4-31 How commitment to OpenStack technology helps HPE customers

Figure 4-31 illustrates how the HPE commitment to OpenStack technology helps customers. Security is a major concern for customers, especially for open-source software.

HPE hired security experts and sponsored several security activities in the OpenStack community. As a result, HPE helped publish *OpenStack Security Guide and Security Notes,* which are best practices for deploying OpenStack technology. HPE also assisted in publishing *OpenStack Security Guidelines*, which are best practices for OpenStack developers. HPE is also participating in the OpenStack Barbican project, which focuses on encryption, and the Kite project, which focuses on Remote Procedure Call(RPC) security.

This level of involvement shows HPE customers that OpenStack can be deployed with enterprise security capabilities and that OpenStack technology is being developed using best practices.

HPE innovation and contributions to OpenStack.org

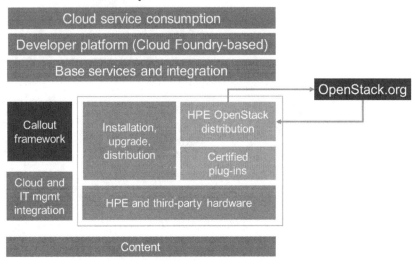

Figure 4-32 HPE innovation and contributions to OpenStack.org

The HPE Helion OpenStack Community makes several contributions to OpenStack.org as shown in Figure 4-32. This support includes blueprints, distribution and release lifecycle management, hardening of the OpenStack code, scaling to enterprise levels, and component certifications for hardware platforms and scale requirements. Customers can also upgrade from the free community edition to Helion OpenStack.

Value-added contributions specific to OpenStack include:

- **Security**—Helion performs security testing of OpenStack releases.

- **Integrated high availability and scalability support**—Helion OpenStack provides cloud failover across all infrastructure services, such as Nova; Cinder; controller services; scale-out reference architecture; and software-defined, large-scale network segmentation via VXLAN overlay networks and block storage scale out.

- **Added lifecycle management capabilities**—Configuration, compatibility, and inventory management support for installation, setup tasks, and upgradeability of releases.

- **Cloud orchestration and automation**—Multitier infrastructure service design and orchestration capabilities leverage other standards-based initiatives aligned to OpenStack.

- **Hybrid cloud support**—Integrated VPN functionality for network fabric connections, resource pool extension across cloud providers, hybrid provisioning, and optional HPE CSA integration support.

- **Specific independent hardware vendor (IHV) and independent software vendor (ISV) plug-ins**—Certified hardware and infrastructure technologies are used in OpenStack deployment (for example, HPE Moonshot). They serve as plug-ins for specific hardware platforms or virtualization technology and are built to match the plug-in architecture and APIs defined in OpenStack.

- **Indemnification**—HPE offers indemnification for Helion OpenStack Community edition customers, subject to basic eligibility requirements, including execution of an indemnification agreement with HPE and a current support contract in place with HPE. The HPE indemnification program has no monetary cap and covers third-party patent, copyright, and trade secret infringement claims directed to OpenStack code alone or in combination with Linux code, whether asserted against direct customers of HPE or an authorized HPE reseller.

- **Added certification**—Helion OpenStack offers certification for enterprises and service providers.

HPE value-added technology specific to base services and integration includes:

- **Call-out framework**—This framework is for integration to external cloud and IT management needs of enterprise and service providers, such as compliance and audit needs or billing integration for chargeback.

- **Extension services**—This functionality is required to support use cases in enterprise and service provider deployments, such as a metadata service (being contributed back to the OpenStack.org community). It is used for tagging cloud assets and entities and also for template repository services for model-based provisioning.

- **Base services**—These are infrastructure extensions as a service, such as DNS as a service (DNSaaS) and load balancing as a service (LBaaS). These are being developed by the community, including new projects and blueprint submissions.

OpenStack projects in Helion OpenStack 2.0

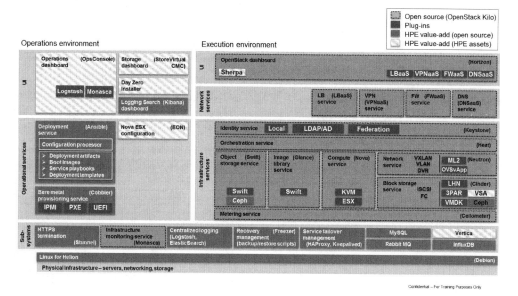

Figure 4-33 HPE, third-party, and open source components in Helion OpenStack

The OpenStack software runs on top of standard hardware, helping customers create an open-source cloud in a software-defined data center. With OpenStack technologies, you can create pools of compute, networking, and storage resources, and then you can create the cloud environment from these pools of resources using either a GUI (such as the OpenStack Horizon dashboard) or applications (using the REST APIs).

As illustrated by Figure 4-33, HPE adds product-level enhancements to provide enterprise-level capabilities not always available in the base OpenStack technologies. For example, the streamlining of the integration of ESXi compute regions with the OpenStack technologies, an operations dashboard (OpsConsole) for managing the cloud infrastructure, a storage dashboard for management of StoreVirtual storage, and a day zero installer that simplifies and speeds up installation and configuration of Helion OpenStack. These additions enhance user capabilities while maintaining portability and avoiding vendor lock-in.

Learning check

1. What is OpenStack?

_____

_____

_____

_____

2. How would you describe the network component of Neutron?

    a. An isolated Layer 2 segment, analogous to a VLAN in the physical networking environment.

    b. A block of IPv4 or IPv6 addresses and the associated configuration state.

    c. A connection point for attaching a single device, such as a NIC of a virtual server, to a virtual network.

    d. The associated network configuration, such as the MAC and IP addresses used on that port.

    e. Compute instances attached to ports.

3. Developers are the only HPE employees who contribute to OpenStack.

    ☐ True

    ☐ False

Learning check answers

1. What is OpenStack?

   - **Open-source cloud computing software platform that controls large pools of compute, storage, and networking resources**

   - **Designed to produce and maintain a massively scalable public and private IaaS cloud on standard hardware**

   - **Receives ongoing, significant code contributions from industry leaders, including HPE**

   - **Free, open-source software (Apache license) governed by the nonprofit OpenStack Foundation**

2. How would you describe the network component of Neutron?

   a. **An isolated Layer 2 segment, analogous to a VLAN in the physical networking environment**

   b. A block of IPv4 or IPv6 addresses and the associated configuration state.

   c. A connection point for attaching a single device, such as a NIC of a virtual server, to a virtual network.

   d. The associated network configuration, such as the MAC and IP addresses used on that port.

   e. Compute instances attached to ports.

3. Developers are the only HPE employees who contribute to OpenStack

   a. True

   b. **False**

Summary

- Open-source software is rapidly becoming the new infrastructure to resolve customers' current challenges. The OpenStack software provides stability, REST APIs, interoperability, and global community support.

- OpenStack is an open-source cloud computing software platform that controls large pools of compute, storage, and networking resources. It is designed to produce and maintain a massively scalable public and private IaaS cloud on standard hardware.

- OpenStack provides an IaaS solution through a set of interrelated services. Each service offers an API that facilitates this integration. Depending on the business need, users can install some or all services. The primary core components of OpenStack are interconnected.

- Cloud storage types include ephemeral (on-instance), block, and object storage. Each type serves its own purposes, and each should be analyzed for each customer based on use case and workload requirements.

- As the only company with dedicated resources to OpenStack, HPE is a leading contributor to OpenStack projects. OpenStack activities include board members, committee members, code contributions and reviews, and Helion value-added contributions specific to OpenStack.

5 A Closer Look at Infrastructure Management

WHAT IS IN THIS CHAPTER FOR YOU?

After completing this chapter, you should be able to:

✓ Explain how converged infrastructure solutions can help customers enable agile IT operations

✓ Describe the benefits of Hewlett Packard Enterprise (HPE) OneView

✓ List the latest advancements provided with HPE OneView 2.0

✓ Identify the benefits of:

  ✓ HPE OneView for Microsoft System Center

  ✓ HPE OneView for VMware vCenter

  ✓ HPE Insight Control server provisioning

Assessment activity

Before proceeding with this section assess your existing knowledge about the topics covered in this chapter by answering the following questions.

1. What is your experience working with management tools such as VCM, VCEM, and HP Systems Insight Manager (HP SIM), which are being replaced by HPE OneView?

   _____

   _____

   _____

   _____

2. What do you already know about HPE OneView? Which parts of HPE OneView are you most interested in learning about? Which parts of the product do you know the least about?

_____

_____

_____

_____

Phases of transformation to on-demand IT infrastructure

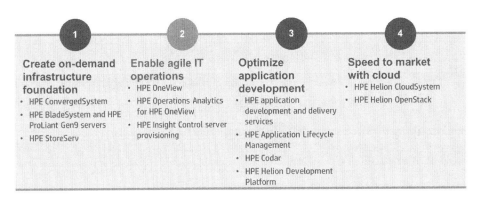

Figure 5-1 Transformation phases

At the beginning of this book, four transformation areas were introduced:

- Protecting the digital enterprise

- Empowering the data-driven organization

- Enabling workplace productivity

- Transforming to a hybrid infrastructure

Each transformation area includes an action plan and associated phases. As shown in Figure 5-1, transforming to a hybrid infrastructure consists of these four phases:

- Creating an on-demand infrastructure foundation, which is supported primarily by HPE ConvergedSystem, HPE BladeSystem and HPE ProLiant Gen9 servers, and HPE StoreServ

- Enabling agile IT operations, which is supported mainly by HPE OneView, HPE Operations Analytics for HPE OneView, and HPE Insight Control server provisioning

- Optimizing application development, enabled by HPE application development and delivery services, HPE Application Lifecycle Management, HPE Codar, and HPE Helion Development Platform

- Speeding to market with cloud with the HPE Helion portfolio of products and solutions

This chapter takes a closer look at the second phase, enabling agile IT operations, and incorporates converged infrastructure management tools from HPE, namely HPE OneView and ICsp.

HPE OneView

Software-defined controls are the key to speed and agility in the strategic use of IT resources in business. A software-defined data center (SDDC) does not replace the need for a cloud solution in most organizations; a cloud is a visible delivery model, whereas the SDDC and converged infra-structure are not. Although the cloud has become the user interface for some organizations, the solutions enabling the SDDC and converged infrastructure support both cloud and noncloud IT processes.

HPE OneView is one solution customers can use for their SDDCs. HPE OneView is a software-defined solution that uses a template-based, collaborative operation model and embodies best prac-tices. The template is designed to capture the most common configuration attributes and functions necessary to deploy a storage, network, server, or power management component.

Software-defined data center

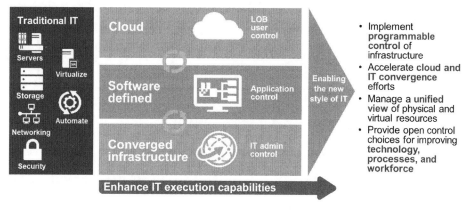

Figure 5-2 Software-defined data center

A SDDC refers to infrastructure that extends the use of virtualization technology by abstracting, pooling, and automating all of the physical data center resources. An SDDC can be defined as sys-tems and procedures that enable infrastructure resources to be controlled at the software level in response to changing business conditions.

The most typical response to changing business conditions is to burst out to additional virtual machines (VMs) using the hybrid cloud model. This can be a useful step, but it is one-dimensional. What if the network conditions or storage requirements change? That is why businesses need to progress toward an SDDC, where computing resources can be fully adapted and conformed to the changing characteristics of business activities as shown in Figure 5-2.

Essentially, implementing an SDDC delivers an IT-as-a-Service (ITaaS) solution. Various infrastructure elements, including network, storage, compute, and security resources, are virtualized and delivered as a service. Although ITaaS might be an SDDC outcome, the SDDC solution is designed to benefit data center architects and IT staff rather than the consumers of the resources. Software abstraction in the data center infrastructure is not visible to the consumers.

The SDDC includes many concepts and data center infrastructure components. Each component can be provisioned, operated, and managed through a programmatic user interface. The core architectural components of a given vendor's SDDC solution might include:

- Compute virtualization, which is a software implementation of a computer's processor, memory, and I/O resources

- Software-defined networking (SDN) or network virtualization, which might involve provisioning virtual LANs (VLANs) on a switch, Ethernet ports operating as a single or aggregated link, ports supporting access or VLAN trunking, security settings, and so forth

- Software-defined storage or storage virtualization, which might involve provisioning storage logical unit numbers (LUNs) on a storage array and host bus adapter (HBA) zoning on a SAN switch

- Management and automation software that enables an administrator to provision, control, and manage all SDDC components

An SDDC is not the same thing as a private cloud because a private cloud only offers a VM self-service solution. Within the private cloud, the IT administrators could use traditional provisioning and management interfaces. Instead, the SDDC could potentially support private, public, and hybrid cloud offerings.

Some of the commonly cited benefits of an SDDC include:

- Improved efficiencies by extending virtualization across all resources

- Increased agility by provisioning resources for business applications more quickly

- Improved control over application availability and security through policy-based definitions

- The flexibility to run new and existing applications in multiple platforms and clouds

- The potential to reduce

    - Energy usage by enabling servers and other data center hardware to run at decreased power levels (or be turned on and off as needed)

    - Costs for data center hardware, challenging traditional hardware vendors to develop new ways to differentiate their products through software and services

Data center resources are becoming increasingly accessible. These resources will require new control options, which suggests that software-defined solutions will be needed to meet those needs.

Overview of HPE OneView

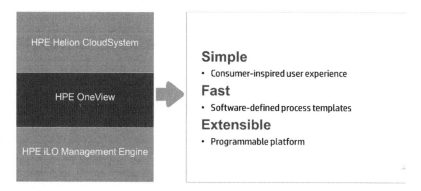

Figure 5-3 HPE OneView

A common complaint from vendors is that they end up with a collection of management tools that look and operate differently. There are typically separate tools, and therefore very different user interfaces, for configuring servers, network devices, and storage systems, and for installing applications. This can cause confusion across tools.

HPE OneView provides an enhanced approach to infrastructure lifecycle management. Its modern architecture is designed to manage converged infrastructure including servers, storage, and networking. This single tool focuses on managing converged infrastructure by incorporating the innovative design of the HPE ProLiant Gen8 and later server platform. The HPE OneView server profile encompasses a wide variety of configuration options and captures a customer's best practices, including BIOS settings, the local HPE Smart Array controller configuration, a firmware baseline policy, and the traditional edge connectivity and identity found in HPE Virtual Connect.

Figure 5-3 shows how HPE OneView fits within the infrastructure lifecycle management stack. HPE OneView is replacing HP SIM, Insight Control, HPE Virtual Connect Manager (VCM) (embedded), and Virtual Connect Enterprise Manager (VCEM). The HPE Integrated Lights-Out (iLO) software and management processor enables HPE OneView and other applications to manage and monitor ProLiant server blade and rack server resources.

Designed for simplicity

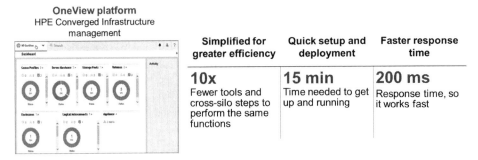

Figure 5-4 HPE Converged Infrastructure management with HPE OneView

The HPE OneView GUI, shown in Figure 5-4, is designed and built for management of an HPE Converged Infrastructure environment. This means fewer tools are involved, which makes the user interface easier to use. HPE OneView is built on an open application programming interface (API) known as the Representational State Transfer (REST) API, which is a central component enabling extensibility for HPE and third-party products.

A key performance target of HPE OneView is that all user interface interactions should take no longer than 200 milliseconds. Those that take longer than 200 milliseconds to process become asynchronous tasks. This means such tasks can continue processing in the background as the administrator moves on to other activities.

HPE OneView supports access through a browser interface and eliminates the requirement for browser plug-ins such as Adobe Flash and Java.

Open integration with existing tools and processes

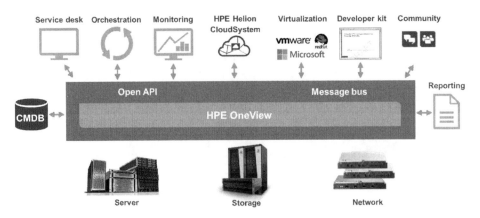

Figure 5-5 Open integration

As illustrated by Figure 5-5, HPE OneView is designed for open integration with existing tools and processes. It enables you to pursue your organization's IT objectives more effectively and to control and fully capitalize on the benefits of a converged infrastructure.

Many commonly used tools, applications, and products surround the HPE OneView environment. These include service desk, reporting, monitoring, and configuration management database (CMDB) tools; HPE Helion CloudSystem software; and Microsoft, VMware, and Red Hat Enterprise Linux hypervisor solutions.

 Note

> A CMDB is a repository that acts as a data warehouse for an IT organization. It contains information describing a collection of IT assets such as software and hardware products and personnel.

RESTful API

REST is a simple stateless architecture that generally runs over HTTP/HTTPS, although other transports can be used. You can use the HPE OneView user interface graphically, and the REST API can be used programmatically. For example, a scripting language such as Microsoft Windows PowerShell can be used to perform any of the tasks you typically might do through the web-based user interface by using RESTful API calls.

Message bus

HPE OneView uses a message bus as a communication channel between application processes and the underlying systems that it supports for management. This message bus is named the State Change Message Bus (SCMB). The SCMB is an interface that uses asynchronous messaging to notify subscribers of changes to managed resources, both logical and physical. For example, you can program applications to receive notifications when new server hardware is added to the managed environment or when the health status of physical resources changes—without having to poll the appliance continuously for status using the RESTful API.

The integrated messaging platform enables dynamic and responsive integrations with the entire IT ecosystem such as Insight Control for VMware vCenter Server and OpenStack for Helion CloudSystem. For instance, when a new enclosure or a rack of equipment is brought online, notifications are automatically sent to the custom application. With HPE OneView management, IT can be more responsive and operate efficiently and predictably.

Single, integrated platform

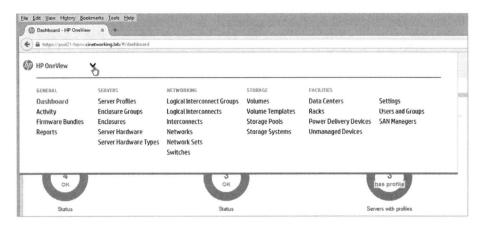

Figure 5-6 Single integrated platform

A key benefit of HPE OneView is that it is implemented as an appliance and is a single tool that uses one data set to present a single view to the administrator. It combines complex and interdependent data center provisioning and management into a simple interface.

The layout of the HPE OneView main menu, shown in Figure 5-6, has been designed for ease of use. There are five commonly used categories:

- General

- Servers

- Networking

- Storage

- Facilities

Within each category, the menu options are listed based on use. The options used most often are at the top of a category's list, and those used less often are placed closer to the bottom of the list.

HPE OneView can be used to:

- **Provision the data center**—The appliance provides several software-defined resources, such as enclosure groups, logical interconnects, network sets, and server profiles, to enable users to capture best practices for implementation across networking, storage, hardware configuration, and operating system build and configuration. By using role-based access control and the various configuration elements in the form of groups, sets, and server profiles, system administrators can provision and manage several hundred servers without involving experts in every server deployment.

- **Manage and maintain firmware and configuration changes**—The appliance provides simple firmware management across the data center. When a resource is added to the appliance, the resource firmware is automatically updated to the minimum version required. This ensures compatibility and seamless operation. An HPE firmware bundle, such as HPE Service Pack for ProLiant (SPP), provides an update package for firmware, drivers, and utilities. Firmware bundles enable you to update firmware on server blades and the infrastructure (enclosures and interconnects). An on-appliance firmware repository allows you to upload SPP firmware bundles and deploy them across the environment according to best practices.

- **Monitor the data center and respond to issues**—The same interface that is used to provision resources is also used to monitor the data center. There are no additional tools or interfaces to learn. When resources are added to the appliance, they are automatically configured for monitoring, and the appliance is automatically registered to receive SNMP traps. You can monitor resources immediately without performing additional configuration or discovery steps. All monitoring and management of ProLiant Gen8 (and later) servers is agentless and done in an out-of-band manner for increased security and reliability.

HPE OneView provides core enterprise management capabilities, including:

- Availability features

- Security features

- Graphical and programmatic interfaces

- Integration with other HPE management software

Consumer-inspired user experience

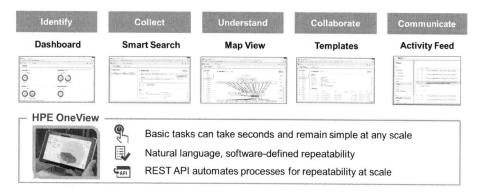

Figure 5-7 HPE OneView key features to help with the five key data center tasks

HPE OneView was developed to ensure that each step in the customer's experience runs efficiently, even when an organization is scaling to accommodate a large number of systems.

HPE OneView has key features to help with the five key data center tasks (Identify, Collect, Understand, Collaborate, and Communicate) as shown in Figure 5-7. These elements can improve device, management, and collaborative processes. These features include:

- **Dashboard**—Provides a variety of capacity and health status information that is well organized and easily accessible. The dashboard offers a high-level overview of the status of the converged infrastructure components. Clicking an object's status briefly summarizes the event.

- **Smart Search**—Enables the administrator to quickly locate configured objects and device information. For instance, users can locate or search for devices based on physical media access control (MAC) addresses and World Wide Names (WWNs).

- **Map View**—Allows the administrator to visualize the relationship between the devices and the objects that are representing them. A "follow the red" status methodology bridges the logical objects to the physical systems, which is especially useful for support calls.

- **Templates**—Used to design boilerplates for the underlying network, server, and storage objects that manage systems in a converged infrastructure. In general, a template is used to define best practices. Templates imply a many-to-one relationship such as the relationship between an enclosure group and its constituent logical interconnect groups. This part of the architecture helps support the needs for documentation and compliance through consistency.

- **Activity Feed**—Allows the administrator to quickly receive alerts and other messages as conditions arise. Activity Feed functions similarly to Twitter in terms of collaboration and communication. An administrator can add notations to events and assign them to an appropriate user. It provides a single place for IT administrators to work together and stay coordinated with alerts, jobs, and projects in real time.

Complete converged infrastructure management

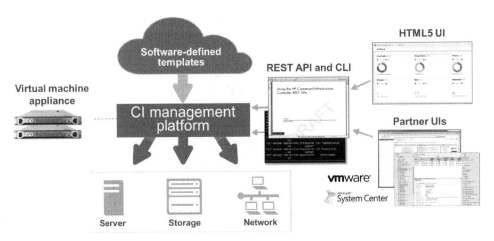

Figure 5-8 HPE OneView delivery

HPE OneView is delivered as a virtual appliance running in a hypervisor VM. The HPE OneView virtual appliance requires a host that is running VMware vSphere ESXi (5.0 and later), Microsoft Hyper-V Server 2012, or Hyper-V Server 2012 R2. Figure 5-8 shows the virtual machine appliance managing HPE Converged Infrastructure (servers, storage, and networking) using the RESTful API, CLI, HTML5 browser-based user interface, and VMware/Microsoft partner user interfaces.

 Note

> With VMware, a licensed version of VMware vSphere is required for the HPE OneView virtual appliance. The free VMware vSphere license is not supported on ProLiant hardware and, therefore, is not supported for HPE OneView.

For VMware vSphere, HPE OneView software is packaged in Open Virtualization Format (OVF) in the form of an Open Virtualization Appliance (OVA), a single file with its contents stored in a tape archive or tar format. OVF is an open standard. The OVA package includes the virtual machine disk (VMDK) and OVF files. For Hyper-V, the HPE OneView software is packaged as a zip file. The zip file includes the VHDX and various XML files.

Customers that deploy or manage large-scale environments require an API to extend the management capabilities of HPE OneView for specific needs in their organization. Through the use of the RESTful API, an organization has a relatively easy solution for automating integration with functions such as enterprise monitoring, CMDB, and service desk applications.

Case study—Activity

Apply the information from this case study to see how HPE OneView can benefit a customer.

Consider the situation of a Fortune 20 company that is having equipment delivered to one of its data centers. Like many organizations, this company does not have a software-defined infrastructure, which can create several challenges.

Without a software-defined infrastructure, the technical personnel must deal with a relatively slow configuration process that involves using disparate tools. Some of the tools are based on a CLI and others employ a GUI. The necessary configuration efforts cannot start until power, networking, and structured cabling is complete.

Realistically, this process does not give the company any opportunities to premodel and test the final deployment configurations. Without a software-defined infrastructure, configuration cannot start until power, networking, and cabling is complete, as the customer waits.

How does HPE OneView help this customer get up and running faster? What benefits can this customer expect? Record your answers here before reading further.

Note

To answer these questions, use the material covered in this chapter or access the HPE converged infrastructure webpage by scanning this QR code or entering the URL into your browser.

https://www.hpe.com/us/en/integrated-systems/software.html

Case study—The solution

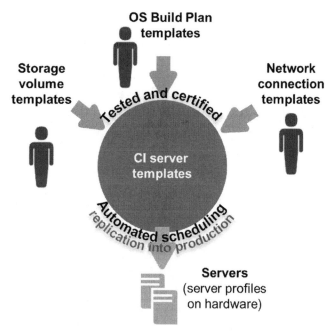

Figure 5-9 HPE OneView collaboration

As illustrated by Figure 5-9, HPE OneView enables collaboration among the different storage, network, and server teams, which can eliminate process serialization and develop standard templates for storage, network, and server objects. Through the development of templates, best practices can be built into the templates to reduce the potential for human error. This allows the HPE OneView administrator to provision servers easily and repeatedly. It also reduces the time needed to build out the infrastructure.

Additionally, because HPE OneView supports the commonly used RESTful API, administrators can automate the deployment of the infrastructure in a programmatic manner without necessarily using the GUI. Combining the use of the RESTful API with tools such as Windows PowerShell or other scripting languages speeds larger scale out of the infrastructure. The RESTful API can also be used to augment HPE OneView support of external systems such as third-party storage systems.

Case study—The results

With HPE OneView

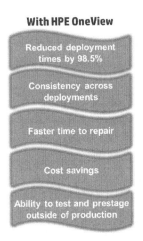

Reduced deployment times by 98.5%

Consistency across deployments

Faster time to repair

Cost savings

Ability to test and prestage outside of production

Call recording technology infrastructure refresh

Before
- Could only build two sites at a time
- Needed 11 days to finish two sites

Today
- Built 12 sites in one night

Instead of taking 66 days, the refresh took a single day

Figure 5-10 Case study results

As shown in Figure 5-10, the Fortune 20 financial company described in the case study benefitted from reduced deployment times, consistency across deployments, faster time to repair, greater cost savings, and the ability to test and prestage the systems before implementing them at their data center site.

HPE OneView 2.0 features

Figure 5-11 HPE OneView 2.0

HPE OneView 2.0, shown in Figure 5-11, introduces these major features:

- Server profile templates

- Driver and firmware updates

- Profile mobility

- Storage snapshots and clones

- SAN health and diagnostics

- Virtual Connect enhancements

Server profile templates

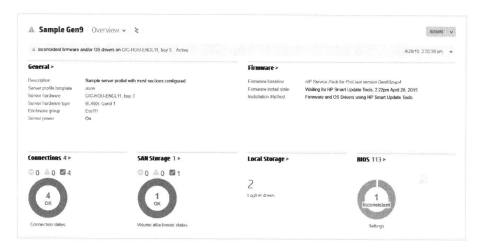

Figure 5-12 Server profile templates

Server profile templates provide automated change management across multiple systems. You make the update once at the template level and automate updates to configurations or apply new system software baselines.

Server profile templates are at the center of software-defined policies and solutions. HPE OneView provides a display of server connections, SAN storage, direct-attached storage (DAS), and BIOS compliance with the profile.

Figure 5-12 shows the warning message an administrator would see if a server was not compliant with the driver and firmware definition in the profile. With HPE OneView, administrators can quickly drill down into dashboard panels to identify issues or troubleshoot problems. By clicking the connection panel, the administrator can easily move from the high-level status down to the connections summary. Individual connections can be expanded to reveal complete connection configuration details. Expanding and collapsing functions allow details for all connections to be revealed or hidden with one click.

Driver and firmware updates

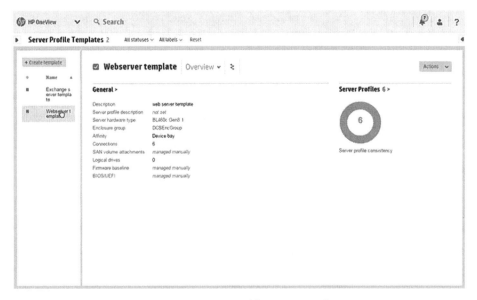

Figure 5-13 Driver and firmware updates

As shown in Figure 5-13, BIOS settings and firmware and driver updates can be made within an HPE OneView template and then propagated out to the server profiles created from that template. HPE OneView templates provide a monitor and flag model. Profiles created from the template are monitored for compliance with the desired configuration. When inconsistencies are detected, the profile is flagged as no longer compliant with the template. When a new update is made at the template level, all profiles parented to that template are flagged as not compliant. From there, the administrator can bring individual or multiple nodes into compliance with the template.

Items that can be updated from template include:

- Firmware baseline
- BIOS settings
- Local RAID settings
- Boot order
- Network and shared storage configurations

The profile can be brought into compliance from the graphic user interface (GUI) or via scripting by using tools such as Windows PowerShell, Python, or RESTful API.

Note

Some updates such as firmware changes require a server reboot.

Profile compliance with the template is evaluated every time the profile or the template is modified; a notification is generated automatically when a compliance issue is detected. The IT administrator has full control over remediation and can choose to update the profile from the template, resolve the inconsistency by editing the server profile directly or dismiss the compliance warning.

Profile mobility

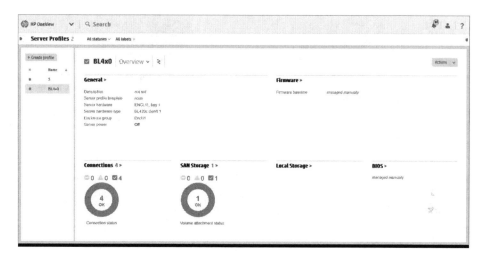

Figure 5-14 Profile mobility

Server profiles in HPE OneView 1.x provided limited mobility across the same server hardware type and enclosure groups.

In version 2.0, shown in Figure 5-14, HPE OneView provides profile mobility across:

- Different adapters
- Different generations
- Different server blade models

These profiles can also be migrated across enclosure groups.

Storage snapshots and clones

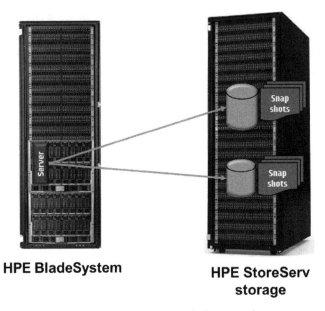

HPE BladeSystem

**HPE StoreServ
storage**

Figure 5-15 Storage snapshots and clones with StoreServ

HPE OneView 1.20 was the industry's first converged and automated management solution that included BladeSystems, ProLiant servers, and StoreServ storage. Advanced automation enables an IT generalist to define and provision storage volumes, automatically zone the SAN as part of the provisioning process, and attach the volumes to server profiles. After they are rolled out, storage and SAN resources are immediately part of the converged infrastructure and monitored by HPE OneView. This means a server administrator can access the power of StoreServ without skilled storage resources and without delays.

HPE OneView 2.0 provides additional enhancements that extend the level of storage automation, making businesses more responsive, secure, and efficient.

StoreServ storage is fully integrated with HPE OneView server profiles for automated, policy-driven rollout of enterprise class storage resources. After the storage has been rolled out, a user can select a StoreServ volume in HPE OneView and create a snapshot from that volume as depicted in Figure 5-15.

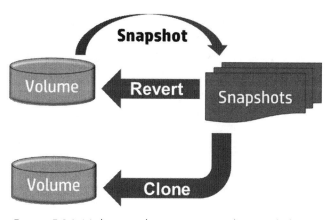

Figure 5-16 Making and reverting snapshots and clones

Snapshots in HPE OneView enable copy and provisioning access to nonstorage professionals such as database administrators, software developers, and test engineers working with systems. Users can safely and easily restore their own copies of test data in seconds without relying on a storage administrator as depicted in Figure 5-16. They can easily replace and restore copies of their volume by copying, promoting, and attaching to server profiles. This enables users to update specific snapshots with more recent snapshots, resulting in faster turnaround times for developers who need to have their snapshots refreshed and alleviating workload for storage administrators.

SAN health and diagnostics

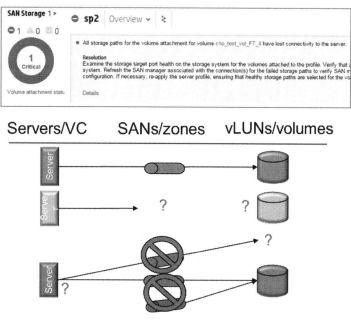

Figure 5-17 SAN health and diagnostics

StoreServ storage is fully integrated with HPE OneView server profiles for automated, policy-driven rollout of enterprise class storage resources. Using StoreServ storage within HPE OneView is as simple as selecting a storage template and a server profile. HPE OneView automation carves out the storage volume, zones the Fibre Channel SAN, and attaches the storage to the server profile.

After they are rolled out, the SAN resources are immediately exposed in the topology map. This includes multihop Fibre Channel and Fibre Channel over Ethernet (FCoE) architectures. As illustrated by Figure 5-17, in HPE OneView 2.0, proactive alerts are provided when the expected and actual connectivity and states differ or when SAN health issues are immediately visible in the topology map. Another feature of HPE OneView 2.0 is the addition of SAN configuration reports. These alert reports include guidance to make the SAN more efficient and to help resolve potential SAN issues before there is a business impact.

Virtual Connect and other enhancements

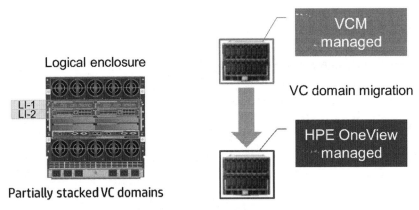

Figure 5-18 Virtual Connect

Virtual Connect interconnects continue to play an integral role in the success of HPE OneView. The HPE OneView team is committed to continue supporting Virtual Connect. Key Virtual Connect features of HPE OneView, some of which are shown in Figure 5-18, include:

- **Partially stacked Virtual Connect domains**—This feature provides air-gap separation between Ethernet networks and enhanced active/active configuration with up to 1000 networks for the active/active pair of connections (instead of the 500 limit with the previous version of Virtual Connect). It also removes the one-to-one relationship between the physical enclosure and the logical interconnect and eliminates the need to stack all interconnects within the enclosure.

Note

Partially stacked domains are not supported for migration from Virtual Connect Manager, but can be configured directly in HPE OneView.

- **Enhanced migration from VCM domains**—Migrating from VCM to HPE OneView 1.1 or earlier was a manual process that required a service outage. HPE OneView 1.2 contained a migration wizard that improved this process. HPE OneView 2.0 contains an embedded VCM migration feature that automates Virtual Connect domain migration with a single push of a button and greatly reduces downtime.

- **Virtual Connect dual-hop FCoE parity support**—This feature allows FCoE traffic out of the enclosure to an external bridge device, which will handle the conversion of FCoE to Fibre Channel traffic. It also provides benefits including cable consolidation, reduction in utilization of the upstream switch ports, and consolidation in management and number of adapters and interconnects required. This feature also supports up to 32 FCoE networks (32 VLANs) and 40 Gb FCoE uplinks of the HPE Virtual Connect FlexFabric-20/40 F8 module. With enhancements around automated storage provisioning on FCoE SANs, HPE OneView can define and provision storage volumes, automatic SAN zoning, and volume attachment to server profiles.

- **Virtual Connect Quality of Service (QoS) priority queuing**—QoS is used to provide different priorities for designated networking traffic flows and guarantee a certain level of performance through resource reservation. The QoS feature enables traffic queues to be configured for different priority network traffic, categorize and prioritize ingress traffic, and adjust DOT1P (IEEE P802.1p) priority settings on egress traffic. Administrators can use these settings to ensure that important traffic receives the highest priority handling and less important traffic is handled at a lower priority.

Other HPE OneView 2.0 features include enhanced monitoring support for Cisco Nexus top-of-rack switches and Cisco Fabric Extender Technology interconnects. Refer to the HPE OneView 2.0 Support Matrix for the list of supported models.

HPE OneView for Microsoft System Center

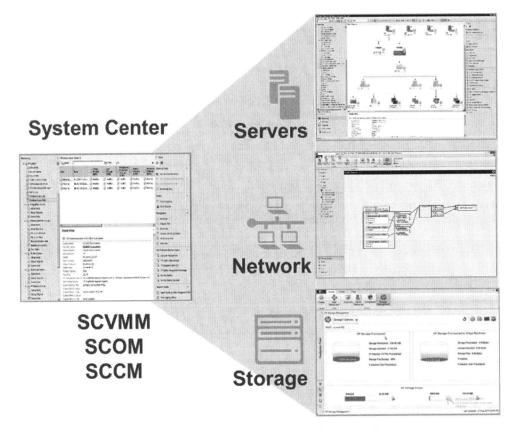

Figure 5-19 HPE OneView for Microsoft System Center

HPE OneView for Microsoft System Center provides seamless integration of unique ProLiant and BladeSystem manageability features into the System Center consoles, as shown in Figure 5-19. Licensed as part of HPE OneView, these extensions deliver:

- Comprehensive system health and alerting

- Configuration management

- Reliable operating system deployment

- Remote control

- HPE Virtual Connect fabric visualization

- Deep levels of HPE hardware inventory

By integrating the server management features of ProLiant and BladeSystem into Microsoft System Center consoles, administrators can gain greater control of their technology environments without having to use multiple interfaces. This provides consistent software deployment and updates. It also speeds the response time in the event of server failure, reducing the risk of downtime.

HPE OneView for Microsoft System Center is an essential infrastructure management platform that helps you:

- Consistently and reliably deploy Hyper-V clusters and take advantage of the automation engine of HPE OneView

- Significantly reduce remediation time when problems arise with easily available solutions

- View physical and virtual relationships with a fabric view from the VM to the network edge

- Decrease administrator time spent on common maintenance tasks by automating updates of ProLiant and BladeSystem servers

Backed by HPE service and support, HPE OneView for Microsoft System Center delivers a superior hardware management experience for customers who have standardized on a Microsoft System Center management platform. It also brings the native manageability of HPE hardware to System Center environments.

 Note

For additional information on HPE OneView for Microsoft System Center, scan this QR code or enter the URL into your browser.

http://www8.hp.com/us/en/products/server-software/product-detail. html?oid=5390822

System Center Virtual Machine Manager integrations

Core server integrations	Enhanced integrations for HPE OneView	HPE storage integrations
• End-to-end Virtual Connect networking view • Operating system deployment (base) • Server firmware and driver updates via automated, rotating workflow	• HPE OneView profile-based deployment (including networking) • Updated end-to-end Virtual Connect networking view • Cluster configuration view • Existing cluster expansion	• Relationship visualization between the Hyper-V VMs, hosts servers, and StoreServ storage • Active management such as create/expand/delete StoreServ volumes • Deployment of HPE StoreVirtual VSA, enabling software-defined storage capabilities
Licensed as part of HPE OneView Advanced or HPE Insight Control	Licensed only as part of HPE OneView Advanced	Free to use with HPE Storage

Figure 5-20 System Center Virtual Machine Manager integrations

HPE OneView for Microsoft System Center provides end-to-end HPE fabric visualization for virtualized environments using HPE Virtual Connect. This view extends from the VM to the edge of the network configuration. As shown in Figure 5-20, it also:

- Provides enhanced provisioning using HPE OneView server profiles to deploy Hyper-V hosts consistently and reliably, including configuration of Windows networking, Virtual Connect, and shared SAN storage

- Facilitates consistency and improves uptime with simplified driver and firmware updates via a rotating, automated workflow for Hyper-V clusters using the ProLiant Updates Catalog

- Identifies mismatched cluster node configurations in the System Center Virtual Machine Manager (SCVMM), including both Windows networking and Virtual Connect

System Center Operations Manager integrations

Core server integrations	Enhanced integrations for HPE OneView	HPE Storage integrations

• Health monitoring and alerting for:
  – Virtual Connect
  – BladeSystem enclosures
  – ProLiant Agentless Management (Gen8 and later)
  – ProLiant Linux
  – ProLiant Windows
	• Consolidated health monitoring view reflects HPE OneView model, showing relationships between devices
	  – StoreServ storage
	  – Virtual Connect
	  – BladeSystem enclosures
		• Events, alerts, and topology view for HPE Storage

Licensed as part of HPE OneView Advanced or HPE Insight Control	Licensed only as part of HPE OneView Advanced	Included with HPE OneView for Microsoft System Center

Figure 5-21 System Center Operations Manager integrations

HPE OneView for Microsoft System Center prevents problems from occurring by proactively monitoring hardware health and intelligently responding to hardware events on servers running Windows and Linux, as well as BladeSystem enclosures, Virtual Connect, and HPE Storage. As shown in Figure 5-21, you can use HPE OneView for Microsoft System Center to:

- Manage the health of ProLiant Gen8 and Gen9 servers without needing to load SNMP agents based on operating systems or WBEM providers

- Monitor the health of:

  - Servers that do not have an operating system loaded

  - ProLiant Gen8 and Gen9 servers running any OS build plans that has a supported agentless monitoring service (such as VMware vSphere ESXi)

HPE OneView integrates with HPE StoreFront Manager for System Center Operations Manager (SCOM), which allows users to monitor and manage HPE Storage for events, alerts, capacity and health dashboards, and detailed virtual infrastructure. SCOM is fully integrated into System Center, so there is one less tool to learn and manage.

Video: Deploying a Hyper-V cluster with HPE OneView for Microsoft System Center

Watch this five-minute demonstration that shows the ease of deploying a Hyper-V cluster using HPE OneView for Microsoft System Center.

 Note

To start the video, scan this QR code or enter the URL into your browser.

https://vrp.glb.itcs.hpe.com/SDP/Content/ContentDetails. aspx?ID=4266&PortalID=1

HPE OneView for VMware vCenter

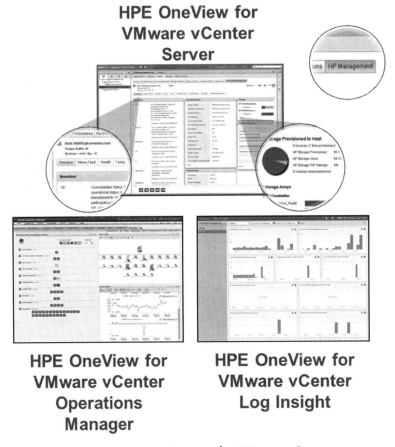

Figure 5-22 HPE OneView for VMware vCenter

As shown in Figure 5-22, HPE OneView for VMware vCenter seamlessly integrates the manageability features of HPE OneView, ProLiant servers, BladeSystem, HPE Networking, and HPE Storage into the VMware vCenter console. The integration of HPE OneView with VMware vCenter allows virtualization administrators to automate control of HPE compute, storage, and networking resources without detailed knowledge of each device. By integrating HPE Converged Infrastructure management features directly into VMware vCenter, administrators can use a familiar VMware management tool to provision, monitor, update, and scale HPE compute, storage, and network resources. This integration simplifies everyday management tasks. Using wizards, administrators can deploy a VMware vSphere cluster in five easy steps and create storage volumes and vSphere data stores. HPE OneView makes the delivery and maintenance of IT services fast, cost-effective, and reliable. The visual mapping of virtualized workloads to physical resources makes it possible to troubleshoot network problems quickly.

Features of HPE OneView for VMware vCenter

Figure 5-23 HPE OneView for VMware vCenter features

With HPE OneView for VMware vCenter, servers licensed by HPE OneView and Insight Control are managed along with HPE Storage under VMware vCenter as shown in Figure 5-23. HPE OneView for VMware vCenter also supports:

- The latest release of HPE OneView

- ProLiant Gen9 servers, including deployment based on server profiles and the Grow Cluster functionality

- Automated deployment of HPE StoreOnce VSA, providing an industry-leading secure backup and recovery solution in your VMware environment

- Root cause and impact analysis via integration with VMware vCenter Operations Manager

- Log analysis via integration with VMware vCenter Log Insight

---

 Note

Customers who purchase HPE OneView are entitled to use the VMware vCenter Operations Manager integration with the standard edition of HPE OneView (without having to upgrade to the advanced edition).

HPE OneView for VMware vCenter Operations Manager

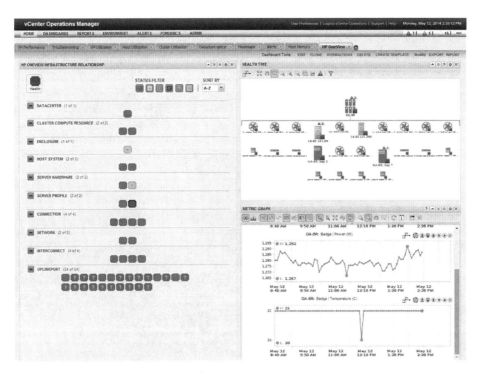

Figure 5-24 HPE OneView for VMware vCenter Operations Manager

HPE OneView for VMware vCenter Operations Manager, shown in Figure 5-24, provides integrated and highly automated performance, capacity, configuration compliance, and cost management tools to the custom GUI within the vCenter Operations Manager. The software uses the vCenter Operations Manager analytics engine that analyzes what is running at acceptable performance levels and then applies those norms to a dynamic server environment.

When the HPE OneView for VMware vCenter Operations Manager is installed, the custom HPE OneView Dashboard is added to the vCenter Operations Manager custom GUI. The HPE OneView dashboard allows administrators to monitor resources in a vCenter environment, including resource health, power and temperature, and system alerts. The analytics engine allows for proactive monitoring of the HPE OneView resource environment. If a problem occurs, an alert is triggered and displayed. The analytics engine also predicts when a resource will reach a critical level.

HPE OneView for VMware vCenter Log Insight

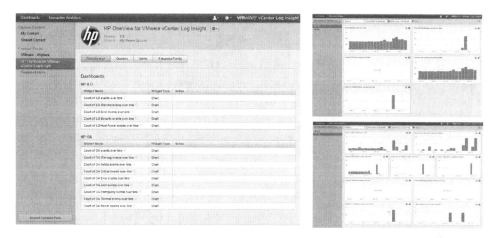

Figure 5-25 HPE OneView for VMware vCenter Log Insight

HPE OneView for VMware vCenter Log Insight, shown in Figure 5-25, provides log aggregation and indexing with search and analytics capabilities. It collects, imports, and analyzes logs to provide information related to systems, services, and applications.

Video: HPE OneView integration with VMware vCenter

Watch this six-minute demonstration that shows the integration of HPE OneView with VMware vCenter.

 Note

To view this and other HPE OneView demonstrations, scan this QR code or enter the URL into your browser. Look under **Resources** for the HPE OneView Integration with VMware vCenter video.

https://www.hpe.com/us/en/integrated-systems/software.html

Video demos—using on-premises management with HPE OneView

Watch this seven-minute HPE OneView demonstration that shows how to deploy and configure the HPE OneView appliance.

 Note

> To view this HPE OneView demonstration, scan this QR code or enter the URL into your browser.

https://youtu.be/aiwWRYj1FaU

Watch this 15-minute HPE OneView demonstration that shows how to create SDDC resources (networks, network sets, Logical Interconnect Groups, and Enclosure Groups) using the OneView appliance.

 Note

> To view this HPE OneView demonstration, scan this QR code or enter the URL into your browser.

https://youtu.be/fiAS1M_cpFo

Watch this eight-minute HPE OneView demonstration that shows how to add an enclosure and update firmware using the OneView appliance.

 Note

> To view this HPE OneView demonstration, scan this QR code or enter the URL into your browser.

https://youtu.be/Art0YMcHcgU

Watch this 20-minute HPE OneView demonstration that shows how to create and manage server profiles and templates in OneView.

 Note

To view this HPE OneView demonstration, scan this QR code or enter the URL into your browser.

https://youtu.be/T2cit2ud5K4

Watch this 13-minute HPE OneView demonstration that shows how to manage storage systems in OneView.

 Note

To view this HPE OneView demonstration, scan this QR code or enter the URL into your browser.

https://youtu.be/4wFM091JdT0

HPE Insight Control server provisioning

Assessment activity

Before proceeding with this section, assess your existing knowledge about HPE Insight Control server provisioning.

1. When using HPE Insight Control server provisioning with ProLiant Gen8 and later servers, what should you use to perform a PXE-free installation of the operating system?

    _____

    _____

    _____

    _____

2. Where are the software components to be provisioned to target servers, such as OS files, captured images, and firmware and driver updates, stored?

_____

3. Which steps are optional in the HPE Insight Control server provisioning installation process? Circle or highlight the optional steps in Figure 5-26.

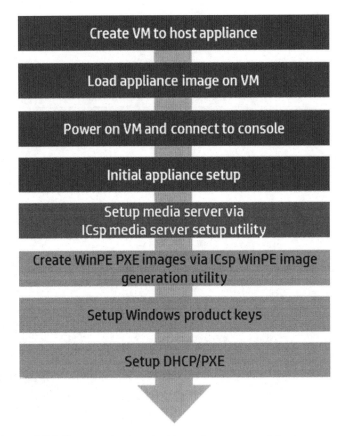

Figure 5-26 Steps in the Insight Control server provisioning process

Assessment activity answers

1. When using HPE Insight Control server provisioning with HPE ProLiant Gen8 and later servers, what should you use to perform a PXE-free installation of the operating system?

   - **HPE Intelligent Provisioning**

2. Where are software components to be provisioned to target servers, such as OS files, captured images, and firmware and driver updates, stored?

   - **On the media server**

3. Which steps are optional in the HPE Insight Control server provisioning installation process?

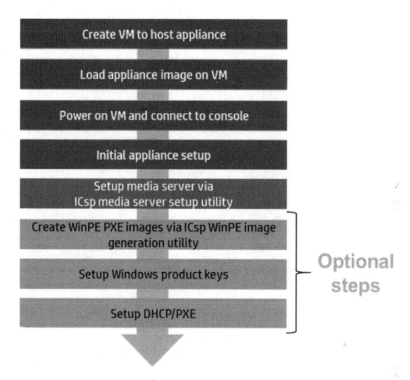

Figure 5-27 Insight Control server provisioning process

HPE Insight Control server provisioning features

HPE Insight Control server provisioning (ICsp) is a tool, packaged as a virtual appliance, used to install and configure ProLiant servers. It is optimized for ProLiant Gen8 and later servers (server blades and rack-mount servers). ICsp uses resources such as OS build plans and scripts to run deployment jobs.

ICsp allows you to:

- Install Windows, Linux, and ESXi on ProLiant servers (bare-metal)

- Deploy operating systems to VMs

- Update drivers, utilities, and firmware on ProLiant servers using the HPE Service Pack for ProLiant (SPP)

- Configure ProLiant system hardware, iLO, BIOS, HPE Smart Array, and FC HBA

- Deploy to target servers without using PXE (for ProLiant Gen8 and Gen9 servers) using the embedded HPE Intelligent Provisioning feature

- Run deployment jobs on multiple servers simultaneously

- Customize ProLiant deployments via an easy-to-use, browser-based interface

- Create and run customized build plans to perform additional configuration tasks either before or after OS deployment

- Migrate from Insight Control server deployment (Rapid Deployment Pack [RDP]) to ICsp

High-level architecture

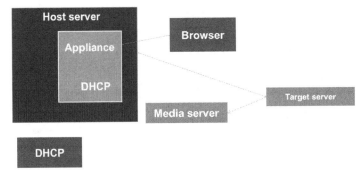

Figure 5-28 ICsp components

As illustrated by Figure 5-28, ICsp includes these components:

- **Appliance**—The ICsp product, delivered as a VM optimized to run the application.

- **DHCP server**—The ICsp appliance comes with an embedded DHCP server. Depending on your environment, you may configure this server for use or disable it using the appliance settings.

- **Appliance IP address** (not shown in the Figure)—The IP address assigned to the appliance. Use this IP address to browse to the appliance via a supported browser or when making REST calls to perform specialized functions.

- **Deployment IP address** (not shown in the Figure)—The IP address used for all deployment operations and target server communications.

- **Target server**—A server managed by ICsp. Each managed server runs an agent. The agent is used for software installation and removal, software and hardware configuration, and status reporting.

- **Media server**—A server that contains vendor-supplied operating system media used during OS provisioning. This server stores captured images. It may also contain media for other purposes, such as firmware and driver updates. The media server is a separate server from the ICsp appliance and is not included as part of the appliance backup and restore actions.

Setup steps

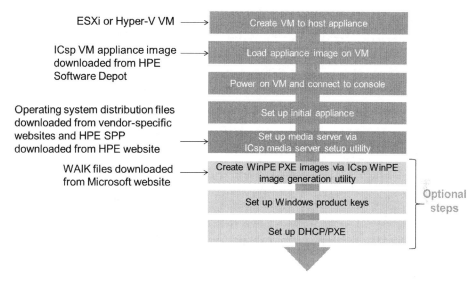

ESXi or Hyper-V VM ──→ Create VM to host appliance

ICsp VM appliance image downloaded from HPE Software Depot ──→ Load appliance image on VM

Power on VM and connect to console

Operating system distribution files downloaded from vendor-specific websites and HPE SPP downloaded from HPE website ──→ Set up initial appliance

Set up media server via ICsp media server setup utility

WAIK files downloaded from Microsoft website ──→ Create WinPE PXE images via ICsp WinPE image generation utility

Set up Windows product keys ⎱ Optional steps

Set up DHCP/PXE

Figure 5-29 ICsp setup steps

As shown in Figure 5-29, the initial installation of the ICsp appliance involves these high-level steps:

1. Create a VM, based on either VMware ESXi or Microsoft Hyper-V, to host the ICsp appliance.

2. Obtain the ICsp software and load it onto the VM. The ICsp VM can be downloaded from the web as a ZIP file, and it is also available on a USB flash drive if you order the media kit. An ESXi VM uses the OVF template; Hyper-V uses image ZIP file.

3. Power on the VM, connect to the console, and log in.

4. Perform the initial appliance setup. The initial appliance configuration includes:

   • Assigning an administrator password

   • Configuring the appliance networking (the appliance and deployment IP addresses)

   • Accessing the appliance from a browser for the first time

   • Activating ICsp

5. Set up the media server. OS build plans use software on the media server for many provisioning functions, including OS and SPP installations as well as firmware and driver updates. Software (media) on the media server can include OS distribution files from HPE or a third-party vendor, captured images, firmware and driver updates, such as HPE SPP, and any custom software or data your build plan may need to access.

Because ICsp is hosted on a virtual appliance optimized to run the ICsp application, there is not enough storage on the appliance to host deployment software. For this reason, a separate media server is required in order to provision servers.

To set up your media server, it is recommended to first download or have media for the actual OS installations. This includes the distribution files from the respective OS supplier, VMware ESXi, and the latest SPP from HPE. When set up, the media server provides a file share that is used for serving the deployment software.

6. Generate and upload Windows PE (WinPE) to the appliance. A version of WinPE needs to be built and uploaded to the appliance for most people performing Windows installations or operations.

7. Set up Microsoft Windows product keys. If you have Microsoft volume license keys, you can enter them, and they will automatically be applied during a Microsoft installation.

8. Set up DHCP and PXE. Although deployments can be performed without PXE when using ProLiant Gen8 and later servers, DHCP is still required for ICsp in all cases. The DHCP server is used to provide

   - IP addresses to servers as they boot into the maintenance OS (even PXE free)

   - Boot information for servers that are PXE booting

   - IP addresses to servers that were not configured with static IP addresses after operating system installation

ICsp has a DHCP server internal to the appliance you may use, or you can set up your own DHCP server external to the appliance.

Note

For more information about the ICsp installation and configuration, see the Insight Control server provisioning installation guide. To access this guide, scan this QR code or enter the URL into your browser.

http://h10032.www1.hp.com/ctg/Manual/c04455190

OS build plan

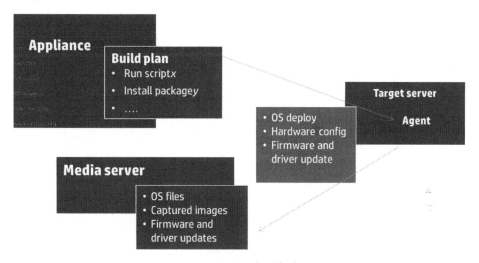

Figure 5-30 OS build plans

OS build plans can be used to perform tasks in ICsp. They are used to cause actions to be performed against servers, for example, installing a server, resetting a server, or updating firmware. An OS build plan is a collection of ordered steps and related parameters to help you perform actions as illustrated by Figure 5-30. ICsp ships with sample build plans and steps that are designed to work right out of the box.

Sample plans demonstrate the steps required to perform the most common deployment-related operations. Most users will modify one of the provided samples to perform the functions they need.

Although build plans are referred to as OS build plans, they do much more than OS deployment. For example, build plans can also be used to:

- Configure a target server's hardware

- Capture a target server's hardware configuration so that the same configuration can later be applied to other servers

- Update the firmware on a target server

- Install software on a target server with a running operating system

ICsp provides four types of sample build plans:

- **ProLiant Hardware**—Build plans labeled with *ProLiant HW* perform hardware-related functions on target servers such as booting the target server to the proper service OS or capturing and configuring hardware settings.

- **ProLiant Operating System**—Build plans labeled with *ProLiant OS* deploy an OS to target servers either via scripted or image installation.

- **ProLiant Software**—Build plans labeled with *ProLiant SW* perform functions on target servers to update the firmware or install and update software on target servers that are running a production operating system.

- **ProLiant Combination**—Beginning with ICsp release 7.2.2, build plans labeled with *ProLiant COMBO* perform a combination of functions on target servers, such as hardware-related configurations, deploying an operating system, and installing software.

HPE OneView and HPE ICsp integration

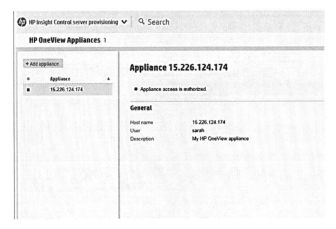

Figure 5-31 HPE OneView and HPE ICsp integration

HPE OneView gives you the right to use ICsp. There are a few things to consider when you are using both management tools together. ICsp is aware of HPE OneView appliances as shown in Figure 5-31. Before running an OS build plan, ICsp checks that the build plan is not accidentally executed and conflicting with certain functions of the target server that are managed by HPE OneView. These functions include the following:

- Configuration and management of BIOS settings

- Management of firmware

- Configuration of SAN boot settings

If HPE OneView is already managing these functions on the target servers, you do not want ICsp to run OS build plans performing the same functions. As a user of ICsp, you can identify the HPE OneView appliances by entering them into the ICsp user interface. You also need to set the appropriate OS build plan type (which is already set for build plans provided by HPE).

When running an OS build plan, ICsp automatically validates the build plan type and the target server against the registered HPE OneView appliances. If there are conflicts, they are displayed in a confirmation dialog box. If needed, the user can override these conflicts and force the execution.

OS build plans integrated with HPE OneView

Type	For scripts this could be OGFS, Python, Unix, Windows .BAT, Windows VBScript, Windows .PS1, etc. See About scripts for additional information. For packages, this is Install ZIP. For configuration files, this is Config File.
History	
Date	Date and time the Action occurred.
Action	
Description: **Examples:**	Describes what happened with the Build Plan. • Create OS Build Plan • Run OS Build Plans • Update OS Build Plan
Custom Attributes	
Name	
Description:	Name of the name/value custom attribute pair associated with this specific Build Plan.
Value	This is the value of the name/value custom attribute pair associated with this Build Plan. The value is not initially shown, but can be displayed by expanding the entry.

Figure 5-32 OS build plans integrated with HPE OneView

HPE ICsp build plans use the Type field to identify what services they manage, as shown in Figure 5-32. This Type field is used when integrating with HPE OneView to identify where the services managed by an OS build plan might overlap services of HPE OneView appliances.

When ICsp runs an OS build plan, it checks to see if you have identified HPE OneView appliances and if the Type is one of these three types:

● HW—SAN Configuration

● HW—BIOS Configuration

● SW—Firmware

ICsp then compares the list of servers and build plans selected with the HPE OneView appliances listed to see if there is a conflict. If a conflict is found, a notification appears. At that point, you can cancel and make corrections or choose to force run the build plan.

 Note

You can find more information about notifications of conflicts when running a build plan by scanning this QR code or entering the URL into your browser.

http://h17007.www1.hp.com/docs/enterprise/servers/icsp/webhelp/ content/s_oneview_notifications.html

Build plans provided by HPE already have the Type set, and therefore no additional configuration is required. Examples of HPE OS build plans that already have the Type field set include:

- The **ProLiant HW—FC HBA Configure Boot Device** and **ProLiant HW—FC HBA Display Configuration** build plans have the Type field set to `HW—SAN Configuration`

- The **System ROM capture**, **System ROM enable BFS**, and **Erase server** build plans have the Type field set to `HW—BIOS Configuration`

- The **Firmware Update** build plan has the Type field set to `SW—Firmware`

If you are integrating with HPE OneView and you create custom build plans, you must set the Type correctly so that ICsp identifies HPE OneView conflicts as it does for build plans provided by HPE. For build plans you create that configure BIOS settings, manage firmware updates, or configure SAN boot, you must specify one of the plan Types. If one or more of the HPE OneView appliances is unavailable at the time an OS Build Plan runs, the situation will be treated as a conflict with HPE OneView and the same options will be provided.

You can check the HPE OneView Appliances screen in the ICsp UI to see the status of appliances. The status indicates whether ICsp can communicate with the HPE OneView appliances.

Notification of conflicts when running a build plan

When you have configured the ICsp integration with HPE OneView and you run a build plan, if conflicts are identified, a notification is displayed on the Run OS build plan screen. The notifications include the following:

- If an HPE OneView server cannot be reached or has other communication errors, an error message appears with instructions to verify that all HPE OneView appliances are running. Then you can try again or force the run and ignore the conflicts. You can select **Yes, force** or **Cancel**.

- If the build plan and HPE OneView appliances have conflicts on specific servers, an error message appears with instructions to correct the conflicts or force the run and ignore the conflicts. You can select **Yes, force** or **Cancel**.

- If checking for conflicts might take a significant amount of time, a message appears with a time estimate, giving you the option to select **Yes, proceed with the checks** or **Run without checking** or **Cancel** and retry with fewer servers.

- If checking for conflicts is estimated to exceed the 10-minute time limit, a message appears giving you the option to cancel and retry with fewer servers or force the run without checking for conflicts. You can select **Yes, force** or **Cancel**.

HPE OneView with ICsp Installation and Startup Service

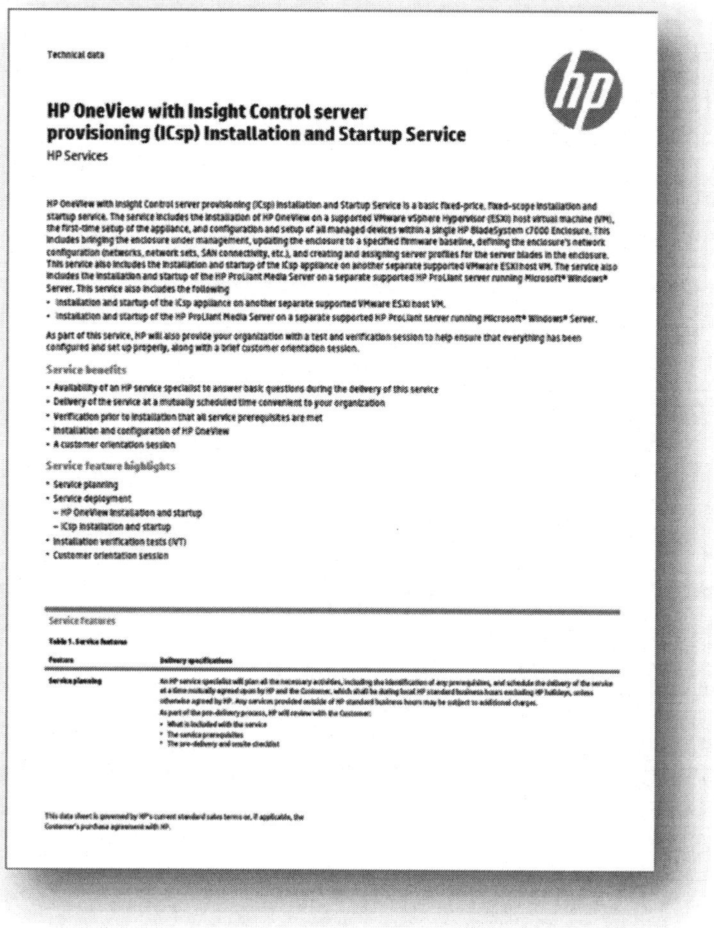

Figure 5-33 HPE OneView with ICsp Installation and Startup Service

HPE OneView with Insight Control server provisioning Installation and Startup Service, the datasheet for which is shown in Figure 5-33, is a basic, fixed-price, fixed-scope implementation service. It includes the installation of HPE OneView on a supported VMware vSphere hypervisor (ESXi) host VM, the initial setup of the appliance, and configuration and setup of all managed devices within a single BladeSystem c7000 enclosure. This includes bringing the enclosure under management, updating it to a specified firmware baseline, defining its network configuration (networks, network sets, SAN connectivity, and so forth), and creating and assigning server profiles for its server blades.

This service also includes the following functions:

- Verification before installation that all service prerequisites are met

- A customer orientation session

- Installation and startup of the ICsp appliance on a separately supported VMware ESXi host VM

- Installation and startup of the ProLiant media server on a separately supported ProLiant server running Microsoft Windows Server

- Installation and configuration of HPE OneView

- A test and verification session to ensure that everything has been configured and set up properly

- Availability of an HPE service specialist to answer basic questions during the delivery of this service

---

 Note

To read the technical datasheet titled *OneView with Insight Control server provisioning (ICsp) Installation and Startup Service,* scan this QR code or enter the URL into your browser.

http://www8.hp.com/h20195/v2/GetPDF.aspx/4AA5-0792ENW.pdf

HPE StoreServ Management Console

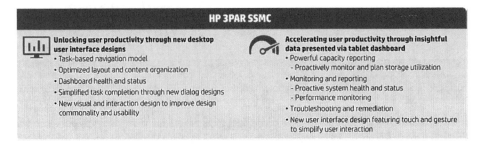

Figure 5-34 HPE StoreServ Management Console

The StoreServ Management Console (SSMC) is the management and reporting console that offers converged management of file and block storage on StoreServ arrays (Figure 5-34).

SSMC offers a consistent feel and a common interface and language with HPE OneView. Using HPE OneView as guidance, SSMC user tasks are accomplished in seconds. Automation combined with a consumer-inspired user experience simplifies basic tasks and everyday processes on StoreServ arrays.

Designed to use the latest API and user interface technologies, SSMC centralizes all StoreServ management under a single console. It offers converged management and reporting for both file and block storage.

StoreServ storage delivers the efficiency and agility required by virtual, cloud, and ITaaS environments through a software portfolio managed by SSMC. As shown in Figure 5-34, key features include the following:

- StoreServ provisioning technologies offer efficiency benefits for primary storage that can significantly reduce both capital and operational costs. Thin provisioning has achieved widespread adoption because it dramatically increases capacity efficiencies. Deduplication is also an essential consideration when looking into deploying workloads onto a flash tier or an all-flash array. Thin technologies can vary widely in how they are implemented.

- HPE StoreServ Remote Copy software brings a rich set of features that can be used to design disaster-tolerant solutions that cost-effectively address disaster recovery challenges.

- HPE StoreServ Adaptive Flash Cache (AFC) allows SSDs to act as Level 2 read cache holding random read data for spinning media that has aged out of DRAM read cache. AFC reduces application response time for read-intensive I/O workloads and can improve write throughput in mixed-workload environment. AFC effectively increases the amount of random read data cached on high-speed media on a node.

- The HPE StoreServ File Persona Software Suite can be enabled on a StoreServ node pair with an optional license. This software suite extends the spectrum of primary storage workloads natively addressed by HPE 3PAR StoreServ from virtualization, databases, and applications via the Block Persona to include client workloads such as home directory consolidation, group and department shares, and corporate shares via the File Persona.

- HPE StoreServ System Reporter is fully integrated into SSMC and offers reporting templates, scheduled reports, and threshold alerts.

HPE StoreServ Replication Software Suite

The StoreServ Replication Software Suite protects data and applications from the unpredictable by delivering simple, fast, and economical data protection and disaster solutions. Replication Software Suite offers autonomic replication, transparent failover, and point-in-time copies enabling seamless disaster recovery for critical data. The intelligence and automation of the software suites offerings reduce complexity, human resource requirements, and potential errors at critical moments.

To help protect, share, and freely move data across data centers without impacting business applications, this software suite bundles:

- **HPE StoreServ Virtual Copy**—Reservation-less, nonduplicative, copy-on-write software that consumes capacity only for changed data in fine-grained increments and without ever duplicating changed data within a snapshot tree. Virtual Copy helps users to protect and share data flexibly and affordably from any application without sacrificing performance, availability, or versatility.

- **HPE StoreServ Remote Copy**—Protects data across different sites using simple but powerful replication technology. It provides enterprise and cloud data centers autonomic disaster recovery and replication technology to protect and share data from any application.

- **HPE StoreServ Peer Persistence**—Offers automated transparent failover across metropolitan distances. It federates a storage infrastructure across data centers unconstrained by their physical boundaries and uses them effectively as one single storage resource. Peer Persistence software helps companies to federate their StoreServ Storage systems present at geographically separated data centers. This inter-site federation of storage allows customers to use their data centers more effectively by allowing them to move applications from one site to another as per their business need and without any application downtime.

- **HPE StoreServ Cluster Extension (CLX)**—Provides protection against system downtime with automatic failover of application services and read/write enabling of remotely mirrored StoreServ storage systems. CLX adapts in real time, to real-life situations, providing protection via rapid site recovery. CLX requires no server reboots or logical unit number (LUN) presentation/mapping changes during failover. This is of benefit if the storage administrator is unaware of the outage, unable to respond, or simply not present, for true hands-free failover/failback decision-making.

HPE StoreServ Data Optimization Software Suite

The StoreServ Data Optimization Software Suite enables you to configure, optimize, and reconfigure StoreServ storage for improved control, efficiency, and effectiveness without disturbing users or applications. The Data Optimization suite is designed specifically for storage requirements that are constantly changing. To help you react quickly to constant fluctuations in allocation, capacity management, and application performance, this software suite bundles:

- **HPE StoreServ Adaptive Optimization**—Helps deliver an increase in IOPS and reduced latency at a lower storage cost than traditional storage

- **HPE StoreServ Dynamic Optimization**—Analyzes how volumes use physical disks and automatically makes intelligent, nondisruptive adjustments to help ensure optimal volume performance and capacity utilization

- **HPE StoreServ Priority Optimization**—Applies priority policies to ensure that the mission-critical applications within the StoreServ system get the required service level and performance without contention

- **HPE StoreServ Peer Motion**—Enables nondisruptive data mobility across systems providing the flexibility to load balance at will, refresh technology seamlessly, reduce asset lifecycle management costs, and lower technology refresh

Together the software suites provide the flexibility to react quickly to changing application and infrastructure requirements all while balancing cost and performance.

Video demo—using on-premises management with HPE ICsp

Watch this seven-minute HPE ICsp demonstration that shows how to deploy an OS to a physical server blade.

 Note

To view this HPE ICsp demonstration, scan this QR code or enter the URL into your browser.

https://youtu.be/ReUDYQoJitw

Resources

 Note

For more information, scan these QR codes or enter the URLs into your browser.

HPE OneView demonstrations

https://vrp.glb.itcs.hpe.com/SDP/Content/Booth.aspx?ID=131&tag=1061

HPE OneView homepage

**http://www8.hp.com/us/en/business-solutions/converged-systems/
oneview.html**

Insight Control server provisioning homepage

**http://www8.hp.com/us/en/products/servers/management/
insight-control/provisioning-migration-server.html**

Learning check

1. Which phase of transforming to a hybrid infrastructure involves converged infrastructure management tools from HPE?

   a. Creating an on-demand infrastructure foundation

   b. Enabling agile IT operations

   c. Optimizing application development

   d. Speeding to market with cloud

2. How can a SDDC help customers enable agile IT operations?

   _____

   _____

   _____

   _____

3. Name a few tasks that can be performed using HPE OneView.

   _____

   _____

   _____

   _____

4. List the benefits specific to HPE OneView 2.0.

   _____

   _____

   _____

   _____

5. HPE OneView for Microsoft System Center allows users to consistently and reliably deploy Hyper-V clusters and take advantage of the automation engine of HPE OneView.

   ☐ True

   ☐ False

6. What are the capabilities and functions of HPE OneView for VMware vCenter Log Insight?

_____

_____

_____

_____

7. What are the types of sample build plans provided by HPE Insight Control server provisioning? (Select four.)

   a. ProLiant Hardware

   b. ProLiant Target Servers

   c. ProLiant Scripting

   d. ProLiant Operating System

   e. ProLiant Software

   f. ProLiant Deploy

   g. ProLiant Combination

8. StoreServ Peer Persistence is the management and reporting console that offers converged management of file and block storage on StoreServ arrays.

   ☐ True

   ☐ False

Learning check answers

1. Which phase of transforming to a hybrid infrastructure involves converged infrastructure management tools from HPE?

   a. Creating an on-demand infrastructure foundation

   b. Enabling agile IT operations

   c. Optimizing application development

   d. Speeding to market with cloud

2. How can a SDDC help customers enable agile IT operations?

   The SDDC solution provides:

   - **Improved efficiencies by extending virtualization across all resources**

   - **Increased agility to provision resources for business applications more quickly**

- **Improved control over application availability and security through policy-based** definitions
- **The flexibility to run new and existing applications in multiple platforms and clouds**
- **The potential to reduce energy usage and costs for data center hardware**

3. Name a few tasks that can be performed using HPE OneView.

   HPE OneView can be used to:

   - **Provision the data center**
   - **Manage and maintain firmware and configuration changes**
   - **Monitor the data center and respond to issues**

4. List the benefits specific to HPE OneView 2.0.

   - **Server profile templates**
   - **Driver and firmware updates**
   - **Profile mobility**
   - **Storage snapshots and clones**
   - **SAN health and diagnostics**
   - **Virtual Connect enhancements**

5. HPE OneView for Microsoft System Center allows users to consistently and reliably deploy Hyper-V clusters and take advantage of the automation engine of HPE OneView.

   ☐ **True**
   ☐ False

6. What are the capabilities and functions of HPE OneView for VMware vCenter Log Insight?

   - **It provides log aggregation and indexing with search and analytics capabilities.**
   - **It collects, imports, and analyzes logs to provide information related to systems, services, and applications.**

7. What are the types of sample build plans provided by HPE Insight Control server provisioning? (Select four.)

   a. ProLiant Hardware

   b. ProLiant Target Servers

   c. ProLiant Scripting

   d. ProLiant Operating System

   e. ProLiant Software

   f. ProLiant Deploy

   g. ProLiant Combination

8. StoreServ Peer Persistence is the management and reporting console that offers converged management of file and block storage on StoreServ arrays.

   ☐ True

   ☐ **False**

Summary

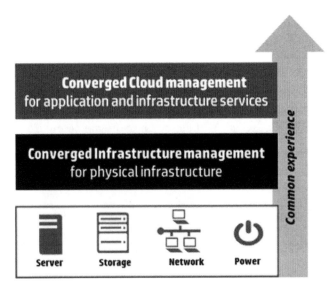

Figure 5-35 Transforming to a hybrid infrastructure

- One of the phases of transforming to a hybrid infrastructure is enabling agile IT operations (Figure 5-35). This phase incorporates converged infrastructure management tools from HPE, namely HPE OneView and ICsp. For customers that are moving toward cloud solutions, HPE

Helion CloudSystem adds applications and an infrastructure services management layer that provides:

- Self-service portal

- Multi-tiered services

- Physical and virtual provisioning

- HPE OneView is a software-defined solution that can be used to provision and monitor the data center, along with managing firmware and configuration changes.

- HPE OneView 2.0 introduces server profile template, driver and firmware updates, profile mobility, storage snapshots and clones, SAN health and diagnostics, Virtual Connect enhancements.

- HPE OneView capabilities include the following:

  - HPE OneView for Microsoft System Center provides seamless integration of unique HPE ProLiant and HPE BladeSystem manageability features into the Microsoft System Center consoles.

  - HPE OneView for VMware vCenter seamlessly integrates the manageability features of HPE OneView, ProLiant servers, BladeSystem, HPE Networking, and HPE Storage into the VMware vCenter console.

  - HPE Insight Control server provisioning is a virtual appliance used to install and configure HPE ProLiant servers. It is optimized for ProLiant Gen8 and later servers.

- The SSMC offers a common interface and language with HPE OneView. SSMC for StoreServ storage delivers the efficiency and agility required by virtual, cloud, and ITaaS environments with a software portfolio managed by SSMC.

Summary of previous HPE OneView releases
HPE OneView 1.1 (June 2014)

- Server profile support for SAN storage provisioning

- Automated SAN zoning

- Configuration support for Density Line (DL) servers

- Automation of enclosure configuration

- Support for non-Virtual Connect environments (within the BladeSystem enclosures)

- Hyper-V appliance as a host

- Support for native Virtual Connect modules based on Fibre Channel

- Integration with VMware vCenter, Microsoft System Center, Red Hat Enterprise Virtualization
- Instant-on, factory-embedded, and enterprise (Activation Key Agreement) licensing

HPE OneView 1.2 (December 2014)

- Licensing changes—free standard offering and for-fee advanced offering
- ProLiant Gen9 support
- Predefined reports, exportable to CSV and Microsoft Excel
- Automated zoning for the HPE 5900CP switches
- Ephemeral StoreServ volume support
- Monitoring of Cisco Nexus 6001 switches
- HPE Operations Analytics for HPE OneView
- Updated third-party monitoring tool integration (VMware vCenter, Microsoft System Center, and Red Hat Enterprise Virtualization)

6 Practice Test

Introduction

The HPE ATP Navigating the Journey to the Cloud certification exam is designed for candidates with "on-the-job" experience in a presales consultative role and have the skill and knowledge on how to identify, describe, and position the HPE Converged Infrastructure and HPE Helion Cloud solutions based on customer requirements. The candidate should also be able to demonstrate the key benefits and functionality of the HPE Converged Infrastructure and HPE cloud technologies.

The intent of this book is to set expectations about the context of the exam and to help candidates prepare for it. Recommended training to prepare for this exam can be found at the HP Certification and Learning website (http://certification-learning.hpe.com), as well as in books like this one. It is important to note that although training is recommended for exam preparation, successful completion of the training alone does not guarantee that you will pass the exam. In addition to training, exam items are based on knowledge gained from on-the-job experience and application, as well as other supplemental reference material that may be specified in this guide.

Who should take this exam?

Most successful candidates have worked in the IT industry for at least 1–2 years and have foundational HPE server, storage, networking, and management product knowledge and have taken the recommended training. The candidate is able to articulate the HPE Converged Infrastructure solutions strategy; including the purpose, benefits, and components; determine high-level customer requirements; generate a basic configuration; and demonstrate the key features of the solutions.

Exam details

The following are details about the exam:

- **Exam ID:** HPE0-D33
- **Number of items:** 65
- **Item types:** matching, multiple choice (single-response), multiple choice (multiple-response), point, and click

- **Exam time**: 1 hour 45 minutes
- **Passing score**: 66%
- **Reference material**: No online or hard copy reference material will be allowed at the testing site.

HPE0-D33 testing objectives

20%—Fundamental Converged Infrastructure and Cloud architectures and technologies

- Identify and describe functional components of a converged infrastructure and the tools that are used to manage the converged infrastructure resources.
- Compare and contrast the types of service models (IaaS, PaaS, SaaS, and XaaS).
- Describe the types of cloud delivery models.
- Identify and describe converged infrastructure and cloud enabling technologies and components.
- Explain business benefits/value and risks/costs associated with converged infrastructure and cloud implementations.

25%—HPE Converged Infrastructure and HPE Helion cloud products, solutions, and warranty/service offerings

- Explain the HPE Helion cloud strategy and Transformation Areas (TAs)
- Identify, describe, and differentiate the functionality of HP Converged Infrastructure software
- Describe the positioning and goals of the HPE Helion CloudSystem software portfolio
- Identify the HPE Security solutions in the context of Transformation Areas
- Identify and position HPE ConvergedSystems
- Identify and position HPE service and warranty offerings for Converged Infrastructure and HPE Helion
- Select and use the appropriate HPE proactive resources to support the customer Converged Infrastructure requirements
- Describe HPE Reference Architecture
- Describe the HPE strategy for converged infrastructure management

12%—Solution planning and design

- Plan and design a Converged Infrastructure solution

- Size a solution

17%—Solution implementation (install, configure, setup, customize, and integrate)

- Perform the predelivery tasks

- Implement the delivery tasks

- Implement postdelivery tasks

11%—Solution enhancement (performance-tune, optimize, and upgrade)

- Compare the existing solution design to the customer requirements and document differences

- Upgrade or expand the solution

15%—Solution management

- Maintain the management infrastructure

- Monitor and manage the HPE Converged Infrastructure resources (physical and virtual)

- Create, maintain, and perform backup and recovery processes, as applicable to HPE Converged Infrastructure and HPE Helion cloud

Test preparation questions and answers

The following questions will help you measure your understanding of the material presented in this book. Read all of the choices carefully, as there may be more than one correct answer. Choose all correct answers for each question.

Questions

1. What do today's businesses need to do in order to thrive in the "idea economy"? (Select two.)

   a. Rapidly compose services

   b. Efficiently store data

c. Rely on silver-bullet technologies

d. Be contextually aware and predictive

e. Migrate data to public cloud

2. Which transformation area is supported by HPE products such as HPE Haven, HPE Vertica, and HPE ControlPoint?

a. Protect the digital enterprise

b. Empower the data-driven organization

c. Enable workplace connectivity

d. Transform to a hybrid infrastructure

3. HPE OneView and HPE Operations Analytics for HPE OneView enable which phase of transformation to hybrid infrastructure and on-demand IT?

a. Create on-demand infrastructure foundation

b. Enable agile IT operations

c. Optimize application development

d. Speed to market with cloud

4. What is the primary goal of implementing a converged infrastructure and cloud?

a. Broker cloud services across multiple clouds

b. Using siloed IT framework

c. Adoption of private cloud technologies

d. Aligning IT to applications, workloads, and business

5. A customer is looking for a packaged solution for their remote office virtual desktop implementation that meets these requirements:

- Has a small footprint
- Is quick and simple to deploy and scale
- Is appliance-based
- Supports VMware

What should you recommend for this customer?

a. HPE ConvergedSystem 250-HC

b. HPE ConvergedSystem 900

c. HPE ConvergedSystem 700x

d. HPE ConvergedSystem 300

6. Where can you find HPE Verified Reference Architectures?

a. HPE SPOCK

b. HPE Information Library

c. HPE Live Network

d. HPE Solution Demonstration Portal

7. What are the benefits of using the Customer Intent Document? (Select three.)

a. Faster time to production

b. Software installation can be executed by customer

c. Systems are preconfigured and ready to deploy

d. Reduced overall installation cost

e. The design and sizing of the solution are part of the CID process

8. Which HPE resource allows customers to access the most current content add-ons and extensions for their HPE IT Performance Suite products?

a. Visio diagrams

b. HPE Factory Express

c. HPE Live Network

d. HPE Information Library

9. What are the characteristics of cloud-native applications? (Select four.)

a. Capable of scaling out

b. Tightly coupled and stateful

c. Having lightweight runtimes

d. Self-configuring

e. Capable of scaling up

f. Administrator-controlled

g. Operating system- and VM-aware

h. Loosely coupled and microservice-based

10. Which cloud service model is described by these statements?

- Provisions processing, storage, networks, and other fundamental computing resources where the consumer is able to deploy and run arbitrary software, which can include operating systems and applications.

- Provides the consumer with control over operating systems, storage, deployed applications, and possibly limited control of select networking components such as host firewalls

- Does not provide the consumer with management capabilities of the underlying cloud infrastructure

    a. IaaS

    b. PaaS

    c. SaaS

    d. ITaaS

11. Which cloud deployment model is described by this definition?

- The cloud infrastructure is provided solely for a single client organization (tenant).

- It is owned and operated by a third-party cloud service provider.

- It can be delivered from a customer's facility or from a facility owned by the service provider.

    a. Public cloud

    b. Hybrid cloud

    c. Managed private cloud

    d. Managed virtual private cloud

12. What is the process of providing access to services from different resource pools, such as from traditional IT or private and public clouds?

    a. Elasticity

    b. Orchestrating

    c. Hosting

    d. Brokering

13. A customer is moving toward a hybrid infrastructure and on-demand IT and is evaluating HPE Helion. One of their requirements is integration with classic applications running in a traditional scale-up environment.

    Which Helion solution meets this customer's requirement?

    a. HPE Helion CloudSystem

    b. HPE Helion Development Platform

    c. HPE Helion OpenStack

    d. HPE ConvergedSystem 900

14. You are helping a customer determine which HPE products meet their hybrid infrastructure and on-demand IT needs. Their requirements are:

    ● Support for traditional and cloud-native workloads

    ● A turnkey hardware and software solution

    ● Complete lifecycle management for application services, including compliance

    Which Helion products should you recommend? (Select two.)

    a. HPE Helion OpenStack

    b. HPE Helion Development Platform

    c. HPE Helion CloudSystem Foundation

    d. HPE Helion CloudSystem Enterprise

    e. HPE Helion Rack

    f. HPE ConvergedSystem 700x

15. Which HPE Helion product includes integration with HPE CloudSystem Matrix and the HPE Matrix Operating Environment, enabling customers to consume resources and services from this legacy cloud environment?

    a. HPE Helion OpenStack

    b. HPE Helion Development Platform

    c. HPE Helion CloudSystem Foundation

    d. HPE Helion CloudSystem Enterprise

16. Which HPE Helion product is an integrated, complete OpenStack Swift-based object storage solution that delivers massive scalability, ease of management, and durability?

    a. HPE Helion Content Depot

    b. HPE Helion Eucalyptus

    c. HPE Helion Rack

    d. HPE Helion CloudSystem

17. How many companies contribute to the global OpenStack community?

    a. 100

    b. Approximately 76

    c. More than 500

    d. More than 200

18. A customer is looking for ways to improve management of devices, networks, and IP addresses in its data center. They require:

    - Rapid provisioning
    - More control
    - Scalability

    Which system meets this customer's requirements?

    a. OpenStack Networking

    b. Object storage

    c. Amazon Elastic Compute Cloud

    d. OpenStack compute

19. Which OpenStack service provides on-demand, self-service persistent block storage for guest virtual machines?

    a. Heat

    b. Swift

    c. Cinder

    d. Ironic

20. Which OpenStack service provisions bare-metal machines instead of virtual machines and is based on the Nova bare-metal driver?

    a. Nova

    b. Ironic

    c. Sahara

    d. Ceilometer

21. Which design principles describe Nova? (Select three.)

    a. Nonrecoverable

    b. Fault tolerant

    c. SDN capable

    d. API compatible

    e. Component based

22. What does Glance use to allow users to query VM image metadata and retrieve the actual images?

    a. RESTful API

    b. Simple file systems

    c. Heat

    d. Messaging bus

23. How does HPE contribute to enhancing security for OpenStack services? (Select three.)

    a. HPE makes financial contributions to OpenStack on a regular basis

    b. HPE sponsored several security activities in the OpenStack community

    c. HPE is participating in the Barbican project, focusing on encryption

    d. HPE led formation of the OpenStack Security Group

    e. HPE protects OpenStack.org from breaches

24. Implementing a software-defined data center enables organizations to deliver which type of service model?

    a. Infrastructure-as-a-Service

    b. Platform-as-a-Service

c. Software-as-a-Service

d. IT-as-a-Service

25. What does HPE OneView use to capture the best practices necessary to deploy a server, storage, network, or power management component?

a. Profile

b. Template

c. ICsp

d. Build plan

26. Which technology enables HPE OneView to monitor and manage ProLiant server blade and rack server resources?

a. Onboard administrator

b. iLO management engine

c. Virtual Connect

d. Insight Online

27. What happens when you issue a task in the HPE OneView interface and it takes longer than 200 milliseconds to complete?

a. The task is suspended with a warning message to the operator.

b. The task becomes asynchronous and continues running.

c. The task gains a priority processing status to finish within the 500-millisecond limit.

d. The task is rescheduled to complete during a predetermined "quiet time."

28. What does HPE Insight Control server provisioning use to perform a PXE-free installation of the Microsoft Windows operating system on bare-metal ProLiant Gen9 server blades?

a. UEFI

b. Intelligent Provisioning

c. Onboard administrator

d. WAIK

29. Where does HPE Insight Control server provisioning store software components such as operating system files, captured images, drivers, and firmware?

    a. On the media server

    b. In the appliance datastore

    c. In a cloud-based object storage repository

    d. On a flash medium connected to the host server

30. Which phase of transforming to a hybrid infrastructure involves HPE application development and delivery services?

    a. Creating an on-demand infrastructure foundation

    b. Optimizing application development

    c. Enabling agile IT operations

    d. Speeding to market with cloud

31. With HPE OneView 2.0, which items can be updated from the template? (Select three.)

    a. Firmware baseline

    b. Python

    c. Local RAID settings

    d. BIOS settings

    e. Enclosure groups

32. You are helping a customer determine which HPE products meet their server management needs. Their requirements are:

    - Manageability features that are compatible with Microsoft System Center consoles

    - Comprehensive system health and alerting

    - HPE Virtual Connect fabric visualization

    Which solution should you recommend?

    a. HPE OneView for Microsoft System Center

    b. HPE OneView for VMware vCenter

    c. HPE Moonshot

33. Which products are integrated within HPE OneView for VMware vCenter? (Select three.)

    a. HPE ProLiant servers

    b. Microsoft Hyper-V

    c. HPE BladeSystem

    d. HPE Networking

    e. Microsoft System Center

Answers

1. ☑ **A** and **D** are correct. To thrive in the "idea economy" customers need to rapidly compose new services from any source to meet the evolving needs of customers and citizens. They also need to harness 100% of data to generate real-time instant insights for continuous improvement, innovation, and learning.

    ☒ **B, C,** and **E** are incorrect. IT must shift focus from just storing and managing data to providing real-time insight and understanding. Many customers struggle with overreliance on silver bullets and public cloud may be the ideal solution for certain workloads and data, but it is not appropriate for all.

    For more information, see Chapter 1.

2. ☑ **B** is correct. HPE Haven enables customers to deploy a big data platform, HPE Vertica helps customers to modernize the enterprise data warehouse, and HPE ControlPoint enables best-in-class data management. These are part of the transformation area *empower the data-driven organization.*

    ☒ **A, C,** and **D** are incorrect. HPE ArcSight, ClearPass, Fortify, and other security products protect the digital enterprise. AirWave from Aruba, HPE WorkSite, AppPulse Mobile, and other workplace connectivity products enable workspace connectivity. ProLiant Gen9 servers, HPE OneView, and other hybrid infrastructure solutions help customers transform to a hybrid infrastructure.

    For more information, see Chapter 1.

3. ☑ **B** is correct. OneView and Operations Analytics transform management of infrastructure and clouds with analytics and automation and enable agile IT operations.

    ☒ **A, C,** and **D** are incorrect ProLiant Gen9 servers, and StoreServ solutions help customers create on-demand infrastructure. HPE application development and delivery services, Application Lifecycle Management, and Codar are used to optimize application development. HPE Helion CloudSystem and Helion OpenStack and Development Platform help customers speed to market with cloud.

    For more information, see Chapter 1.

4. ☑ **D** is correct. Aligning IT to the applications, workloads, and business is the primary goal of a converged infrastructure and cloud.

   ☒ **A, B,** and **C** are incorrect. Brokering cloud services across multiple clouds, using siloed IT framework, and adopting private cloud technologies are not the primary goal of implementing a converged infrastructure and cloud.

   For more information, see Chapter 2.

5. ☑ **A** is correct. HPE ConvergedSystem 250-HC provides an entry-level solution with an easy, open path to hybrid cloud delivery.

   ☒ **B, C,** and **D** are incorrect. HPE ConvergedSystem 900 is designed for SAP HANA scale-out and scale-up configurations, HPE ConvergedSystem 700x provides a private cloud solution and supports multi-hypervisor, multi-OS, and heterogeneous infrastructure environments, and HPE ConvergedSystem 300 is designed for high-scalability and massively parallel processing (MPP) to support high-volume data demands in a Microsoft Analytics Platform environment.

   For more information, see Chapter 2.

6. ☑ **B** is correct. Reference architectures are posted in the HPE Information Library under the Reference Architectures and Related Applications/Software menu.

   ☒ **A, C,** and **D** are incorrect. HPE SPOCK provides detailed information about supported HPE storage product configurations, HPE Live Network provides access to the most current content add-ons and extensions for HPE IT Performance Suite products, and HPE Solution Demonstration Portal provides a central location for all demonstrations, webinars, and supporting collateral that showcase HPE technologies.

   For more information, see Chapter 2.

7. ☑ **A, C,** and **D** are correct. The CID reduces time to utilization and speeds time to production, ultimately helping the customer achieve a faster ROI. It is used by the HPE Factory Express team to build and deliver a complex solution from order to operations in as little as 20 days, meaning that the customer takes delivery of the finished solution rather than the individual parts. The CID also reduces overall cost of the installation due to lower labor rates and better resource utilization.

   ☒ **B** and **E** are incorrect. Software installation can be executed by customer and designing and sizing the solution are not benefits of the CID.

   For more information, see Chapter 2.

8. ☑ **C** is correct. HPE Live Network provides access to the most current content add-ons and extensions for HPE IT Performance Suite products.

   ☒ **A, B,** and **D** are incorrect. Visio diagrams, HPE Factory Express, and HPE Information Library do not allow customers to access the most current content add-ons and extensions for their HPE IT Performance Suite products.

   For more information, see Chapter 2.

9. ☑ **A, C, D,** and **H** are correct. Capable of scaling out, having lightweight runtimes, self-configuring and loosely coupled, and microservice-based are characteristics of cloud-native applications.

   ☒ **B, E, F,** and **G** are incorrect. Tightly coupled and stateful, capable of scaling up, administrator-controlled and operating system- and VM-aware are not characteristics of cloud-native applications.

   For more information, see Chapter 3.

10. ☑ **A** is correct. IaaS provides for the services described by all of the statements.

    ☒ **B, C,** and **D** are incorrect. PaaS enables the consumer to deploy onto the cloud infrastructure consumer-created applications using programming languages and tools supported by the provider, such as Java, Python, or .NET. SaaS enables the consumer to use the provider's applications running on a cloud infrastructure. ITaaS involves changing the business approach from a technology focus to a service focus.

    For more information, see Chapter 3.

11. ☑ **C** is correct. Managed private cloud is a secure, dedicated private cloud that runs on infrastructure that is dedicated to a single client organization (tenant), but managed by a third party.

    ☒ **A, B,** and **D** are incorrect. In a public cloud (also known as an *external cloud*), all cloud assets that underlie the services are owned and operated by a provider, and multiple users access the services over the Internet. A hybrid cloud environment makes use of services sourced from internal and external providers. A managed virtual private cloud is a secure, multitenant private cloud that allows multiple organizations to share the cloud infrastructure.

    For more information, see Chapter 3.

12. ☑ **D** is correct. Brokering enables access to services from different resource pools, such as from traditional IT or private and public clouds.

    ☒ **A, B,** and **C** are incorrect. Elasticity is the ability to dynamically expand or contract a computing resource based on demand. Orchestrating means executing a predefined workflow used in the process of provisioning and managing infrastructure. Hosting is the process of making services available via a network.

    For more information, see Chapter 3.

13. ☑ **A** is correct. HPE Helion CloudSystem integrates with classic applications running in a traditional scale-up environment, for example, Oracle on Linux and SQL Server on Windows.

    ☒ **B, C,** and **D** are incorrect. HPE Helion Development Platform enables customers to develop cloud-native applications. HPE Helion OpenStack is not a good fit for a customer requiring integration with classic applications running in a traditional scale-up environment. HPE ConvergedSystem 900 is designed for SAP HANA scale-out and scale-up configurations.

    For more information, see Chapter 3.

14. ☑ **D** and **F** are correct. HPE Helion CloudSystem Enterprise and HPE ConvergedSystem 700x meet all of the customer requirements.

　☒ **A, B, C,** and **E** are incorrect. HPE Helion OpenStack does not provide a turnkey hardware and software solution. HPE Helion Development Platform, HPE Helion CloudSystem Foundation, and HPE Helion Rack do not provide complete lifecycle management for application services, including compliance.

　For more information, see Chapter 3.

15. ☑ **D** is correct. HPE Helion CloudSystem Enterprise includes integration with HPE CloudSystem Matrix and the HPE Matrix Operating Environment.

　☒ **A, B,** and **C** are incorrect. HPE Helion OpenStack, HPE Helion Development Platform, and HPE Helion CloudSystem Foundation do not include integration with HPE CloudSystem Matrix and the HPE Matrix Operating Environment.

　For more information, see Chapter 3.

16. ☑ **A** is correct. HPE Helion Content Depot is an integrated, complete OpenStack Swift-based object storage solution.

　☒ **B, C,** and **D** are incorrect. HPE Helion Eucalyptus is AWS-compatible private cloud software. HPE Helion Rack is an OpenStack-based private cloud solution. HPE Helion CloudSystem is a fully integrated, end-to-end, private cloud solution built for traditional and cloud-native workloads.

　For more information, see Chapter 3.

17. ☑ **D** is correct. More than 200 companies are involved in the OpenStack ecosystem, including HPE, Rackspace, IBM, NTT, Dell, Canonical, Red Hat, and VMware.

　☒ **A, B,** and **C** are incorrect. More than 200 companies are involved in the OpenStack ecosystem.

　For more information, see Chapter 4.

18. ☑ **A** is correct. OpenStack Networking provides ways to improve management of devices, networks, and IP addresses in a data center.

　☒ **B, C,** and **D** are incorrect. Object storage provides a block storage service. Amazon Elastic Compute Cloud provides IaaS. OpenStack compute provides virtual servers on demand.

　For more information, see Chapter 4.

19. ☑ **C** is correct. Cinder provides persistent block storage to VMs.

　☒ **A, B,** and **D** are incorrect. Heat is a template-based orchestration solution for provisioning the resources required to deploy a cloud. Swift provides object storage. Ironic is an integrated OpenStack service that provisions bare-metal machines instead of virtual machines.

　For more information, see Chapter 4.

20. ☑ **B** is correct. Ironic is an integrated OpenStack service that provisions bare-metal machines instead of virtual machines.

    ☒ **A, C,** and **D** are incorrect. Nova provides virtual servers on demand. Sahara provides capabilities to provision and scale Hadoop clusters in OpenStack. Ceilometer provides utilization metering for OpenStack deployments.

    For more information, see Chapter 4.

21. ☑ **B, D,** and **E** are correct. Nova isolates processes to avoid cascading failures, thereby providing fault tolerance. It offers API compatibility with popular public cloud providers such as Amazon EC2 and is component based allowing quick changes and additions to the functionality.

    ☒ **A** and **C** are incorrect. Nova provides easy diagnosis, debugging, and correction of failures, and it does not provide SDN capability (Neutron does this).

    For more information, see Chapter 4.

22. ☑ **A** is correct. Glance allows users to query VM image metadata and retrieve the actual images via a RESTful API.

    ☒ **B, C,** and **D** are incorrect. Glance is not concerned with file systems. Heat is a template-based orchestration solution for provisioning the resources required to deploy a cloud. The messaging bus enables components of individual OpenStack services to communicate with each other. It does not allow users to query VM image metadata.

    For more information, see Chapter 4.

23. ☑ **B, C,** and **D** are correct. HPE hired security experts and sponsored several security activities in the OpenStack community. HPE is participating in the OpenStack Barbican project, which focuses on encryption. HPE led the formation of the OpenStack Security Group.

    ☒ **A** and **E** are incorrect. HPE does not make financial contributions to OpenStack. HPE does not protect OpenStack.org from breaches.

    For more information, see Chapter 4.

24. ☑ **D** is correct. Implementing an SDDC delivers an IT-as-a-Service solution.

    ☒ **A, B,** and **C** are incorrect. IaaS, PaaS, and SaaS are individual components of ITaaS.

    For more information, see Chapter 5.

25. ☑ **B** is correct. Templates are used to design boilerplates for the underlying network, server, and storage objects that manage systems in a converged infrastructure.

    ☒ **A, C,** and **D** are incorrect. Server profiles define the personality assigned to a server that includes firmware baseline, BIOS settings, network connectivity, boot configuration, Integrated Lights-Out (iLO) settings, and unique IDs. HPE Insight Control server provisioning (ICsp) is a tool, packaged as a virtual appliance, used to install and configure ProLiant servers. Build plans are used in ICsp to cause actions to be performed against servers. For example, installing a server, resetting a server, or updating firmware.

For more information, see Chapter 5.

26. ☑ **B** is correct. OneView communicates with ProLiant servers using the iLO management engine.

   ☒ **A, C,** and **D** are incorrect. OneView does not use the OA to communicate with ProLiant servers, it communicates directly with the servers' iLO. Virtual Connect provides wire-once connectivity for LAN and SAN resources. Insight Online is a feature in the HP Support Center that allows customers to view all of their devices that are being remotely monitored by HP.

For more information, see Chapter 5.

27. ☑ **B** is correct. Tasks that take longer than 200 milliseconds to process become asynchronous tasks. This means such tasks can continue processing in the background as the administrator moves on to other activities.

   ☒ **A, C,** and **D** are incorrect. Tasks that take longer than 200 milliseconds are not suspended, do not gain a priority processing status, and are not rescheduled to complete during a predetermined "quiet time".

For more information, see Chapter 5.

28. ☑ **B** is correct. When using HPE Insight Control server provisioning with HPE ProLiant Gen8 and later servers, Intelligent Provisioning is used to perform a PXE-free installation of the operating system.

   ☒ **A, C,** and **D** are incorrect. UEFI is the preboot environment on ProLiant Gen9 servers. OA provides local and remote administration of HP BladeSystem c-Class enclosures. Windows Automated Installation Kit is a collection of Microsoft tools designed to help deploy Microsoft Windows operating system images to target computers.

For more information, see Chapter 5.

29. ☑ **A** is correct. There is insufficient storage space on the ICsp appliance, so HPE Insight Control server provisioning stores software components such as operating system files, captured images, drivers, and firmware on the media server.

   ☒ **B, C,** and **D** are incorrect. There is insufficient storage space on the ICsp appliance datastore, and ICsp does not store software components in a cloud-based object storage repository or on flash storage connected to the host computer.

For more information, see Chapter 5.

30. ☑ **B** is correct. Optimizing application development is supported by HPE application development and delivery services.

    ☒ **A, C,** and **D** are incorrect. Creating an on-demand infrastructure foundation is supported primarily by HPE ConvergedSystem, HPE BladeSystem and HPE ProLiant Gen9 servers, and HPE StoreServ. Enabling agile IT operations is supported mainly by HPE OneView, HPE Operations Analytics for HPE OneView and HPE Insight Control server provisioning. Speeding to market with cloud is supported mainly by the HPE Helion portfolio of products and solutions.

    For more information, see Chapter 5.

31. ☑ **A, C,** and **D** are correct. OneView 2.0 allows the firmware baseline to be updated from a template and allows local RAID settings and BIOS settings to be updated from a template.

    ☒ **B** and **E** are incorrect. Profiles can be brought into compliance using tools such as Python, but Python is not represented in the template. OneView 2.0 does not allow enclosure groups to be updated from a template.

    For more information, see Chapter 5.

32. ☑ **A** is correct. HPE OneView for Microsoft System Center meets all of the customer needs.

    ☒ **B** and **C** are incorrect OneView for VMware vCenter integrates with VMware vCenter, not Microsoft System Center. HPE Moonshot is an energy-efficient, integrated server system and does not include HPE Virtual Connect.

    For more information, see Chapter 5.

33. ☑ **A, C,** and **D** are correct. HPE OneView for VMware vCenter seamlessly integrates the manageability features of OneView with HPE ProLiant servers, HPE BladeSystem, and HPE Networking into the VMware vCenter console.

    ☒ **B** and **E** are incorrect. HPE OneView for VMware vCenter does not integrate with Microsoft Hyper-V or Microsoft System Center.

    For more information, see Chapter 5.

HPE ASE — Data Center and Cloud Architect V3

7 A Closer Look at HPE Helion OpenStack

WHAT IS IN THIS CHAPTER FOR YOU?

After completing this chapter, you should be able to:

✓ Recognize the main components and benefits of the Hewlett Packard Enterprise (HPE) Helion Platform

✓ Explain how HPE Helion Rack provides value for customers

OpenStack software

Before proceeding with this section, assess your existing experience with Helion OpenStack by answering the following questions.

Assessment activity

1. Do you have any experience with Helion OpenStack?

   _____

   _____

   _____

2. Do you know any customers who have implemented Helion OpenStack? If so, what version? What were the use cases and the customers' experience?

   _____

   _____

   _____

   _____

3. Have any of your customers deployed the Helion Development Platform? If so, what was their experience?

_____

_____

_____

_____

OpenStack logical architecture

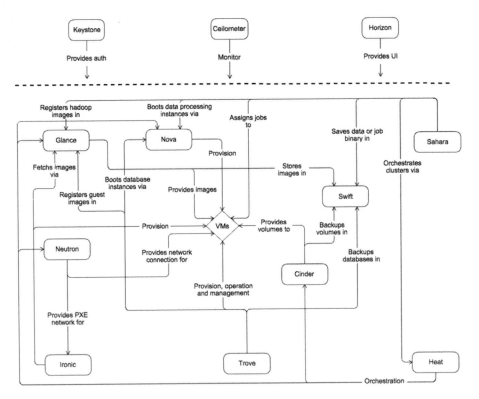

Figure 7-1 OpenStack logical architecture

Figure 7-1 shows how the major OpenStack services are interconnected.

In this architecture:

- Users can interact through a common web interface (Horizon) or directly with each service through an API.

- All services authenticate through a common source (facilitated through Keystone).

- Individual services interact with each other through their public APIs, except where privileged administrator commands are necessary.

Accessing OpenStack services with Horizon

Figure 7-2 Accessing OpenStack services with Horizon

You can log in to Horizon as a cloud user or a cloud administrator as shown in Figure 7-2. Depending on your role, a different dashboard view and different set of functions will be available to you.

 Note

This optional two-minute demonstration of the Horizon dashboard (Juno release) shows the data processing capability (Sahara), which can help run big data solutions such as Hadoop on OpenStack. It also shows the database-as-a-service capability (Trove) for automating database management. To watch the video, scan this QR code or enter the URL into your browser.

https://www.youtube.com/watch?feature=player_embedded&v=TgPTjrf1y0A

User view

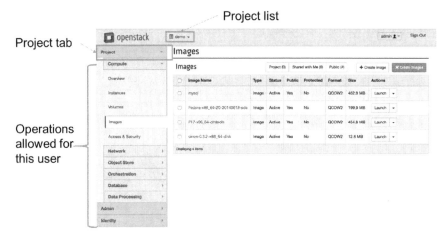

Figure 7-3 User view

The visible tabs and functions in the OpenStack Horizon dashboard depend on the access permissions, or roles, of the user. If you are logged in as an administrative user, the Admin and Projects tabs are displayed. If you are logged in as a non-administrative user, only the Projects tab is displayed. Figure 7-3 shows the dashboard for a user with administrative access. The top right of the window displays the user name.

The Project tab enables you to view and manage the resources for a selected project, including instances and images. You can select a project from the list at the top of the window. The Project tab provides you with access to the Compute, Network, Object Store, and Orchestration menus.

- The Compute menu includes the following:

    - **Overview**—View reports for the project.

    - **Instances**—View, launch, and create a snapshot from stop, pause, or reboot instances (or connect to them through virtual network computing).

    - **Volumes**—View, create, edit, and delete volumes/volume snapshots/images; view instance snapshots; and launch instances from images and snapshots.

    - **Access and Security**—View, create, edit, and delete security groups and security group rules; view, create, edit, import, and delete key pairs; allocate floating IP addresses or release them from a project; and view API endpoints.

- The Network menu includes the following:

    - **Network Topology**—View the network topology.

    - **Networks**—Create and manage public and private networks.

    - **Routers**—Create and manage subnets.

- The Object Store menu includes the following:

  - **Containers**—Create and manage containers and objects.

- The Orchestration menu includes the following:

  - **Stacks**—Orchestrate multiple composite cloud applications.

Note

Depending on the version of OpenStack, additional menu options and services may be available (such as database and data processing, as shown in Figure 7-3).

Administrator view

Figure 7-4 Administrator view

If you are logged in as an administrator, the Projects tab and the Admin tab are displayed as shown in Figure 7-4. The Admin tab offers basic administrator functions. Full capabilities are available from the command line and through the OpenStack API.

Administrators can use the Admin tab to:

- View usage

- Manage:

  - Instances

  - Volumes

  - Flavors

  - Images

  - Projects

  - Users

- Services

- Quotas

Accessing OpenStack services with the API

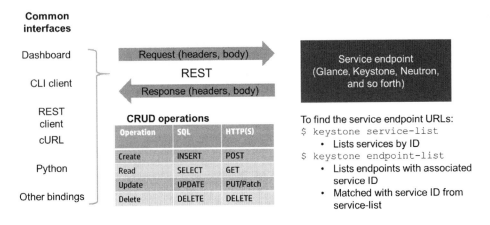

Figure 7-5 Accessing OpenStack services with the API

Horizon is just one way to interact with OpenStack resources. Developers can automate access or build tools to manage their resources using the native OpenStack API or the Amazon EC2 compatibility API.

The left side of Figure 7-5 shows how OpenStack supports several common API interfaces. Through these interfaces, requests can be made and responses can be received from service endpoints of services such as Nova, Glance, Keystone, and Neutron. The interfaces include the following:

- Horizon dashboard

- CLI client

- REST client

- cURL

- Python code

- Other applications, such as database workbenches

The OpenStack API supports the standard create, read, update and delete (CRUD) operations, which are known as post, get, put, and delete from an HTTP/HTTPS perspective.

The right side of Figure 7-5 shows the keystone CLI command being used to list available services and API endpoints.

Example 1: Using the OpenStack CLI client

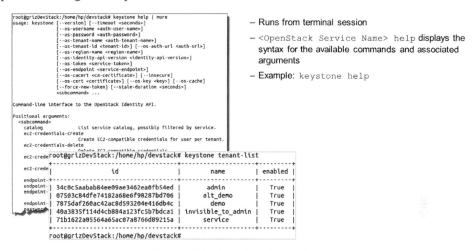

- Runs from terminal session
- `<OpenStack Service Name> help` displays the syntax for the available commands and associated arguments
- Example: `keystone help`

Figure 7-6 OpenStack CLI client

A common method for accessing and using OpenStack is through the OpenStack CLI clients. Commands are executed from a terminal session command line.

OpenStack CLI clients have an extensive help system that is accessible by entering an OpenStack service name followed by the word "help," which displays the syntax for the available commands and any associated arguments. For example, keystone help | more

Figure 7-6 shows the output of the Keystone help command and a list of tenant IDs in Keystone (keystone tenant-list).

Example 2: Using the RESTClient

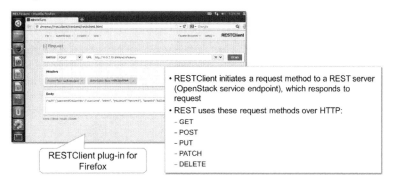

- RESTClient initiates a request method to a REST server (OpenStack service endpoint), which responds to request
- REST uses these request methods over HTTP:
  - GET
  - POST
  - PUT
  - PATCH
  - DELETE

RESTClient plug-in for Firefox

Figure 7-7 RESTClient

The RESTClient plug-in for Firefox, shown in Figure 7-7, allows you to construct and send custom HTTP requests to an OpenStack endpoint service API. This tool is useful when exploring RESTful web services, which use a wide range of HTTP verbs.

RESTClient is a useful alternative to cURL. It provides a user-friendly interface with easy copy-and-paste capability when testing services. This is primarily designed as a developer tool, but it may also be easier to use when demonstrating REST rather than using cURL.

The architectural properties of REST are realized by applying specific interaction constraints, connectors, and data elements. Formal REST constraints include the following:

- **Client/server**—A uniform interface separates clients from servers. This separation means that clients are not concerned with data storage, which remains internal to each server. Therefore, the portability of client code improves. Servers are not concerned with the user interface or user state, so servers can be simple and scalable. Servers and clients can be replaced and developed independently if the interface between them does not change.

- **Stateless**—The client/server communication is further constrained because client context is not stored on the server between requests. Each client request contains all of the information necessary to service the request, and session state is held in the client. It is important to note that the session state can be transferred by the server to another service such as a database to maintain a persistent state for a period of time and allow authentication.

- **Cacheable**—As on the World Wide Web, clients can cache responses. Therefore, responses must, implicitly or explicitly, be defined as cacheable or not cacheable to prevent clients from reusing stale or inappropriate data in response to further requests. Well-managed caching partially or completely eliminates some client/server interactions, further improving scalability and performance.

- **Code on demand (optional)**—Servers can temporarily extend or customize the functionality of a client through the transfer of executable code. Examples of this include compiled components such as Java applets and client-side scripts such as JavaScript. Code on demand is the only optional constraint of the REST architecture.

- **Uniform interface**—The uniform interface between clients and servers simplifies and decouples the architecture, which enables each part to evolve independently.

Applications conforming to these REST constraints are called *RESTful*. If a service violates any of the required constraints, it cannot be considered RESTful. Compliance with these constraints, and thus adherence to the REST architectural style, means that any distributed hypermedia system with desirable properties, such as performance, scalability, simplicity, modifiability, visibility, portability, and reliability, is supported.

Note

These REST constraint details were extracted and modified from *Principled Design of the Modern Web Architecture*, an article published in the Association for Computer Machinery's *Transactions on Internet Technology*. To access the article, scan this QR code or enter the URL into your browser:

https://www.ics.uci.edu/~taylor/documents/2002-REST-TOIT.pdf

Example 3: Using cURL to invoke REST API calls

```
stack@hpdevstack:~/devstack$ curl -i 'http://192.168.5.139:5000/v2.0/tokens' -X POST -H "Con
tent-Type: application/json" -H "Accept: application/json" -d '{"auth": {"tenantName": "adm
in", "passwordCredentials": {"username": "admin", "password": "hpinvent"}}}'
HTTP/1.1 200 OK
Date: Thu, 04 Dec 2014 13:16:48 GMT
Server: Apache/2.4.7 (Ubuntu)
Vary: X-Auth-Token
Content-Length: 5669
Content-Type: application/json

{"access": {"token": {"issued_at": "2014-12-04T13:16:48.540387", "expires": "2014-12-04T14:1
6:48Z", "id": "c1de5b2a750f4b708e12519475be69fb", "tenant": {"description": null, "enabled":
 true, "id": "12e0f9b15e3f4d8ebf0a5d7b48bde070", "name": "admin"}, "audit_ids": ["wx8TZRcpRK
SRE2qe9k6tcw"]}, "serviceCatalog": [{"endpoints": [{"adminURL": "http://192.168.5.139:8004/v
1/12e0f9b15e3f4d8ebf0a5d7b48bde070", "region": "RegionOne", "internalURL": "http://192.168.5
.139:8004/v1/12e0f9b15e3f4d8ebf0a5d7b48bde070", "id": "3499a53d9a6542f5ae8d15efce190341", "p
ublicURL": "http://192.168.5.139:8004/v1/12e0f9b15e3f4d8ebf0a5d7b48bde070"}], "endpoints_lin
ks": [], "type": "orchestration", "name": "heat"}, {"endpoints": [{"adminURL": "http://192.1
68.5.139:8774/v2/12e0f9b15e3f4d8ebf0a5d7b48bde070", "region": "RegionOne", "internalURL": "h
ttp://192.168.5.139:8774/v2/12e0f9b15e3f4d8ebf0a5d7b48bde070", "id": "c00a4821ea864351b6d9c7
```

Figure 7-8 Using cURL

cURL is a useful tool for transferring data to or from a server (an OpenStack service endpoint) using supported protocol services such as the HTTP protocol used by OpenStack APIs. Figure 7-8 shows cURL invoking REST API calls.

Note

To learn more about cURL, scan this QR code or enter the URL into your browser:

https://en.wikipedia.org/wiki/CURL

Example 4: Using Python to invoke OpenStack API

```python
#!/usr/bin/env python
s = { "server": { "name": sname, "imageRef": sImageRef, "flavorRef":
sFlavorRef, "metadata": sMetadata, "personality": sPersonality } }
sj = json.dumps(s)
params = sj
headers = { "X-Auth-Token":apitoken, "Content-type":"application/json" }
conn = httplib.HTTPConnection("localhost:8774")
conn.request("POST", "%s/servers" % apiurlt[2], params, headers)
response = conn.getresponse()
data = response.read()
dd = json.loads(data4)
conn.close()
```

Figure 7-9 Using Python

Many users write automation scripts directly against the OpenStack REST API or shell scripts that invoke the CLI in services such as Keystone or Nova. Python provides an alternative way to write OpenStack automation scripts and an example is shown in Figure 7-9. All of the OpenStack services have native Python APIs that expose the same feature set as the command line tools.

For Python programmers, the Python APIs are much simpler to work with than the command line tools or the REST API.

 Note

To learn more about Python, scan this QR code or enter the URL into your browser.

https://en.wikipedia.org/wiki/Python_(programming_language)

Internal Keystone components

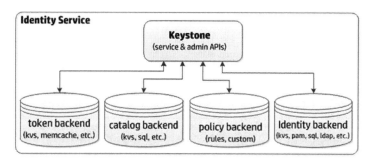

Figure 7-10 Internal Keystone components

Keystone is organized as a group of internal services which are exposed to one or more endpoints. The primary components of Keystone, aside from the Keystone service itself, are the data stores (backends) shown in Figure 7-10:

- **The token backend**—Validates and manages tokens used for authenticating requests after user or tenant credentials have been verified.

- **The catalog backend**—Provides an endpoint registry used for endpoint discovery.

- **The policy backend**—Provides a rule-based affinity authorization engine.

- **The identity backend**—Provides authentication credential validation and data for users, tenants, and roles (as well as any associated metadata).

Example: Keystone authentication and authorization process

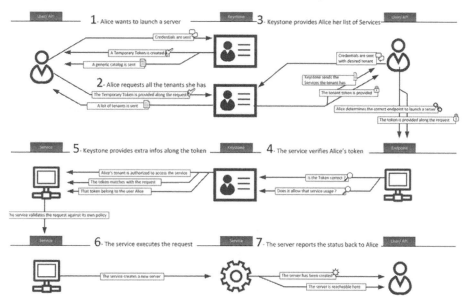

Figure 7-11 Keystone authentication and authorization process

Figure 7-11 shows how Keystone provides authentication and authorization to an OpenStack service request. In this example, a user named Alice wants to launch a server instance in OpenStack. The user's credentials are sent to Keystone, and Keystone responds with a temporary token and a generic catalog of services.

Alice has a list of tenants. She requests those tenants from the server and provides the temporary token. Keystone responds to this request with a list of tenants.

Alice then requests a list of services for a specific tenant, and Keystone provides this list along with a security token for that tenant. She then determines the correct endpoint from which to launch the server and sends the service request along with the tenant token to the service endpoint.

The service endpoint sends a tenant token to Keystone for authentication and authorization. Keystone responds by authenticating the token and verifying that the tenant is authorized to access the service endpoint. The service endpoint verifies from its policy whether the tenant user role is authorized to create a server. The service executes the request. In this case, the request is creating a new server and Nova compute is the service.

The server reports the status of the request execution to Alice and, upon completion of the request, sends information about how the server can be reached.

 Note

For additional information, scan this QR code or enter the URL into your browser.

http://docs.openstack.org/admin-guide-cloud/content/keystone-concepts.html

---

Keystone—Activity

Keystone is an OpenStack project that provides identity, token, catalog, and policy services for use in the OpenStack family.

Use this activity to review some key concepts in Keystone. Mark the correct answers for each question.

1. What in Keystone fits all three of these descriptions?

    - Represents a tenant, a group, or an organization
    - Must be specified when making OpenStack requests

- Consists of users with assigned roles and computing resources available to them
  - ☐ Domain
  - ☐ Role
  - ☐ Project

2. What captures operations that a user can perform in a given project and defines the "personality" of the user?
    - ☐ Domain
    - ☐ Role
    - ☐ Project

3. What in Keystone fits all four of these descriptions?
    - Defines administrative boundaries for the management of identity entities
    - Represents an individual, a company, or an operator-owned space
    - Exposes administrative activities directly to the system users
    - Consists of a collection of projects and users
      - ☐ Domain
      - ☐ Role
      - ☐ Project

Keystone activity—answers

1. What in Keystone fits all three of these descriptions?
    - Represents a tenant, a group, or an organization
    - Must be specified when making OpenStack requests
    - Consists of users with assigned roles and computing resources available to them
      - **Project**

2. What captures operations that a user can perform in a given project and defines the "personality" of the user?
    - **Role**

3. What in Keystone fits all four of these descriptions?

    - Defines administrative boundaries for the management of identity entities

    - Represents an individual, a company, or an operator-owned space

    - Exposes administrative activities directly to the system users

    - Consists of a collection of projects and users

      - **Domain**

Nova architecture

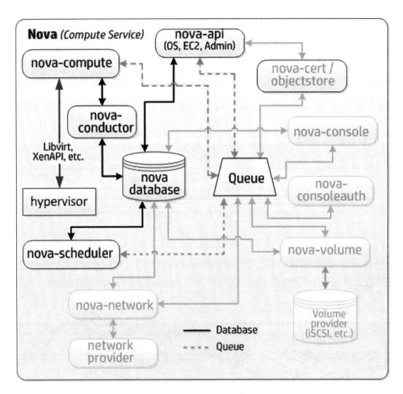

Figure 7-12 Nova architecture

Figure 7-12 shows the common architectural components in Nova, which include:

- **nova-api** that provides these functions:

  - It receives http requests, converts them into commands, and communicates with other components.

- It accepts and responds to user compute API calls. It supports the OpenStack Compute API, Amazon EC2 API, and a special Admin API for privileged users performing administrative actions.

- It initiates most of the orchestration activities, such as running an instance, and enforces some policies.

By default, the nova-api listens on port 8774 for OpenStack API requests. To accept and fulfill these requests, the nova-api initiates most of the orchestration activities such as booting servers and creating flavors. It also enforces some policies (mainly authentication, authorization, and quota checks). For some requests, it fulfills the entire request by querying the database and then returning the answer. For more complicated requests, it passes messages to other daemons through a combination of writing information to the database and adding messages to the queue.

- **nova-compute** manages communication with hypervisors and VMs. It creates and terminates VM instances using the hypervisor APIs, which include XenAPI for XenServer and XCP, libvirt for KVM or Quick EMUlator (QEMU), and VMware API for VMware. This simple process includes accepting actions from the queue and then performing a series of system commands such as launching a KVM instance while updating the state in the database.

- **nova-scheduler** decides which host receives each VM. It takes a VM instance request from the queue and determines specifically which compute server host it should run on.

- **nova-cert** manages x509 certificates.

- **nova-novncproxy** provides a proxy for accessing running instances through a Virtual Network Computing (VNC) connection. It supports browser-based novnc clients.

- **nova-consoleauth** authorizes console access based on the user, project, and role.

- **nova-conductor** mediates interactions between nova-compute and the database.

Instance creation process overview

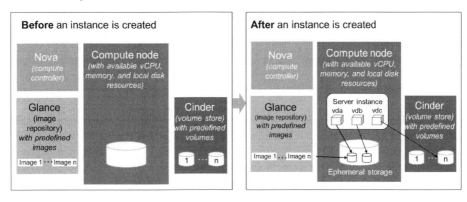

Figure 7-13 Instance creation process

Figure 7-13 shows an overview of the instance creation process.

These criteria must be met before an instance is created:

- The image must exist in Glance.

- The volume, if required, must be available to Cinder.

- The instance flavor must be specified.

- Nova-scheduler must be able to locate a compute node that can accommodate the specified flavor.

In the instance creation process:

- An image is copied from Glance to the local storage of the selected compute node (on vda in the example graphic). If required, a new empty disk image is created to present as the second disk (vdb in the example).

- If block storage is required, the compute node attaches to the requested cinder-volume using iSCSI or Fibre Channel and then maps to the third disk (vdc) as requested.

- The vCPU and memory resources are provisioned, and the instance is booted from the vda volume.

After the instance is created:

- The compute node operating system is operational on vda.

- The vdb node is available for storage.

- The vdc node is connected by an iSCSI or Fibre Channel connection to a Cinder volume.

Instance creation process steps

The instance creation process flow is completed in 11 steps, as shown in Figure 7-14:

1. A request is sent to Keystone to generate and return an authorization token.

2. An API request is sent to nova-api with the authorization token.

3. Nova-api validates the authorization token with Keystone.

4. Nova-api processes the API request.

5. Nova-api makes a remote procedure call (RPC) to nova-scheduler.

6. Nova-scheduler selects a host based on the scheduling policies.

7. Provisioning of the VM begins on the compute node as a result of an RPC request to nova-compute.

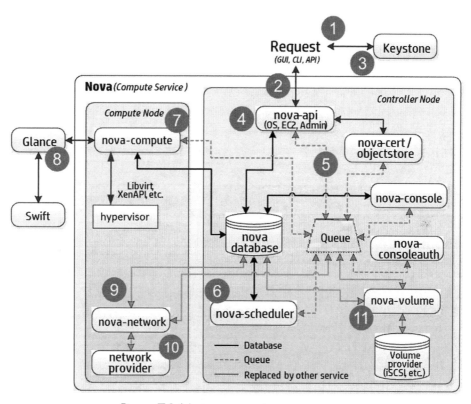

Figure 7-14 Instance creation process steps

- Nova-compute requests an OS image from Glance.
- Glance returns the image URI to nova-compute.

8. The image is downloaded from an image repository such as Swift.

9. Nova-api makes an RPC to allocate a network.

10. Nova-network creates the network and associates it with the VM.

11. A block volume is attached as appropriate (using an RPC cast).

 Note

The VM creation process described shows nova-network being used to create the network for the new VM. Nova-network has been deprecated, and it will be replaced with the OpenStack Neutron networking service in due course.

Nova—Activity

Answer the following questions to review some key concepts in Nova. Mark the correct answers for each question.

1. What in Nova defines an available hardware configuration for a server and specifies a combination of disk space, memory capacity, and priority for CPU time?

   ☐ Server instance

   ☐ Image

   ☐ Flavor

2. Fill in the blank. (Select one option for both sentences.)

   A soft _____ signals the operating system to restart, which allows for a graceful shutdown of all processes.

   A hard _____ simply restarts the system.

   ☐ Rebuild

   ☐ Resize

   ☐ Reboot

3. What is a collection of operating system files provided by cloud operators that are used to create or rebuild a server?

   ☐ Server instance

   ☐ Image

   ☐ Flavor

4. Fill in the blank:

   _____ converts an existing server to a different flavor.

   ☐ Rebuild

   ☐ Resize

   ☐ Reboot

5. What in Nova refers to a VM or bare-metal system onto which an operating system can be installed?

   ☐ Server instance

   ☐ Image

   ☐ Flavor

6. Which activity removes all data on the server and replaces it with the specified image, but leaves the server ID and IP addresses the same?

☐ Rebuild

☐ Resize

☐ Reboot

Nova activity—answers

1. What in Nova defines an available hardware configuration for a server and specifies a combination of disk space, memory capacity, and priority for CPU time?

   • **Flavor**

2. Fill in the blank (Select one):

   • A soft **reboot** signals the operating system to restart, which allows for a graceful shutdown of all processes.

   • A hard **reboot** simply restarts the system.

3. What is a collection of operating system files provided by cloud operators that are used to create or rebuild a server?

   • **Image**

4. Fill in the blank: **Resize** converts an existing server to a different flavor.

5. What in Nova refers to a VM or a bare-metal system onto which an operating system can be installed?

   • **Server instance**

6. Which activity removes all data on the server and replaces it with the specified image, but leaves the server ID and IP addresses the same?

   • **Rebuild**

Cinder architecture

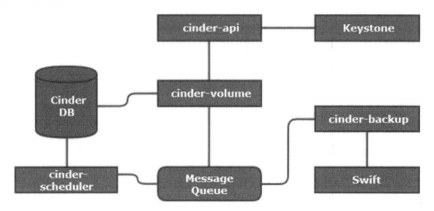

Figure 7-15 Cinder architecture

Cinder provides a Block Storage service and enables the management of volumes, volume snapshots, and volume types. Block storage volumes are analogous to virtual USB hard drives in that they provide persistent storage to VMs and can be attached to and detached from VMs at will. As shown in Figure 7-15, Cinder includes these components:

- **cinder-api** accepts API requests and routes them to the cinder-volume service for action. The cinder-api service is a Web Server Gateway Interface (WSGI) application that authenticates and routes requests throughout the block storage system. It supports the OpenStack APIs only, although a translation can be done using the Nova EC2 interface, which calls the Cinder client.

- **cinder-volume** responds to read and write requests against the Cinder database. It interacts with other processes (such as cinder-scheduler) through a message queue. It can interact with a variety of storage providers as well.

  The cinder-volume service is responsible for managing block storage devices. It provides volumes in RAW format, which have to be formatted with the operating system flavor or disk format. The main difference between the legacy nova-volume service and cinder-volume is that the legacy nova-volume service returns the volume to the instance, whereas the cinder-volume service returns the volume abstraction.

- **cinder-scheduler** selects the optimal block storage provider node on which to create a volume. The cinder-scheduler service is responsible for scheduling and routing requests to the appropriate volume service. Depending on your configuration, the routing could be simple round-robin scheduling to the running volume services, or it could be more sophisticated through the use of the Filter Scheduler. The Filter Scheduler is the default and enables filtering on aspects such as Capacity, Availability Zone, Volume Types, and Capabilities as well as custom filters.

- **cinder-backup** enables you to back up a Cinder volume. The default backup location is the OpenStack object store (Swift), but an NFS export can be used as the backup repository if required.

- **Messaging queue** routes information between the block storage service processes and a database that stores the volume state.

Cinder — Activity

Answer the following questions to review some of the key concepts in Cinder:

1. A detachable block storage device (synonymous with a USB device) that is attachable to one instance at the time

   _____

2. Administrator-defined type of block storage volume (such as SATA, SCSI, or SSD)

   _____

3. A point-in-time copy of the data contained in the volume

   _____

4. A virtual machine that runs in the cloud

   _____

5. A full copy of a volume that is stored by an external object storage service, such as Swift, and is restorable into the same volume or a new volume

   _____

Cinder activity — answers

1. A detachable block storage device (synonymous with a USB device) that is attachable to one instance at the time

   - **Volume**

2. Administrator-defined type of block storage volume (such as SATA, SCSI, or SSD)

   - **Volume type**

3. A point-in-time copy of the data contained in the volume

   - **Snapshot**

4. A virtual machine that runs in the cloud

   ● **Instance**

5. A full copy of a volume that is stored by an external object storage service, such as Swift, and is restorable into the same volume or a new volume

   ● **Backup**

Swift architecture

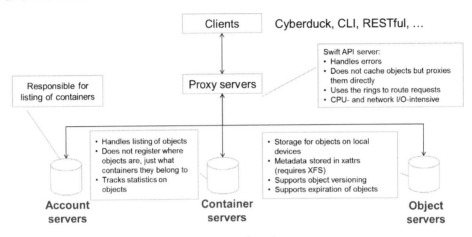

Figure 7-16 Swift architecture

Swift is an object storage service that can be used to store unstructured data and is often used for file sharing and backup purposes. Common examples of object storage services are Dropbox (for sharing and backing up files), Facebook (for sharing photographs), and Spotify (for sharing audio files). The Swift architecture is distributed to prevent any single point of failure (SPOF) and to scale horizontally. As shown in Figure 7-16, it includes these components:

● **Proxy servers (swift-proxy-server)** accept incoming requests through the OpenStack Object API. They accept files to upload, modifications to metadata, or container creation. In addition, they also serve files or container listings to web browsers. The proxy servers can use an optional cache (usually deployed with memcached) to improve performance.

The proxy servers are responsible for integrating the rest of the OpenStack object storage architecture. For each request, they look up the location of the account, container, or object in the ring (the ring is a mapping between the names of objects stored on disk and their physical location) and then route the request accordingly. The public API is also exposed through the proxy servers.

● **Account servers** manage accounts defined with the object storage service.

● **Container servers** manage mapping of containers (such as folders) within the object store service.

- **Object servers** manage actual objects (files) on the storage nodes.

  The object servers are simple Binary Large Object (BLOB) storage servers that can store, retrieve, and delete objects stored on local devices. Objects are stored as binary files on the file system with metadata stored in the extended attributes (xattrs) of the file. This requires that the underlying file system choice for object servers supports xattrs on files. Some file systems, such as ext3, have xattrs turned off by default.

  Each object is stored using a path derived from the object name hash and the operation time-stamp. The last write always wins, which ensures that the latest object version is served. A deletion is also treated as a version of the file. It is a zero byte file ending with ".ts", which stands for tombstone. This ensures that deleted files are replicated correctly and older versions do not reappear after a failure scenario.

A number of periodic processes also run to perform housekeeping tasks on a large datastore. The most important of these processes are the replication services, which ensure consistency and availability through the cluster. Other periodic processes include auditors, updaters, and reapers.

Authentication is handled through the configurable WSGI middleware, which is usually Keystone.

Sample Swift deployment architecture

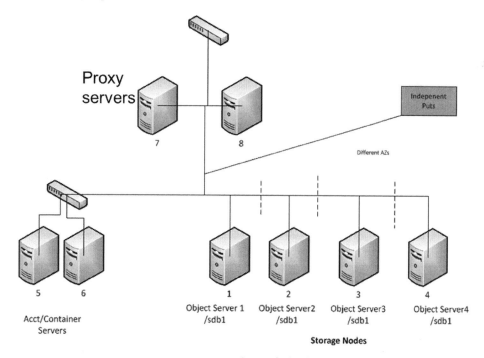

Figure 7-17 Sample Swift deployment

Figure 7-17 shows an implementation in which the proxy servers are load balanced. Multiple account and container servers are configured, as well as several storage nodes.

Swift—Activity

Answer the following questions to review some key concepts in Swift. Mark the correct answers for each question.

1. What in Swift fits all three of these descriptions?

   - Defines a namespace for objects (holds objects)

   - Represents a location (a storage compartment) for the objects (synonymous with folders or directories)

   - Does not support nesting

     ☐ Region

     ☐ Partition

     ☐ Container

2. What in Swift fits all four of these descriptions?

   - Represents the top level of the Swift hierarchy

   - Owns the containers within this hierarchy

   - Is created by the service provider

   - Is synonymous with a project or tenant

     ☐ Object

     ☐ Account

     ☐ Ring

3. What in Swift fits all three of these descriptions?

   - Accepts incoming requests via the OpenStack object API or raw HTTP

   - Accepts files to upload, modifications to metadata, and container creation requests

   - Serves files and container listings to web servers

     ☐ Ring

     ☐ Proxy server

     ☐ Swift controller

4. What in Swift fits all three of these descriptions?

    - Represents the basic storage entity within Swift

    - Stores data content, such as documents, images, videos, and optional custom metadata

    - Does not compress or encrypt the stored data

        ☐ Object

        ☐ Container

        ☐ Zone

Swift activity—answers

1. What in Swift fits all three of these descriptions?

    - Defines a namespace for objects (holds objects)

    - Represents a location (a storage compartment) for the objects (synonymous with folders or directories)

    - Does not support nesting

        - **Container**

2. What in Swift fits all four of these descriptions?

    - Represents the top level of the Swift hierarchy

    - Owns the containers within this hierarchy

    - Is created by the service provider

    - Is synonymous with a project or tenant

        - **Account**

3. What in Swift fits all three of these descriptions?

    - Accepts incoming requests through the OpenStack object API or raw HTTP

    - Accepts files to upload, modifications to metadata, and container creation requests

    - Serves files and container listings to web servers

        - **Proxy server**

Glance architecture

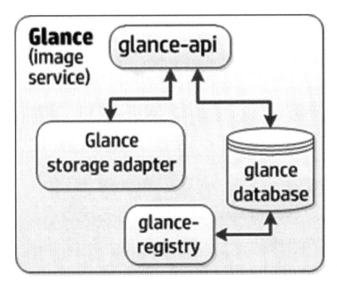

Figure 7-18 Glance architecture

Glance provides image services that enable discovering, registering, and retrieving virtual machine images. Nova compute uses Glance to manage images that are used to boot VMs.

As shown in Figure 7-18, the Glance architecture consists of four primary components:

- **glance-api** enables users to query the VM image metadata and retrieve actual images from the HTTP requests.

- **Glance storage adapter** provides the interface between Glance and vendor-specific devices for saving, retrieving, and deleting images, along with verifying operations.

- **Glance database** stores image-related information such as the location of the image, the image properties, and the image membership.

- **Glance registry** stores and retrieves metadata about images.

Glance operations using Horizon

Figure 7-19 Glance operations using Horizon

Most of the Glance operations can be performed using the OpenStack dashboard (Horizon) UI. As shown in Figure 7-19, clicking the Images option in the navigation pane displays a list of available images. From this list, you can:

- Click the **Image Name** link to view details about the selected image
- View the status of the image, whether it is public, private, or protected, and its disk format

The Actions column of the table enables you to update some of the image information, including its name, description, architecture, disk format, public availability, and protection.

The Horizon UI also enables you to create new images and delete existing images.

Glance—Activity

Answer the following questions to review some key concepts in Glance. Mark the correct answers for each question.

1. Which Glance component provides the interface between Glance and vendor-specific devices for saving, retrieving, and deleting images?

    ☐ Glance domain controller

    ☐ REST API

    ☐ Glance storage adaptor

2. Which Glance component stores and retrieves metadata about images?

☐ Glance registry

☐ Glance store

☐ Client

3. Which Glance component enables users to query the VM image metadata and retrieve actual images?

☐ Database abstraction layer

☐ Glance store

☐ glance-api

4. Which component in Glance stores image-related information such as the location of the image, the image properties, and the image membership?

☐ Glance database

☐ REST API

☐ Registry layer

Glance activity—answers

1. Which Glance component provides the interface between Glance and vendor-specific devices for saving, retrieving, and deleting images?

- **Glance storage adaptor**

2. Which Glance component stores and retrieves metadata about images?

- **Glance registry**

3. Which Glance component enables users to query the VM image metadata and retrieve actual images?

- **glance-api**

4. Which component in Glance stores image-related information such as the location of the image, the image properties, and the image membership?

- **Glance database**

Ceilometer metering

Ceilometer collects and stores utilization data for the physical and virtual resources in OpenStack clouds, and can trigger actions when defined criteria are met. A simple example of a triggered action is when CPU utilization rises above 70%, Ceilometer sends a warning email to an administrator.

To better understand what can be done with the collected data and how the metering process works, the standard billing process commonly used in the telecommunications industry can be divided into three steps.

The three steps are typically defined as follows:

- **Metering**—Collecting information about services that can be billed

- **Rating**—Analyzing a series of tickets according to the business rules

- **Billing**—Assembling the bill line items into a single, per-customer bill

Ceilometer is limited to the metering function. However, you can use Ceilometer data and the Ceilometer API to build your own billing system that does more than metering.

Accessing collected data

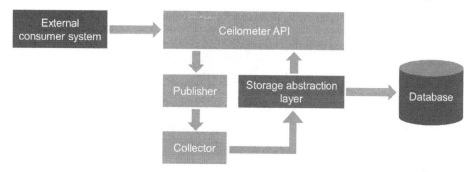

Figure 7-20 Accessing collected data

The data that Ceilometer collects is stored in a database. You can have multiple types of databases through the use of different database plug-ins. In addition, the schema and dictionary of the database can evolve over time, so Ceilometer exposes the REST API that can be used to access the stored metering data. As shown in Figure 7-20, the publisher pushes samples to a message queue, which arethen consumed by the collector. The collector gathers the event and metering data and writes it, via the storage abstraction layer, to a storage repository such as a database, a file, or an HTTP location.

Ceilometer is a part of OpenStack, but it is not tied to the standard definitions of "users" and "tenants." The "source" field of each sample refers to the authority defining the user and tenant associated with the sample. Those who deploy OpenStack can define custom sources through a configuration file and then create agents to collect samples for new meters using those sources.

Users can also send their own application-specific data to the database through the REST API for various use cases, such as to trigger alarms.

Ceilometer alarms

The alarm component of Ceilometer enables you to set alarms based on a threshold evaluation of a collection of samples. An alarm can be set on a single meter or on a combination of meters.

Example

You want an alarm to be triggered when the memory consumption reaches 70% in a given instance, if the instance has been running for more than 10 minutes. To set up this alarm, you could use the Ceilometer CLI command, the Horizon interface, or one of the other supported API access methods and specify the alarm conditions and the action to take.

Architecture—Activity

Watch an 11-minute video about the inner workings of Helion OpenStack, featuring Mark Perreira, chief architect for Helion cloud.

 Note

To watch this video, scan this QR code or enter the URL into your browser:

https://www.youtube.com/watch?v=FTrrARc3hsY

After you watch the video, answer the following questions.

1. What are the two major environments within Helion OpenStack?

2. Which Linux distribution serves as the foundation for Helion OpenStack?

3. Which OpenStack services are part of the Helion OpenStack execution environment?

4. What are the HPE value-added services to OpenStack that are responsible for VMware ESX integration and Cinder configuration?

    _____

    _____

    _____

5. Which UI dashboards are part of the Helion OpenStack operational environment?

    _____

    _____

    _____

Architecture activity—answers

1. Which OpenStack services are part of the Helion OpenStack execution environment?

    - **Keystone—identity**

    - **Heat—orchestration**

    - **Swift—object storage**

    - **Glance—images for VMs**

    - **Nova—compute**

    - **Cinder—block storage**

    - **Neutron—networking**

2. What are the HPE value-added services to OpenStack that are responsible for VMware ESX integration and Cinder configuration?

    - **EON (ESX integration service)**

    - **Sirius (Cinder configuration service)**

3. Which UI dashboards are part of the Helion OpenStack operational environment?

    - **Operational dashboard—based on Horizon, with added views for EON, Sherpa, and Sirius**

    - **IMC/CMC dashboard—HPE value-add for StoreServ or StoreVirtual VSA storage**

    - **Icinga web—monitoring**

    - **Kibana—search capabilities**

Helion OpenStack architecture (based on Kilo release)

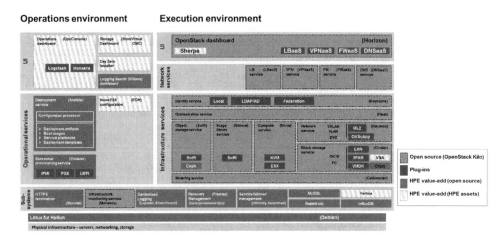

Figure 7-21 Helion OpenStack architecture

Figure 7-21 represents the Helion OpenStack architecture. It is an enhanced version of the architecture that Mark Perreira illustrated in the video from the previous activity. Important considerations regarding Helion OpenStack architecture include the following:

- **High availability**—Almost all services (except for Ceilometer and Sherpa) are deployed in a three-node cluster with active/active high availability.

 Note

Sherpa is the remote web software repository used for updates. The Helion OpenStack Sherpa service provides a link to the remote web catalog of software available for purchase and download into the Helion environment.

- **Security**—All services as well as the overall network design of this distribution adheres to a security review process, requirements, and penetration testing (also known as "pen testing"). Pen testing refers to testing a computer system, network, or web application to find vulnerabilities that an attacker could exploit.

- **Recovery**—Backup and restore processes apply to all operational and execution environment databases.

- **Deployment**—The deployment service for Helion deployments uses the OpenStack Ironic baremetal service for provisioning Helion servers with industry standard IPMI/PXE boot protocols.

- **Patches/updates**—Patch and update processes for all components of Helion OpenStack are delivered through the Helion download network, which is integrated into the Horizon-based operational dashboard (Sherpa).

- **Compute**—Nova KVM hypervisor support via the Debian Linux distribution is delivered as part of the Helion OpenStack distribution.

Introducing the HPE Helion Platform

The HPE Helion Platform is a combined infrastructure-as-a-service (IaaS) and platform-as-a-service (PaaS) solution for cloud-native workloads. The Helion Platform is based on industry-leading, open-source projects in OpenStack and Cloud Foundry. It provides enterprise customers with flexibility in software and hardware. The Helion Platform can simplify the development, deployment, and delivery of applications that have been designed for the cloud.

The Helion Platform offers:

- Accelerated application development and deployment

- Workload portability

- No vendor lock-in

- Easier administration of OpenStack and Cloud Foundry technologies

- Enhanced software delivery with integrated IaaS and PaaS

The platform consists of two main components:

- HPE Helion OpenStack

- HPE Helion Development Platform

HPE Helion OpenStack 2.0

Helion OpenStack 2.0 launched on 28 October 2015. Version 2.0 is the latest OpenStack-based IaaS cloud platform release from HPE. It is based on the OpenStack Kilo release and implements core services and features of Kilo while providing new installation and management features.

Features, benefits, and value

These tables highlight the new features, benefits, value, and target market segments for version 2.0.

Table 7-1 HPE Helion OpenStack new features (1 of 2)

Feature	Benefits	Value	Market segment			
			Private cloud	Managed cloud	NFV (service provider)	NFV (vendor)
Flexible bare-metal provisioning						
Operator can choose to use their existing operating system installation tool	• Enterprise IT and cloud administrators can use their favorite or existing tool • Existing compliance procedures remain intact	Reduce OPEX in adopting new tools	✓			
Customizable OpenStack services (HPE support via professional services)						
Customize compute scheduler filters	Better utilization of compute resources	Reduce CAPEX	✓	✓		
Upstream supported Cinder and Neutron plug-ins	• Reuse existing storage hardware • Use favorite networking gear	Reduce CAPEX	✓			
Upgrade						
Live updates (bug fixes, security patches) and seamless upgrades (major versions)	Highly available and secure cloud	Reduce OPEX and increase revenue by saving maintenance time and meeting SLAs	✓	✓	✓	

Table 7-2 HPE Helion OpenStack new features (2 of 2)

Feature	Benefits	Value	Market segment			
			Private cloud	Managed cloud	NFV (service provider)	NFV (vendor)
Flexible topology						
Flexible control plane	Right footprint for the size and usage profile of cloud	Reduce CAPEX by using fewer servers for control plane	✓		✓	
Flexible network topology (underlay)						
NIC bonding	• High availability • Better performance	Reduce OPEX and increase revenue by saving maintenance time and meeting SLAs	✓		✓	
Network separation (multi-NIC)	• Better performance • Increased security	• Increase revenue by providing better quality of service • Reduce risk and achieve regulatory compliance	✓		✓	
Provider VLAN, L2 VxLAN gateway	Smooth migration of NFV and enterprise workloads to cloud	Reduce risk and CAPEX/OPEX	✓		✓	
Others						
No need for Ubuntu licensing	All support comes from one vendor (HPE)	Reduce CAPEX and OPEX in support costs				✓
Scale out the cloud without downtime	Highly available and scalable cloud	Reduce OPEX and increase revenue by saving maintenance time and meeting SLAs	✓	✓	✓	

 Note

NFV stands for network function virtualization, and it is a networking concept that uses IT virtualization technologies to virtualize network node functions into building blocks that can be connected together to create complex, multitier communication services.

Helion OpenStack Cloud model

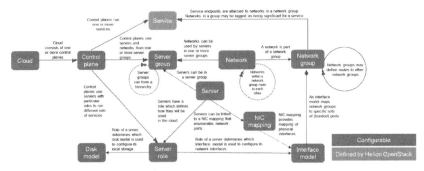

Figure 7-22 OpenStack cloud model

A Helion OpenStack 2.0 cloud is defined by a declarative model that is described in a series of configuration objects. These configuration objects are represented in YAML files, which can be used as configuration templates with this release. These examples can be used nearly unchanged, with the exception of necessary changes to IP addresses and other site and hardware-specific identifiers. Alternatively, the examples may be customized to meet site requirements.

 Note

The term "A Helion OpenStack 2.0 cloud" mentioned in the previous paragraph refers to the underlying servers, storage, and networking resources that make up the so-called undercloud (a single-node, OpenStack installation running on a single physical server used to deploy, test, manage, and update the overcloud servers) and overcloud (the functional cloud on which guest virtual machines are created). It does not refer to the guest VMs that are created by tenants.

 Note

Yet Another Markup Language (YAML) is a machine-parsable data serialization format designed for human readability and interaction with scripting languages such as Perl and Python. YAML is optimized for configuration settings, log files, Internet messaging, and filtering.

The cloud model describes the physical underlay for the cloud (servers, networks, and storage) and how OpenStack and other services map to this infrastructure. This model represents the core interface into the Helion OpenStack lifecycle management.

In Figure 7-22:

- There is one cloud (each instance of the model describes one cloud). A cloud can have one or more control planes.

- A control plane runs one or more services (for example, nova-api). These services expose endpoints.

- Control planes use servers that run different sets of services. For example, a role might be a compute or networking node.

- Servers can be in a server group. Server groups correspond to physical failure domains (for example, data center, rack, or enclosure). Availability zones in Cinder, Nova, and other OpenStack components are specified by server groups. Layer 3 networks are assigned to a server group, and server groups form a hierarchy.

- Servers can be linked to a NIC mapping, which enumerates network ports so that the cloud can communicate with the right interface.

- Servers have a role which is linked to a disk model, controlling how local storage is presented.

- A network is part of a network group. Service endpoints are attached to networks in a network group.

- Network groups may define routes to other network groups.

- Servers also have a role which is linked to the interface model, mapping network groups to specific sets of optionally bonded ports. The NIC mapping ensures communication with the correct NIC.

- A network group has one or more networks associated with it. A network is associated with one or more server groups. Networks within a group must route to each other.

- Control planes use servers and networks from one or more server groups.

Installation workflow

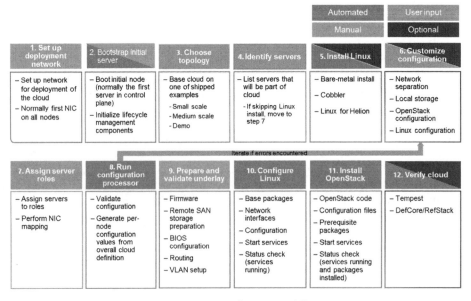

Figure 7-23 Installation workflow

Figure 7-23 shows an example of steps that a cloud administrator can take to perform the first-time installation of Helion OpenStack 2.0:

Note

The solution is flexible and other installation sequences are possible.

1. **Set up the deployment network**—This is the network that will be used to deploy Helion OpenStack for network booting of the servers to install Linux and then for Ansible access (via Secure Shell [SSH]). Ansible is a simple automation platform for configuration and orchestration.

2. **Bootstrap initial server**—Helion OpenStack ships as an ISO file. You have to boot this ISO and install the Helion OpenStack lifecycle management components on the first server of your control plane. You can use USB, CD, or iLO Virtual Media to accomplish this task.

3. **Choose the topology**—Helion OpenStack ships with a number of sample topologies. You can choose one of these on which to base your cloud. Then, you can customize the sample topology to better suit your needs.

4. **Identify the servers**—You need to specify the servers that will make up your cloud.

5. **Install Linux**—PXE boot your servers and install a base Linux operating system (HPE Linux for Helion, a Debian derivative). At the completion of this step, you can connect with SSH to all your servers. This step is optional; if you have your own server imaging solution, then you can use it to install the supplied image and continue with Step 6. However, you still need to specify which servers to use and ensure that they are accessible via SSH from the lifecycle management server.

6. **Customize the configuration**—Change the sample configuration (topology) that you chose in Step 3 to better meet your needs. This configuration describes both the underlay (servers, networks, and storage), how OpenStack services map to the underlay, and how the OpenStack services are configured.

7. **Assign server roles**—Specify which servers should perform which roles. You will have a list of roles that depends on the sample topology chosen in Step 3. Helion OpenStack can assign roles for you if you do not have specialized hardware.

8. **Run the configuration processor**—This step is key to Helion OpenStack lifecycle management. It converts the configuration that you described in Steps 3, 4, 6, and 7 from an easy-to-use, normalized data model into specific values that Ansible will deploy on each server. The configuration processor deals with IP address allocation, resolution and provision of dependencies, and so on. It provides warnings and errors if your model contains mistakes or inconsistencies. You might need to go back to Step 6 in order to correct these.

9. **Prepare and validate the underlay**—The configuration processor outputs information about the requirements for the physical underlay. You need to ensure that your physical underlay works as expected. It is important to ensure that the right VLANs are exposed to the right ports and the necessary routing is in place.

10. **Configure Linux**—Take the base Linux installation performed in Step 5 and configure it for your cloud needs. This is a separate step that provides support for different host operating systems if needed in the future. There is a status check at the end of this step.

11. **Install OpenStack**—Deploy the OpenStack services and start them up.

12. **Verify cloud**—At the end of the deployment, you have the option to verify your cloud using the Tempest tests. These tests allow you to use the new cloud as a user and report whether your cloud conforms to the OpenStack DefCore standard for interoperability between different OpenStack clouds.

Example configurations

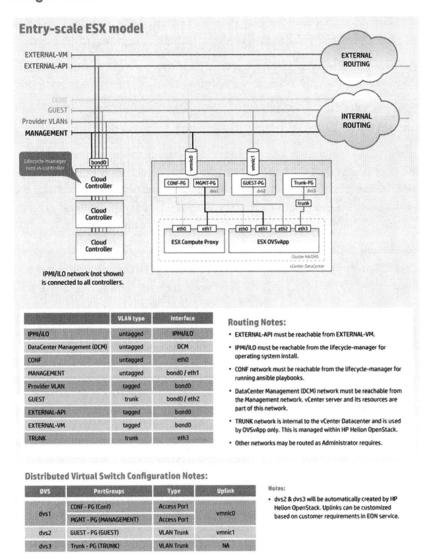

Figure 7-24 Example configuration

Helion OpenStack 2.0 ships with a collection of prequalified example configurations. These are designed to help you get up and run quickly with a minimum number of configuration changes. Figure 7-24 shows one such example configuration.

The Helion OpenStack input model allows a wide variety of configuration parameters that may, at first glance, appear daunting. The example configurations are designed to simplify this process by providing prebuilt and prequalified examples that need only a minimum number of modifications to get started.

Helion OpenStack 2.0 ships with two classes of sample cloud models: *examples* and *tech-preview*. The models in the examples directory have been qualified by the HPE Quality Engineering team, while the tech-preview models are more experimental.

The prequalified examples shipped with Helion OpenStack 2.0 are the following:

- Entry-scale KVM with VSA model (entry-scale-kvm-vsa)

- Entry-scale Swift-only model (entry-scale-swift)

- Entry-scale KVM with Ceph model (entry-scale-kvm-ceph)

- Entry-scale ESX model (entry-scale-esx)

These systems are designed to provide an entry-level solution that can be scaled from a small number of nodes to a moderately high node count (approximately 100 compute nodes, for example).

The tech-preview configuration currently includes the mid-scale KVM with VSA model (mid-scale-kvm-vsa).

In this model, the cloud control plane is subdivided into a number of dedicated service clusters to provide more processing power for individual control plane elements. This enables a greater number of resources to be supported (including compute nodes and Swift object servers). This model also shows how a segmented network can be expressed in the Helion OpenStack model.

First-time installer

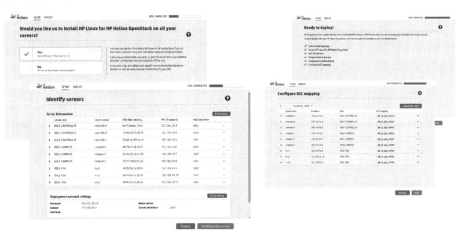

Figure 7-25 First-time installer

Helion OpenStack 2.0 provides a GUI experience for the first-time installation as shown in Figure 7-25. It is accessible from a web browser pointed at the lifecycle manager after the bootstrap step has been completed. The first-time installer guides you through the initial installation and introduces you to some of the concepts in the cloud configuration model.

The first-time installation is a self-contained experience with detailed context-sensitive help screens that enable you to complete your first installation with minimum effort.

The GUI supports most aspects of the underlying cloud configuration model including bare-metal installation, NIC bonding, network traffic separation, local disk partitioning and management, NIC mapping, and server role assignment. It does not support overriding the base OpenStack configuration, which can be done using the CLI.

HPE Helion Development Platform

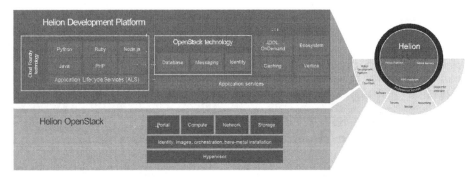

Figure 7-26 HPE Helion Development Platform

The Helion Development Platform, illustrated in Figure 7-26, is an open-source cloud application platform and ecosystem. This solution makes it easy to develop, deploy, and deliver highly available and scalable cloud-native applications. It provides developers with instant access to a variety of language runtimes, frameworks, and services to develop applications quickly. Because the platform is based on the open-source technologies of Cloud Foundry, Docker, and OpenStack software, enterprises can deploy and scale applications across clouds without having to rewrite code.

The Helion Development Platform optimizes Cloud Foundry for Helion OpenStack to take advantage of underlying IaaS high availability and adds highly available application services for improved performance. The platform runs on top of the Helion OpenStack infrastructure. To get started with this platform, there are two options:

- For prototyping and light-duty production, use the Helion Development Platform by downloading Helion OpenStack Community for free.

- For production apps and development/testing environments, a limited trial version of the Helion Development Platform is available for no additional charge beyond the VM cost. The trial enables customers to experience this platform without needing to create their own Helion OpenStack environment.

HPE and Cloud Foundry

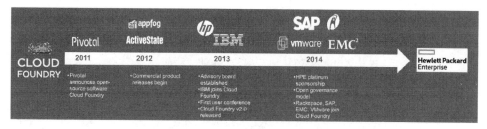

Figure 7-27 HPE and Cloud Foundry

Cloud Foundry is an open-source cloud computing PaaS, originally developed by VMware. As shown in Figure 7-27, it was initially owned by Pivotal Software; in 2011, it separated and become open source. Cloud Foundry is primarily written in Ruby and Go programming languages.

HPE is a platinum sponsor of the Cloud Foundry project. HPE has been working with Cloud Foundry since March 2013. The Cloud Foundry technology has been integrated with Helion OpenStack since that launch.

Use case example: Developer/tester system

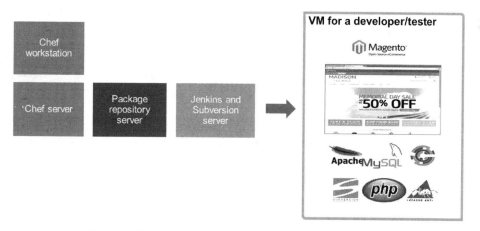

Figure 7-28 Use case example: Developer/tester system

The Helion Development Platform provides an ideal environment for developing and testing cloud applications with an open-source IaaS and PaaS platform that also integrates with other third-party tools, as shown in Figure 7-28, including the following:

- Chef is a configuration management tool used to streamline the maintenance of a company's servers. It integrates with cloud-based platforms such as Rackspace, Amazon Elastic Compute Cloud, Microsoft Azure, OpenStack, and Helion OpenStack.

- Jenkins is an open-source continuous integration tool written in Java. It provides services for software development. Continuous integration is the practice of merging all developer working copies with a shared mainline (also called trunk, baseline, and master) several times a day.

- Apache Subversion is a free, open-source control system. It manages files and directories, along with changes to them over time. Subversion also allows recovery of older versions of data.

- Git is a widely used, open-source distributed version control system for software development. It was designed with an emphasis on speed, data integrity, and support for distributed, nonlinear workflows.

HPE Helion Development Platform—Activity

Two case studies follow. Read through them and answer the questions that follow them.

Case study 1: Using microservices for customer surveys

A customer survey feature is needed for a legacy application with millions of daily requests. Functional requirements include autoscale, third-party integrations, and the selection of a cutting-edge programming technology for rapid prototyping and frequent releases. HPE Helion Development Platform makes it all possible.

Introduction

This case study explores a typical use case scenario for deploying a microservice to extend the functionality of one of your legacy applications. We have provided a realistic business case, which highlights the web scale of this application, how to architect the application, and the next steps for a developer. Please note that the number of requests, unique visitors, and page views referenced in this document show an example of the typical high-volume application requiring support.

Business summary

A legacy high-volume Web application needs to add customer survey functionality. The survey will be presented to every user, once per visit. Current traffic levels are at 1,300,000 unique visitors per day, which generate over 4,000,000 page views. The microservice will be implemented as an AJAX component and embedded into the legacy application to avoid unnecessary modifications to the legacy applications. HPE Helion Development Platform has been chosen to accommodate the scale, release frequency, and core technology choices.

Application scale requirements

A survey completion rate of 7% is expected, meaning 91,000 surveys will be completed per day. Each survey will include the survey data made up of four to eight questions, plus a handful of metadata. Every page view will send data to the survey microservice, which will either set or update a cookie to track user behavior.

Total daily traffic will be around 4,500,000 requests, with distribution of requests peaking in the evening hours. For our application, this means that spike capacity will be required daily and the HPE Helion Development Platform autoscale functionality will address this requirement.

Release frequency

An internal research team coordinates with business groups to identify knowledge gaps, which then become the basis of new surveys. New surveys are added every month. Each new survey requires a release due to the custom survey structures, which gives the research team complete flexibility. Occasionally, more frequent releases will be required in cases when a survey is found to produce flawed data. The command line client for HPE Helion Development Platform makes it possible to push new releases in minutes for development or production.

Data needs

Data structure will change from one survey to the next, requiring an unstructured datastore technology. Due to the high volume of writes during peak usage, the datastore needs to be highly available and scalable. Data also needs to be easily accessible for post processing.

Third-party analytics

Data will also be sanitized and sent to a third-party vendor to identify high-level consumer trends and provide preliminary input to the research and business teams.

Collected survey data requires postprocessing to identify geo data based on IP addresses and market segmentation based on correlation with external consumer databases.

Technologies

High concurrency requirements make the Go programming language a natural fit. Go will further cement RESTful API and stateless best practices. Each application instance will have low memory requirements and have a constant service capacity. HPE Helion Development Platform build-pack technology gives the application team complete control over the runtime environment.

The MongoDB datastore technology has been selected to accommodate the uptime and scale requirements. The MongoDB cluster will be hosted by the Data Services group, who will manage scale and backups. Hadoop will be used for periodic batch processing. The services integration feature in HPE Helion Development Platform allows the application team to retain their complete focus on developing the application while still accessing externally supported datastore and third-party services.

HPE Helion Development Platform provides the flexibility and speed needed by the application team. Autoscale, build-pack technology, and service integration decrease their time to go live, while at the same time decreasing risk and simplifying processes.

Technical implementation

The microservice is self-contained, which makes it easy to embed into the legacy application (or reuse elsewhere). There are three primary functions in the API that makes this embedding possible: render, submit, and extract as shown in Table 7-3.

Table 7-3 Microservice methods for rendering, submitting, and extracting surveys

METHOD	HEADERS	FUNCTION	ARGUMENTS/BODY
GET	Content-Type: text/html	Render	Survey: Which survey should be rendered?
POST	Content-Type: application/x-www-form-urlencoded	Submit	Survey data
GET	Content-Type: text/html	Extract	Survey: Which survey should be rendered?

Figure 7-29 shows the interactions between components.

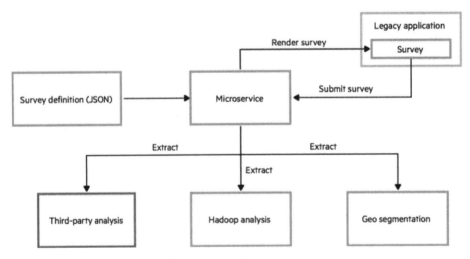

Figure 7-29 Interactions between components

The survey definitions are based on a predefined structure. This makes it easy to collaborate with other groups. In the case of this microservice, the research team responsible for designing new surveys can easily submit those new designs in the form of a regular text file. This also makes them easy to track using revision control.

The diagram then illustrates how the three functions work. The render function includes an argument, which indicates the survey definition to use. This produces HTML that can be embedded directly in the legacy application. Survey submissions come directly back to the microservice submit endpoint, which captures and stores the data. Finally, the extract function makes it possible for external systems to operate on survey data.

Case study 1 questions

1. What are microservices? (Use the Internet if necessary.)

   _____

   _____

   _____

2. What is the business case and background for this study?

   _____

   _____

   _____

3. What are the application scale requirements, release frequency, and data needs?

   _____

   _____

   _____

4. What technologies does the solution include?

   _____

   _____

   _____

Case study 2: Mendix (rapid app development)

Mendix and HPE Helion deliver rapid app development in the cloud.

Objective

Seamless integration between Mendix and HPE Helion Development Platform to foster fast cloud application development.

Approach

Partner with HPE to provide an integrated development experience for rapid development and deployment of open cloud applications.

IT Matters

- Applications run as a native cloud application built without coding.

- Visual/data application modeling enables the rapid building, integration, deployment, and sharing of multidevice business applications.

- Portability allows applications to run across browsers, smartphones, and tablets.

Business Matters

- Companies achieve ROI in days versus months.

- Collaboration between HPE, Mendix, and the open source community fuels greater innovation.

- Partnership advances a New Style of IT to solve challenges across cloud, social, mobile, and Big Data technology.

Mendix no-code rapid application development experience

Many enterprises have a mix of developers—from a traditional developer who uses a variety of languages to a rapid developer who prefers not to code and instead seeks visual development tools to quickly create applications. Mendix addresses the needs of the rapid developer by providing a visual application modeling solution that enables companies to build, integrate, and deploy multidevice business applications six times faster than traditional programming languages.

Mendix applications are cloud native and can be built without coding. Plus, they run across browsers, smartphones, tablets, and anywhere else the application is displayed. Using visual, model-driven development, one-click deployment, social collaboration, and a central app governance all-in-one complete solution, companies can drive ROI in a matter of days versus months. Mendix is ideal for delivering differentiated and innovative multichannel, enterprise-class systems. This is a natural addition to the developer tooling and experience—a key feature of the HPE Helion Development Platform, complementing deeper languages and frameworks used by developers such as Java, Node. js, Perl, Python, Ruby, and Go.

Mendix seamlessly integrates with the HPE Helion Development Platform, an open, interoperable platform that enables application development and portability across traditional, private, and public clouds. Based on Cloud Foundry and integrated with HPE Helion OpenStack to allow businesses to quickly develop, deploy, and deliver cloud-native applications, Mendix and the HPE Helion Development Platform support the complete application lifecycle:

- **Design**—Capture, refine, and prioritize user stories; estimate, plan, and monitor sprints; manage releases, and so forth

- **Build**—Build multichannel, multidevice applications with visual, model-driven development to promote communication, productivity, quality, and short iterations

- **Deploy**—Provision and manage apps in the cloud, including one-click, auto-scale deployment; manage test, acceptance, and production environments

- **Manage**—Manage all applications from a single dashboard, delivering availability, security, performance, and scalability

- **Iterate**—Gather feedback, including end-user input, and use in the next cycle of capture-develop-deploy-iterate

- **Integrate**—Seamlessly integrate new applications with existing systems and processes across the enterprise, including the HPE Helion portfolio of products

Shadow IT creates barriers for all companies

With the speed of business today, IT organizations are discovering that new cloud applications are not being delivered fast enough using traditional application development methods. And when backlogs inevitably occur, the trend is clear: LOB managers will go with "good enough" apps and services from SaaS and public cloud providers. In addition to creating silos and pockets of technology, such shadow IT practices introduce control, manageability, security, and governance concerns. This presents a huge challenge for IT organizations that are striving to modernize data centers and implement a converged cloud infrastructure that will meet the business and technical needs of their entire organizations.

According to Mark Rogers, Mendix VP of Business Development,

> CIOs see cloud solutions as a way to transform their business, but the path to get there is not always well defined. They will turn to off-shore development as a solution only to find an additional layer of complexity. When it comes to application portfolio management, CIOs have many questions, such as, "How do I move existing apps to the cloud, or is it better to rewrite them instead?"

Mendix and HPE Helion integration speeds app delivery

The equation is simple. If you can empower a company's business analysts or domain experts to create and deliver new apps faster, especially in an environment supported and governed by IT, then you can divest yourself of the complications that come with shadow IT, off-shoring, and portfolio management. You can then address important technology initiatives such as establishing and evolving your private or hybrid clouds.

What makes the Mendix and HPE Helion solution different? HPE is working with Mendix to provide a simple, integrated development experience for rapidly developing and deploying applications to an open cloud solution. Mendix applications run in the cloud as a native cloud application (Cloud Foundry apps) and can be developed without the need for coding. The HPE Helion Development Platform uses open source technologies such as Cloud Foundry software and Docker containers and is integrated with HPE Helion OpenStack, providing developers and IT operations with an open solution for agility, portability, and scalability. Mendix and HPE help to accelerate and simplify the deployment of enterprise-ready Cloud Foundry-based applications, as well as fuel collaboration and innovation of the HPE Helion ecosystem for cloud development by combining the Mendix model-driven development approach with one-click, auto-scale deployment to the HPE Helion Development Platform. HPE and Mendix are collaborating together with a network of system integrators and service providers to enable fast industry-specific application development capabilities and templates that can be used by business-oriented developers and users.

What HPE Helion and Mendix mean for customers

Through the HPE Helion Ready Program, HPE and Mendix will help CIOs advance the New Style of IT across social, mobile, cloud, and Big Data challenges by offering a better way to deliver new, migrated, or modernized multichannel applications. The partnership will provide an integrated experience to simplify the development and deployment of enterprise-ready Cloud Foundry-based applications. Advanced features, including security model integration and application lifecycle management, will further accelerate and produce cloud-native development and deployment for professional and rapid developers alike.

In summary, Rogers states,

> Our joint customers will use the HPE Helion Development Platform combined with Mendix to build a wide range of multichannel apps across industries and use cases—including insurance and loan processing, clinical systems, logistics workflows, multichannel order management, and customer and partner portals. This will certainly be a win-win situation for customers and both HPE and Mendix.

Customer at a glance

HPE Helion Cloud solution

- Private cloud

Software

- Mendix App Platform
- HPE Helion Development Platform

Case study 2 questions

1. What is Mendix? (Use the Internet if necessary.)

   _____

   _____

   _____

2. What experience does Mendix have with no-code rapid application development?

   _____

   _____

   _____

3. How are Mendix apps integrated with the Helion Development Platform?

_____

_____

_____

HPE Helion Development Platform activity—answers

Case study 1—Using microservices for customer surveys

1. What are microservices?

   - **Microservice is a software architecture style in which complex applications are composed of small, independent processes communicating with each other using language-agnostic APIs. These services are small, highly decoupled and focus on doing a small task, facilitating a modular approach to system-building**

2. What is the business case and background for this study?

   - **A legacy, high-volume web application needs a customer survey functionality added to it. This survey needs to be presented to every user, once per visit. The current traffic levels are at 1,300,000 unique visitors per day, which generate over 4,000,000 page views. The peak traffic is in the evenings**

3. What are the application scale requirements, release frequency, and data needs?

   - **The customer survey microservice must be embedded into the legacy web application and not require modifications to this application. It must automatically scale up to 4,500,000 daily request, peaking in the evening hours. A survey completion of seven percent of unique visitors is expected each day, which yields 91,000 surveys per day (1,300,000 * 7%). Each survey must consist of four to eight questions, plus metadata**

   - **A monthly release cadence is required to support new surveys from an internal research team. When there are flaws in the collected survey data, this release frequency might increase.**

   - **Because the data structure will change from one survey to another, the solution must support unstructured datastore technology. This datastore must be highly available and scalable. The stored data must also be easily accessible for postprocessing**

4. What technologies does the solution include?

   - **Go programming language—Supports the high-concurrency requirements, RESTful API, and stateless applications; it also has low memory requirements**

   - **MongoDB datastore—Provides the uptime and scale requirements**

   - **Hadoop—Is used for batch processing**

- Helion Development Platform—Provides flexibility and speed required by the development team, auto-scale to improve resource utilization and application responsiveness, build-pack technology, and service integration

Case study 2—Mendix (rapid app development)

1. What is Mendix?

   - Mendix applications are cloud native and can be built without coding. They run across browsers, smartphones, and tablets. This is a natural extension of the Helion Development Platform, which provides the developer tooling, deeper languages, and integrated ecosystem. The solution is a seamless integration between Mendix and the Helion Development Platform to foster rapid application development in the cloud and an integrated development experience.

2. What experience does Mendix have with no-code rapid app development?

   - Rapid developers utilizing the Mendix and Helion Development Platform solution can build, integrate, and deploy multidevice applications six times faster than using traditional programming languages. They can deliver major applications in weeks or days. The applications run as native cloud applications that are portable across browsers, smartphones, and tablets.

3. How are Mendix apps integrated with the Helion Development Platform?

   - Mendix is part of the HPE Helion ecosystem through the HPE Helion Ready Program for Solution Providers.

Benefits to customers

Benefits of the Helion Development Platform include:

- **Agility and flexibility**—Enables accelerated development of cloud-native applications with instant access to a variety of runtimes, frameworks, and services.

- **Resilient, highly available services**—Allows services to be leveraged inside applications.

- **Modern developer experience**—Frees up developers to focus on coding instead of managing resources.

- **Easy administration**—Uses a common platform across development and operations and automates processes to reduce administration overhead.

- **Workload portability**—Prevents lock in with the ability to move applications from a private or a public cloud without having to change the code (the cloud workloads are hardware-agnostic).

- **Low TCO**—Provides an open-source software solution and a favorable pricing structure.

- **Enterprise-grade security**—Allows all components to support SSL connections.

The Helion Development Platform can support customers in multiple ways. The platform:

- Handles the automatic configuration of the language runtime, web server, application dependencies, databases, and other services

- Supports languages such as Java, Node.js, ActivePerl, PHP, ActivePython, Ruby, Erlang, Scala, and Clojure

- Supports web servers such as Nginx, Apache, and Apache TomEE

Getting started

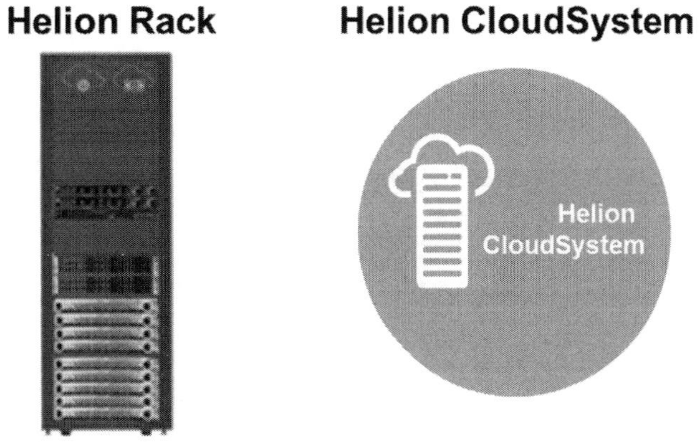

Figure 7-30 Two ways to start using the Helion Development Platform

As shown in Figure 7-30, customers can start using the Helion Development Platform in two ways:

- **Helion Rack**—Helion Rack is a complete private cloud solution with the versatility and performance of ProLiant servers. It is optimized for Helion OpenStack for enterprise-grade security, manageability, and reliability, as well as full integration with the Helion Development Platform.

- **Helion CloudSystem 9.0**—This release of Helion CloudSystem includes the Helion Development Platform.

Stackato

Stackato is a PaaS product for enterprise IT departments and developers based on various open-source projects, including Cloud Foundry and Docker. It is a commercially supported and extended version of Cloud Foundry that runs on any IaaS (VMware, Microsoft Azure, Amazon Web Services, OpenStack, and others).

Note

HPE acquired Stackato from ActiveState on 28 July 2015 to complement the Helion Development Platform.

HPE Helion Rack

HPE Helion Rack overview—Activity

Watch a two-minute video on YouTube about Helion Rack.

 Note

To watch this video, scan this QR code or enter the URL into your browser:

https://www.youtube.com/watch?v=knvCzlPihXl

Then watch a three-minute video about Helion Rack, featuring Ken Won, director of cloud solutions marketing, and Lisa-Marie Namphy, Helion and OpenStack Marketing.

After you watch the videos, answer these questions.

1. What is Helion Rack?

    _____

    _____

    _____

2. What are the use cases for Helion Rack?

    _____

    _____

    _____

3. Which HPE servers are included in Helion Rack?

   a. BladeSystem c7000 enclosures and ProLiant BL460c server blades

   b. ProLiant DL360 and DL380 servers

   c. ProLiant SL4500 servers

   d. Moonshot 1500 and ProLiant m800 server cartridges

4. Which software components are included in Helion Rack? (Select three.)

   a. Helion Development Platform

   b. Cloud Service Automation

   c. Operations Orchestration

   d. Helion OpenStack

   e. OneView

HPE Helion Rack overview activity—answers

1. What is Helion Rack?

   - **An optimized cloud solution—a predefined, prebuilt, preconfigured private cloud solution based on OpenStack and Cloud Foundry technologies**

2. What are the use cases for Helion Rack?

   1. **Testing/development environments for cloud-native applications (development and deployment)**

   2. **Demanding workloads such as analytics and high-performance databases**

   3. **Web tier applications that need security, compliance, and performance**

3. Which HPE servers are included in Helion Rack?

   a. BladeSystem c7000 enclosures and ProLiant BL460c server blades.

   b. **ProLiant DL360 and DL380 servers**

   c. ProLiant SL4500 servers.

   d. Moonshot 1500 and ProLiant m800 server cartridges.

4. Which software components are included in Helion Rack? (Select two.)

   a. Helion Development Platform

   b. Cloud Service Automation

   c. Operations Orchestration

   d. Helion OpenStack

   e. OneView

HPE Helion Rack configuration

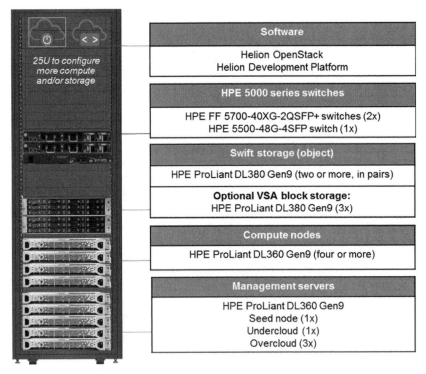

Figure 7-31 HPE Helion Rack configuration

The Helion Rack configuration, shown in Figure 7-31, provides everything customers need with options to scale out. ProLiant Gen9 servers power the Helion Rack, providing performance in a high-density compute, storage, and networking environment. ProLiant DL360 Gen9 servers are used for compute, and ProLiant DL380 Gen9 servers are used for storage. They are supported by HPE Networking 5700 and 5500 switches.

The base configuration of 17 units includes four compute nodes and two Swift storage nodes. Storage nodes must be added in units of two. The configuration can be scaled up to fill the remaining 25U in the 42U rack with any mix of compute or Swift storage as desired. A three-year hardware warranty is included.

Helion Rack software consists of Helion OpenStack and Helion Development Platform. Helion OpenStack Accelerator Service is included with Helion Rack, meaning that HPE specialists complete the onsite installation after the hardware is delivered. This offers the fastest possible installation and accelerates time to value for customers by leveraging the full value of HPE cloud solutions and OpenStack technology expertise and leadership. The base configuration also includes Helion compute and storage annual licenses with 24×7 support to ensure that a solution runs at its optimal performance. A 9×5 support service is also available.

Optional training and consulting services are available from Helion Professional Services to extend cloud and OpenStack skills and cloud operational capabilities as customers' business needs grow.

HPE interactive technology demonstration — Activity

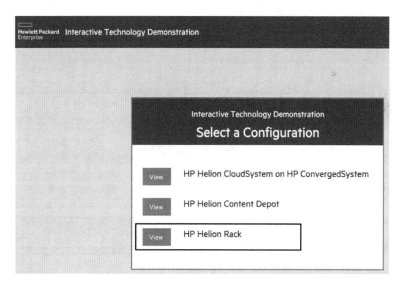

Figure 7-32 HPE Helion Rack demonstration

Take a few minutes to explore the interactive technology demonstration of Helion Rack by selecting HP Helion Rack as shown in Figure 7-32.

Note

To access the demonstration, scan this QR code or enter the URL into your browser:

https://hpcloud.xperionportal.com/desktop#/menu/707

 Note

Other interactive demonstrations on this site include Helion CloudSystem on ConvergedSystem and Helion Content Depot.

After exploring the demo, answer the following questions:

1. What type of internal disk drives are in the default configuration for the compute nodes, and how many drives are included?

_____

_____

_____

2. How many Swift starter or object store nodes are part of the default configuration?

_____

_____

_____

3. What are the type and number of internal disk drives for the overcloud controller?

_____

_____

_____

4. Does the configuration include a console kit? If so, which model is it?

_____

_____

_____

5. How many internal drives are used by the VSA block storage servers?

_____

_____

_____

HPE interactive technology demonstration activity—answers

1. What type of internal disk drives are in the default configuration for the compute nodes, and how many drives are included?

   - **Eight SFF**

2. How many Swift starter or object store nodes are part of the default configuration?

   - **Two**

3. What are the type and number of internal disk drives for the overcloud controller?

   - **Four LFF**

4. Does the configuration include a console kit? If so, which model is it?

   - **Yes, HPE LCD8500 1U Rackmount Console Kit**

5. How many and what type of internal drives are used by the VSA block storage servers?

   - **24 SFF**

Learning check

1. What is the Helion Platform?

   _____

   _____

   _____

2. What are the benefits of Helion OpenStack? (Select three.)

   a. Flexible deployment across cloud models

   b. Microservices

   c. Enterprise-grade security

   d. Web tier applications

   e. Global support and professional services

3. What is the Helion Development Platform?

   _____

   _____

   _____

4. HPE Integrity servers power the Helion Rack, providing flexibility and performance in a high-density compute, storage, and networking environment.

   ☐ True

   ☐ False

Learning check answers

1. What is the Helion Platform?

   - **It is a combined IaaS and PaaS platform based on OpenStack and Cloud Foundry**

   - **It consists of Helion OpenStack and Helion Development Platform**

   - **It provides accelerated application development and deployment, application portability, and no vendor lock in**

2. What are the benefits of Helion OpenStack? (Select three.)

   a. **Flexible deployment across cloud models**

   b. Microservices

   c. **Enterprise-grade security**

   d. Web tier applications

   e. **Global support and professional services**

3. What is the Helion Development Platform?

   • **An open-source cloud application platform and ecosystem that makes it easier to develop, deploy, and deliver highly available and scalable cloud-native applications**

   • **Provides developers with instant access to a variety of language runtime, frameworks, and services to quickly develop applications**

   • **Based on the open-source technologies of Cloud Foundry, Docker, and OpenStack software**

4. HPE Integrity servers power the Helion Rack, providing flexibility and performance in a high-density compute, storage, and networking environment

   ☐ True

   ☐ **False**

Resources

To access these resources, scan the QR codes or enter the URLs into your browser:

 Note

Helion OpenStack and Helion Development Platform overviews.

http://www.hp.com/helion/hphelionopenstack

Helion OpenStack solution brief

**https://www.hpe.com/h20195/v2/GetPDF.aspx/
4AA5-6770ENW.pdf**

Helion technical documentation

http://docs.hpcloud.com/

Helion Rack overview

http://www.hp.com/helion/helionrack

Summary

Open-source software is rapidly becoming the new infrastructure to resolve customers' current challenges. These solutions provide:

- No vendor lock in with access to a rich ecosystem offering breadth and choice

- Transparency of governance, road map, blueprints, and development

- Easier migration from traditional data centers to cloud environments

- Shorter development process

- Rapid innovation through contributions from subject matters experts

The HPE Helion Platform is a combined IaaS and PaaS solution based on OpenStack and Cloud Foundry. It provides accelerated application development and deployment along with easy administration of OpenStack and Cloud Foundry technologies. Two main components are Helion OpenStack and Helion Development Platform.

Helion Rack is a predefined, prebuilt, and preconfigured private cloud solution based on OpenStack and Cloud Foundry. The use cases for Helion Rack include:

- Test/dev environments for cloud-native applications (development and deployment)

- Demanding workloads such as analytics and high-performance databases

- Web tier applications that need security, compliance, and performance

8 A Closer Look at CloudSystem Foundation – Architecture and Installation

---

WHAT IS IN THIS CHAPTER FOR YOU?

After completing this chapter, you should be able to:

✓ Describe HPE Helion CloudSystem

✓ List components of the architecture

✓ Explain networking in CloudSystem 9.0

✓ Identify best practices to ensure security in the CloudSystem environment

✓ Outline the steps for installing CloudSystem 9.0

HPE Helion CloudSystem

Before proceeding with this section, assess your existing knowledge of HPE Helion CloudSystem by answering the following questions.

Assessment activity

1. Have you had previous exposure to Helion CloudSystem 8.x?

_____

_____

_____

2. Do you have customers who are currently using CloudSystem 8.x? If so, what type of environment have they deployed this solution into?

_____

_____

_____

_____

3. What do you or your customers like about this solution? Which functionality do you or your customers wish it had?

_____

_____

_____

_____

Enterprise-ready high availability architecture

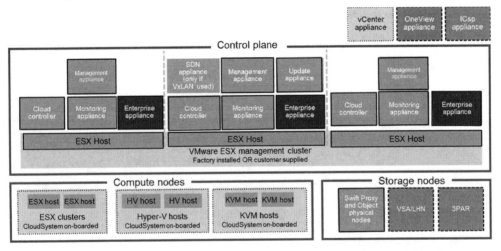

Figure 8-1 Helion CloudSystem 9.0 architecture

Figure 8-1 represents the Helion CloudSystem 9.0 enterprise-ready, high availability architecture and the distribution of CloudSystem appliances and managed resources in a deployed cloud. In this example, CloudSystem is deployed across a cluster of three VMware ESX hosts.

CloudSystem appliances are represented by the boxes bounded by solid lines (management appliance, cloud controller (CC), monitoring appliance, enterprise appliance (EA), update appliance, ESX hosts, Swift proxy, and Object physical nodes). Boxes bounded by dashed lines (OneView appliance, ICsp appliance, VSA/LHN, and 3PAR) are other HPE appliances or storage components that integrate with CloudSystem. Boxes bounded by dotted lines (vCenter appliance, ESXclusters, Hyper-V hosts, and KVM hosts) represent third-party partner components that are provided by the customer. The Software-defined networking (SDN) appliance is optional and only used with virtual extensible LAN (VxLAN).

Figure 8-1 shows appliances that belong to CloudSystem Foundation:

- Management appliance

- Monitoring appliance

- Cloud controller

- SDN appliance (optional and used only with VxLAN)

- Update appliance

CloudSystem Enterprise builds on the CloudSystem Foundation solution by adding:

- Enterprise appliance

- OneView appliance

- Insight Control server provisioning (ICsp) appliance

 Note

The control plane is also supported on Red Hat KVM.

Architecture

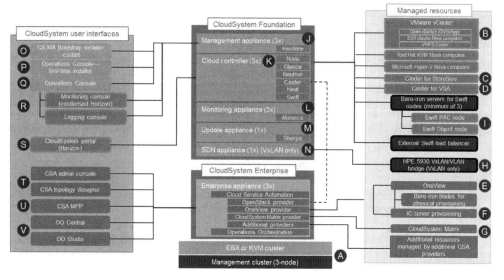

Figure 8-2 Management hypervisors and managed resources

The Helion CloudSystem 9.0 solution components fall into four categories:

- Management hypervisors and managed resources
- CloudSystem virtual appliances
- CloudSystem user interfaces
- CloudSystem storage

Management hypervisors and managed resources

Figure 8-2 shows management hypervisors and managed resources. Descriptions and corresponding diagram labels include the following:

- **Management hypervisors** (Label A) host the virtual appliances that compose the CloudSystem solution. These hypervisors are arranged as a three-node configuration of ESXi clusters or KVM hosts.

- **VMware vCenter** (Label B) acts as a central administrator for ESXi clusters that are connected on a network. VMware vCenter allows users to pool and manage the resources of multiple hosts, as well as monitor and manage the physical and virtual infrastructure. Customers can activate ESXi clusters in the CloudSystem Operations Console after registering a connection with vCenter. ESXi instance security is provided by the Virtual Cloud Networking Open vSwitch vApp (OVSvApp) appliance. This appliance is automatically installed on each ESXi compute hypervisor during activation after an OVSvApp image is loaded in the customer's data store.

- **HPE StoreServ storage system** (Label C) provides a method of carving storage for KVM and Hyper-V compute nodes. StoreServ block storage drivers are registered in the Operations Console.

- **StoreVirtual VSA** (Label D) provides block storage for KVM and Hyper-V compute nodes. StoreVirtual VSA block storage drivers are registered in the Operations Console.

 Note

A Fibre Channel (FC) SAN, iSCSI, or Flat SAN network connects the StoreServ storage system to compute nodes or ESXi clusters. An iSCSI connection is required for StoreVirtual VSA storage.

- **OneView** (Label E) manages the converged infrastructure and supports key scenarios such as deploying bare-metal servers, performing ongoing hardware maintenance, and responding to alerts. It is designed for the physical infrastructure needed to support virtualization, cloud computing, big data, and mixed computing environments.

- **ICsp** (Label F) deploys operating systems on ProLiant bare-metal servers; updates drivers, utilities, and firmware; and configures system hardware.

- **CloudSystem Matrix** (Label G) can be configured as an additional provider in CSA (CSA is provided with CloudSystem Enterprise).

- **FlexFabric 5930 Switch Series** (Label H) is a family of high-performance and ultra-low-latency 40 GbE top-of-rack (ToR) data center switches.

- **Swift Proxy Account Container (PAC) and Swift Object servers** (Label I) support object storage. These servers are not configured as part of the initial CloudSystem installation.

The object storage networks (OBS) must be configured appropriately and you must install the OpenStack Swift CLI on the management appliance to manage scale-out object storage. An external load balancer is also required.

CloudSystem Foundation virtual appliances

Helion CloudSystem supports a three-node KVM management host or VMware ESXi management cluster, hosting these virtual appliances in a high availability configuration.

- **Management appliance (3×)** (Label J) is responsible for standing up and managing CloudSystem virtual appliances. The Operations Console is the administrative interface for this appliance.

- **Cloud controller (3×)** (Label K) contains the majority of OpenStack services used in CloudSystem. The OpenStack Horizon user portal is the cloud user interface for this appliance. A data volume containing the Glance repository is part of the cloud appliance trio. Glance is configured to use a local disk as its image store location (this is not a shared disk).

- **Monitoring appliance (3×)** (Label L) contains the monitoring services that are used to monitor the performance and health of CloudSystem virtual appliances and compute nodes.

- **Update appliance (1×)** (Label M) manages patches and upgrades to the CloudSystem environment.

- **SDN controller appliance (1×)** (Label N) is only deployed in environments configured to support VxLAN for Tenant and Provider networks. It manages the Layer 2 (L2) gateway to bridge the cloud VxLAN network and the legacy data center virtual LAN (VLAN) network.

CloudSystem Enterprise virtual appliances

CloudSystem Enterprise adds a trio of appliances to the Foundation installation. These appliances provide the core functionality of the Enterprise offering. They include CSA, the Marketplace Portal, the Topology Designer, and the Sequential Designer. Operations Orchestration (OO) Central and OO Studio are also embedded in the Enterprise appliance.

- **CSA Cloud Service Management Console** is the administrative portal for the Enterprise appliance. Designs are provisioned as offerings in the CSA console.

- **Marketplace Portal** displays offerings that cloud users can purchase as a subscription.

- **CSA Topology Designer** is an easy-to-use solution for infrastructure provisioning designs.

- **CSA Sequential Designer** handles more complex application provisioning designs.

- **OO Central** provides the ability to run scripted workflows in CSA.

- **OO Studio** provides the ability to create and customize new workflows and debug and edit existing workflows. OO Studio is installed separately, using an executable file included in CloudSystem Enterprise.

CloudSystem user interfaces

Table 8-1 Helion CloudSystem user interfaces for administrators and cloud users

Label	UI	How to access	Virtual appliance hosting the UI
O	Management appliance installer	Launch the csstartgui.bat file included in the CloudSystem release package from a staging server	N/A (installer is run from a staging server)
P	First-time installer	Launch the Operations Console for the first time and the installer launches automatically	Management appliance
Q	Operations Console	http://<MA_DCM_vip_address>	Management appliance
R	OpenStack monitoring portal	https://<MA_DCM_vip_address>:9090	Management appliance
S	OpenStack user portal (Horizon)	https://<CC_CAN_vip_address>	Cloud controller
T	CSA Management Console	https://<EA_CAN_vip_address>:8444/csa	Enterprise appliance
U	Marketplace Portal	https://<EA_CAN_vip_address>:8089	Enterprise appliance
V	OO	http://<EA_CAN_vip_address>:9090/oo	Enterprise appliance

Label	UI	Used to	Credentials
O	Management appliance installer	Install the management appliance, create the Data Center Management Network and prepare the Cloud Management Network	N/A
P	First-time installer	Install the remaining virtual appliances and create the rest of the network infrastructure	N/A
Q	Operations Console	Manage the cloud environment	Set during first-time installation
R	OpenStack monitoring portal	Manage monitoring services	Set during first-time installation
S	OpenStack user portal (Horizon)	Create, launch, and manage virtual machine instances	Set during first-time installation
T	CSA Management Console	Create and manage service offerings and service catalogs	admin/cloud
U	Marketplace Portal	Select cloud services from a catalog and monitor and manage existing services with subscription pricing	consumer/cloud
V	OO	Attach workflows to server lifecycle actions or schedule workflows for regular execution	Set during first-time installation

Note

In Table 8-1 MA stands for management appliance, DCM is the data center management network, CC is the cloud controller, CAN is the consumer access network, and EA is the enterprise appliance.

CloudSystem storage

CloudSystem storage is provided through a variety of solutions:

- Block storage

  - **VMware Virtual Machine File System (VMFS)**—Provides boot-from-volume functionality for ESXi compute hosts.

  - **StoreServ FC**—Provides instance data storage for KVM compute hosts.

  - **StoreServ iSCSI**—Provides boot and instance data storage for KVM and Hyper-V compute hosts.

  - **StoreVirtual VSA**—Provides virtual storage for instances on ESXi and Hyper-V compute hosts.

- Object storage (Swift)

  - **Scale-out object storage** is provided by four dedicated servers: two Swift PAC servers and two Swift Object servers. The object proxy network (OPN) and OBS support object storage. Users can manage object storage using the Swift CLI on the management appliance.

- File storage

  - **Ephemeral** (assigned to a VM instance when the instance is created and released when the instance is deleted) storage for ESXi, KVM, and Hyper-V compute hosts.

  - **Management hypervisor storage** for CloudSystem virtual appliances.

Networking

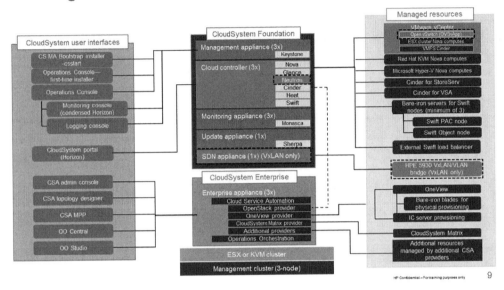

Figure 8-3 Neutron networking in CloudSystem 9.0

As shown by the boxes bounded by dashed lines in Figure 8-3, Helion OpenStack Neutron networking components in CloudSystem are present in both the management cluster and the compute nodes (labeled managed resources in Figure 8-3):

- **Management cluster**

  - OpenStack Neutron services are incorporated into the cloud controllers.

  - The SDN controller is deployed in the management cluster for VxLAN configurations.

- **Compute node**—OVSvApp L2 agent is automatically deployed on a compute node when the compute node is activated in the CloudSystem environment.

 Note

For more detailed information about Neutron and using the OVSvApp solution in ESX deployments, scan this QR code or enter the URL into your browser:

https://wiki.openstack.org/wiki/Neutron/Networking-vSphere

Networking options

Tables 8-2 and 8-3 show which network deployment options are supported by Helion CloudSystem 9.0.

 Note

Distributed Virtual Routing (DVR) and Centralized Virtual Routing (CVR) are discussed on the next page.

Table 8-2 Management hypervisor networking options

Hypervisor type	VLAN	VxLAN	CVR	DVR
ESXi	✓	✓	✓	✓1
KVM	✓	✓	✓	✓2

 NOTES

1 SDN Controller is launched to manage compute security.

2 SDN Controller is **not** launched.

Table 8-3 Compute node networking options

Compute node type	VLAN	VxLAN	CVR	DVR
ESXi	✓	✓	✓	✓
KVM	✓	✓	✓	✓
Hyper-V	✓		✓	

 Important

Networking options and choice of hypervisors must be decided prior to CloudSystem installation. They cannot be changed later.

OpenStack Distributed Virtual Routing vs. Centralized Virtual Routing

OpenStack introduced DVR in the Juno release. When installing CloudSystem, customers can choose to implement CloudSystem with DVR or with traditional CVR.

With DVR, L3 forwarding and NAT functions are moved from the network nodes to the compute nodes. DVR provides greater east/west load distribution compared to CVR and better network scalability.

VLAN vs. VxLAN

VxLAN is designed to enhance L2 network services beyond what IEEE 802.1Q VLAN currently supports. The VxLAN specification was created through a partnership with VMware, Arista Networks, and Cisco. VxLAN offers several benefits over VLAN:

- VxLAN provides high scalability, allowing environments to scale beyond the 802.1Q limit of 4094 VLANs.

- VxLAN technology provides cloud service providers with multitenancy (meaning duplicate IP) options.

- VxLAN allows for greater virtual machine (VM) instance mobility across the entire data center.

Network decision points

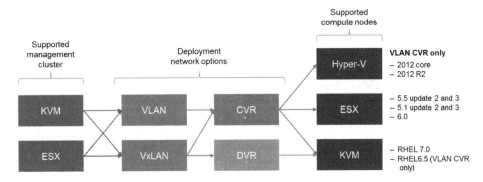

Figure 8-4 Network decision points

Figure 8-4 depicts the network decision points with respect to the supported management cluster (ESX and KVM), the deployment network options (CVR, DVR, VLAN, and VxLAN), and the supported compute nodes.

Consider the following example where a customer is using KVM with VLAN. Their only choice for routing is CVR (there is no arrow going from the VLAN box to the DVR box). Consider another example where a customer is using KVM with VxLAN. They could choose CVR or DVR, but choosing DVR would limit their choice of compute nodes to KVM. Choosing CVR would allow them to choose Hyper-V, ESX, or KVM compute nodes.

CloudSystem networks

Table 8-4 CloudSystem network organization

Trunk	Network	Required?	Type
Management	Data center management (DCM) network	Yes	L3
	Cloud management (CLM) network	Yes	L2
	Consumer access network (CAN)	Yes	L3
	External network (EXT)	Yes	L3
	PXE network	No	L3
Cloud data (either VLAN trunk or VxLAN underlay)	Provider networks	No	L3
	Tenant networks	No	L2
	Tenant underlay network (TUL)	No	L2
	Service network for Helion Development Platform	No	L3
Storage	Block storage network (BLS)	No	L2
	Object proxy network (OBP)	No	L2
	Object storage network (OBS)	No	L2

As shown in Table 8-4, Helion CloudSystem 9.0 networks are organized into three network trunks—management trunk, cloud data trunk, and storage trunk:

- **Management trunk**—Requires a virtual distributed switch (VDS), which connects the trunk to the CloudSystem control plane and contains these networks:

  - Data Center Management (DCM) Network

  - Cloud Management (CLM) Network

  - Consumer Access Network (CAN)

  - External network (EXT)

  - PXE network (EXT)

- **Cloud data trunk**—Contains these networks:

  - Provider networks

  - Tenant networks

  - Tenant Underlay (TUL) Network

  - Service network for Helion Development Platform

Important

When deploying CloudSystem, you must choose a VLAN or a VxLAN network underlay. You cannot change this after CloudSystem is deployed. Future migration operations might not support migration of a cloud environment configured to use VxLAN.

- **Storage trunk**—Contains these networks:
  - Block storage network (BLS)
  - Object proxy network (OBP)
  - Object storage network (OBS)

Management trunk—Learner activity

Use the *Helion CloudSystem 9.0 Network Planning Guide*, available on the HPE Information Library, to assign the correct definitions to the Management trunk networks.

 Note

To access the Enterprise Information Library, scan this QR code or enter the URL into your browser:

http://www.hp.com/go/CloudSystem/docs

Fill in the blanks that correctly describe the bullet points underneath each line.

- _____:

  - A public network that allows cloud users to access the OpenStack user portal (Horizon), CSA, the Marketplace Portal, and OpenStack and Marketplace Portal APIs
  - Created during CloudSystem deployment by the first-time installer
  - Used by providers

- _____:

  - Connects virtual appliances to StoreServ arrays, StoreVirtual VSA, VMware vCenter Server, and enclosures
  - Provides access to administrator roles only

- Is created during CloudSystem deployment by the Cloud Management Appliance Installer
- Used by the Operations Console and OpenStack API admin URLs and the CSA APIs
- Used by OpenStack Keystone for the admin endpoints for all services
- Used to transmit sensitive information such as CloudSystem backup files
- Does not support native VLAN

- _____:

- Enables routing of VM instances on Tenant networks out from the CloudSystem private cloud to the data center, the corporate intranet, or the Internet
- Allows cloud users to attach public IP addresses to their provisioned VM instances
- Supports and requires floating IP addresses to connect instances to this network
- Has its VLAN ID identified during CloudSystem deployment by the first-time installer

- _____:

- An untagged (native VLAN) network with DHCP and used to boot object storage (Swift) servers
- Not connected until you deploy scale-out Swift
- Attached to the first management appliance
- Connected to the object storage nodes exclusively at eth0 (or the first NIC of the machine)—no other networks can be connected to this NIC
- Created when CloudSystem is deployed by the first-time installer

- _____:

- Used by the management appliances to communicate with other CloudSystem appliances
- Used when operating system and security updates are applied to the Enterprise appliance
- Provides solution components with a dedicated cloud management network
- Supports all appliance-to-appliance and appliance-to-compute node traffic, HA heartbeats, and OpenStack APIs (internal URLs)
- Created during CloudSystem deployment, initially by the Cloud Management Appliance Installer

Management trunk learner activity—answers

- **Consumer access network:**

  - A public network that allows cloud users to access the OpenStack user portal (Horizon), CSA, the Marketplace Portal, and OpenStack and Marketplace Portal APIs

  - Created during CloudSystem deployment by the first-time installer

  - Used by providers

- **Data center management network:**

  - Connects virtual appliances to StoreServ arrays, StoreVirtual VSA, VMware vCenter Server, and enclosures

  - Provides access to administrator roles only

  - Is created during CloudSystem deployment by the Cloud Management Appliance Installer

  - Used by the Operations Console and OpenStack API admin URLs and the CSA APIs

  - Used by OpenStack Keystone for the admin endpoints for all services

  - Used to transmit sensitive information such as CloudSystem backup files

  - Does not support native VLAN

- **External network:**

  - Enables routing of VM instances on Tenant networks out from the CloudSystem private cloud to the data center, the corporate intranet, or the Internet

  - Allows cloud users to attach public IP addresses to their provisioned VM instances

  - Supports and requires floating IP addresses to connect instances to this network

  - Has its VLAN ID identified during CloudSystem deployment by the first-time installer

- **PXE network:**

  - An untagged (native VLAN) network with DHCP and used to boot object storage (Swift) servers

  - Not connected until you deploy scale-out Swift

  - Attached to the first management appliance

  - Connected to the object storage nodes exclusively at eth0 (or the first NIC of the machine)— no other networks can be connected to this NIC

  - Created when CloudSystem is deployed by the first-time installer

- **CLM network:**

    - Used by the management appliances to communicate with other CloudSystem appliances

    - Used when operating system and security updates are applied to the Enterprise appliance

    - Provides solution components with a dedicated cloud management network

    - Supports all appliance-to-appliance and appliance-to-compute node traffic, HA heartbeats, and OpenStack APIs (internal URLs)

    - Created during CloudSystem deployment, initially by the Cloud Management Appliance Installer

Management trunk networks

As shown in Figure 8-5, the DCM network connects virtual appliances (management appliance, cloud controller, Enterprise appliance, and update appliance) to StoreServ 3PAR arrays, StoreVirtual VSA, and compute nodes (KVM, Hyper-V, and ESXi).

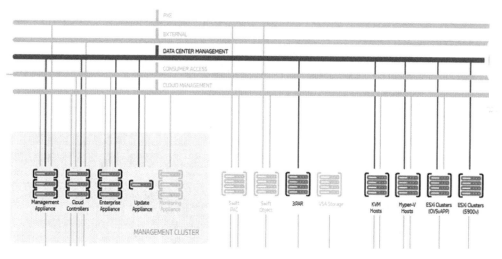

Figure 8-5 DCM network

As shown in Figure 8-6, the CLM network is used by the management appliances to communicate with other CloudSystem appliances, with Swift storage nodes, and with compute hosts.

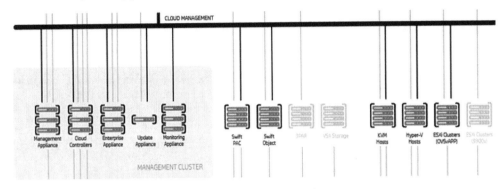

Figure 8-6 CLM network

As shown in Figure 8-7, the CAN is a public network connected to an external load balancer that allows cloud users to access Horizon, CSA, the Marketplace Portal, and OpenStack and Marketplace Portal APIs served by the CCs and Enterprise Appliances.

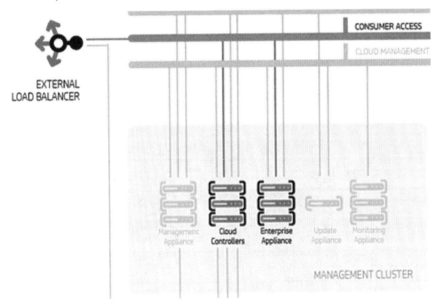

Figure 8-7 Consumer access network (CAN)

As shown in Figure 8-8, the external network (EXT) is only connected to the CCs. It routes VM instances on Tenant networks out from the CloudSystem private cloud to the data center, the corporate intranet, or the Internet. This network enables cloud users to attach public IP addresses to their provisioned virtual machine instances.

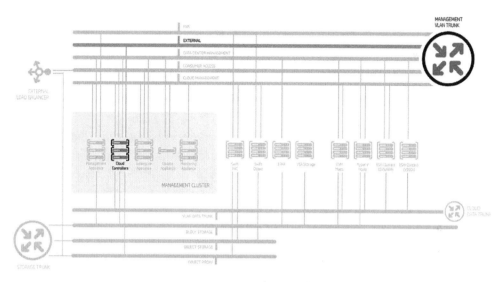

Figure 8-8 External network (EXT)

The PXE network shown in Figure 8-9 connects the management appliance to the Swift PAC and Object servers and is used to boot these storage servers.

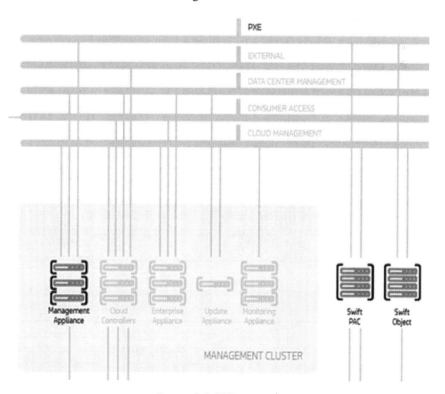

Figure 8-9 PXE network

Cloud data trunk—Learner activity

Use the *Helion CloudSystem 9.0 Network Planning Guide*, available on the Enterprise Information Library, to assign the correct definitions to the cloud data trunk networks.

 Note

To access the Enterprise Information Library, scan this QR code or enter the URL into your browser:

http://www.hp.com/go/CloudSystem/docs

Fill in the blanks that correctly describe the bullets underneath each line.

● _____:

- A single network VLAN that encapsulates and carries Tenant and Provider networks as VxLANs

- Only available in VxLAN configurations

- Is an alternative to the Cloud Data Trunk and VLANs for the Tenant and Provider networks (because it becomes the "trunk" for Tenant and Provider VxLANs)

- Configured when the first-time installer deploys CloudSystem with the VxLAN option selected

 Note

The SDN controller virtual appliance is automatically configured when this network is created and the VxLAN option is selected during the first-time installation.

● _____:

- Data center networks that are routed through the existing data center infrastructure and used to provision VM instances

- Are created for cloud tenants in the Operations Console by cloud admins

- Can be shared among tenants or assigned to a specific tenant

- Mapped or routed to an existing physical network in the data center so that cloud instances can communicate with legacy data center resources

Note

If the cloud network type is VLAN, the Provider network could be part of the Cloud Data Trunk or routed with the Cloud Data Trunk. If the type is VxLAN, the Provider network requires an SDN controller and at least one physical HPE FlexFabric 5930 switch (L2 hardware gateway) to bridge communication with legacy data center networks (external VLANs).

- _____:

  - Are restricted and can be accessed only by VM instances assigned to the network—subnets must be defined in the OpenStack user portal before using this network

  - Have segmentation ID ranges created in the Operations Console by the cloud administrators

  - Created by cloud users in the OpenStack user portal

Cloud data trunk learner activity—answers

- **Tenant underlay network:**

  - A single network VLAN that encapsulates and carries Tenant and Provider networks as VxLANs

  - Only available in VxLAN configurations

  - Is an alternative to the Cloud Data Trunk and VLANs for the Tenant and Provider networks (because it becomes the "trunk" for Tenant and Provider VxLANs)

  - Configured when the first-time installer deploys CloudSystem with the VxLAN option selected

- **Provider network:**

  - Data center networks that are routed through the existing data center infrastructure and used to provision VM instances

  - Are created for cloud tenants in the Operations Console by cloud admins

  - Can be shared among tenants or assigned to a specific tenant

  - Mapped or routed to an existing physical network in the data center so that cloud instances can communicate with legacy data center resources

- **Tenant network:**

  - Are restricted and can be accessed only by VM instances assigned to that network—subnets must be defined in the OpenStack user portal before using this network

- Have segmentation ID ranges created in the Operations Console by the cloud administrators

- Created by cloud users in the OpenStack user portal

Cloud data trunk networks

The cloud data trunk with VLAN shown in Figure 8-10 connects the CCs with the hypervisor hosts. It carries network traffic to and from the VM instances running on the compute hosts.

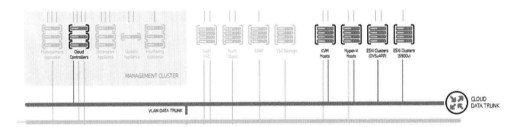

Figure 8-10 Cloud data trunk with VLAN

The cloud data trunk with VxLAN shown in Figure 8-11 connects the CCs and the management appliance with the hypervisor hosts. This network encapsulates and carries network traffic to and from the VM instances running on the compute hosts.

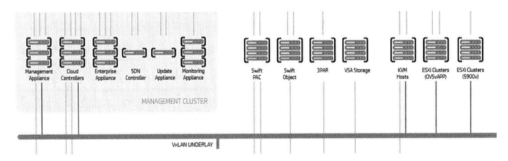

Figure 8-11 Cloud data trunk with VxLAN

Storage trunk—Learner activity

Use the *Helion CloudSystem 9.0 Network Planning Guide*, available on the Enterprise Information Library, to assign the correct definitions to the storage trunk networks.

Note

To access the Enterprise Information Library, scan this QR code or enter the URL into your browser:

http://www.hp.com/go/CloudSystem/docs

Fill in the blanks that correctly describe the bullets underneath each line.

● _____:

　● An iSCSI network used to integrate Cinder for VSA and iSCSI into CloudSystem

　● Supports block storage for StoreServ and VSA storage devices

● _____:

　● A load-balancing network that connects the control plane with external Object storage PAC and Object storage nodes

　● Completely private for communication among proxies, containers, or objects

　● Can be connected to a load balancer

● _____:

　● Supports traffic between Object storage (Swift) PAC and Object storage nodes

　● Is outside of the CloudSystem-defined networks and must be created manually when configuring object storage

　● Hosts all account, container, and object services

Storage trunk learner activity—answers

● **Block storage network**:

　● An iSCSI network used to integrate Cinder for VSA and iSCSI into CloudSystem

　● Supports block storage for StoreServ and VSA storage devices

● **OPN**:

　● A load-balancing network that connects the control plane with external Object storage PAC and Object storage nodes

- Completely private for communication among proxies, containers, or objects

- Can be connected to a load balancer

- **OBS**:

  - Supports traffic between Object storage (Swift) PAC and Object storage nodes

  - Is outside of the CloudSystem-defined networks and must be created manually when configuring object storage

  - Hosts all account, container, and object services

Storage trunk networks

The block storage network (BLS) shown in Figure 8-12 is an iSCSI network that integrates Cinder for VSA and/or iSCSI into CloudSystem. Compute nodes use this network to connect block storage volumes to virtual machines.

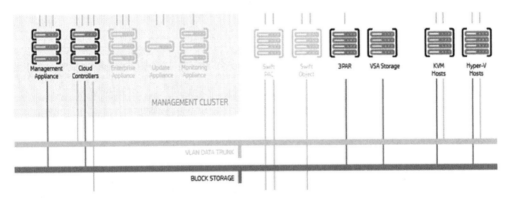

Figure 8-12 Block storage network (BLS)

The OPN shown in Figure 8-13 connects the control plane with external Object storage PAC and Object storage nodes. This network can be connected via a load balancer to the CAN. It carries traffic between Swift PAC and Swift object nodes.

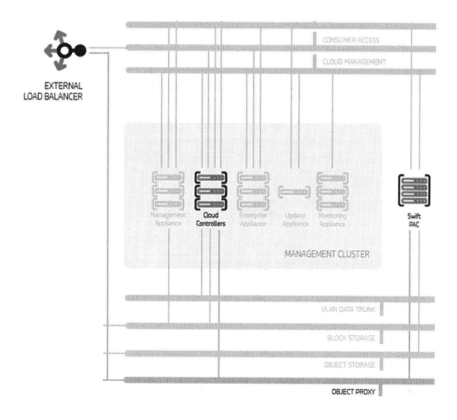

Figure 8-13 Object proxy network (OPN)

The OBS shown in Figure 8-14 carries traffic between Swift object storage PAC and object storage nodes.

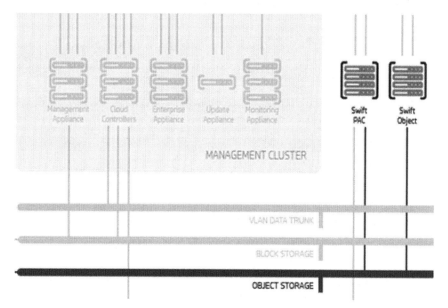

Figure 8-14 Object storage network (OBS)

Deployment scenarios

Figures 8-15 and 8-16 show the initial network choices when the first-time installer launches, along with the final network settings upon completion.

HP Helion CloudSystem Enterprise

HP Helion CloudSystem Enterprise contains the HP CSA Cloud Service Management Console, the Mar to access these features.

Network Settings: Data Trunk

VLAN with Centralized Virtual Routing is supported in both ESXi and KVM management environment

VxLAN with Centralized Virtual Routing is supported in both ESXi and KVM management environmen for VxLAN with KVM compute nodes.

TENANT NETWORK TYPE:

◉ VLAN   ○ VxLAN

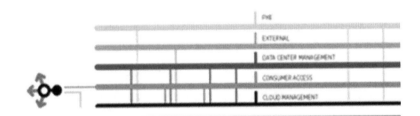

Figure 8-15 Initial network choices in First-Time Installer (FTI)

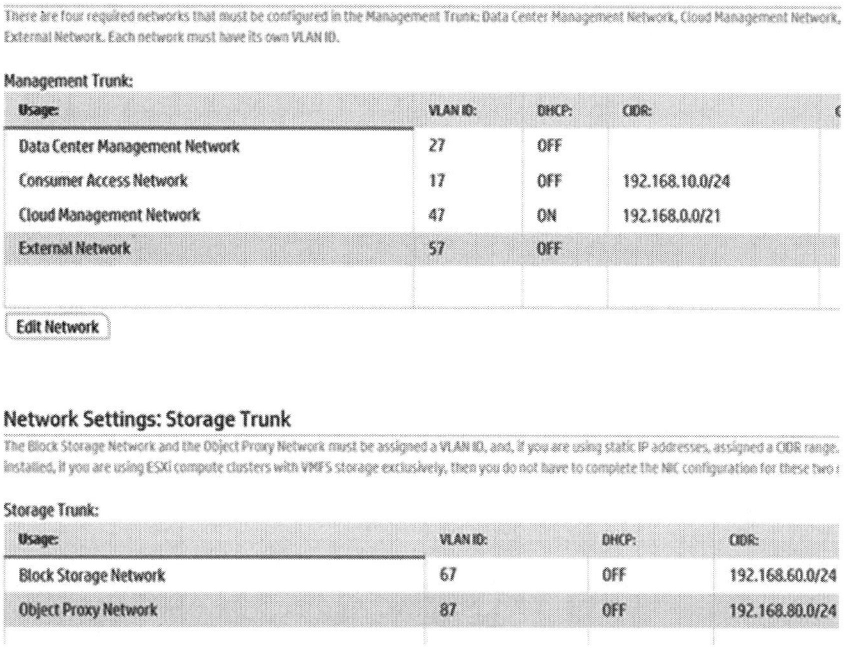

There are four required networks that must be configured in the Management Trunk: Data Center Management Network, Cloud Management Network, External Network. Each network must have its own VLAN ID.

Management Trunk:

Usage:	VLAN ID:	DHCP:	CIDR:	
Data Center Management Network	27	OFF		
Consumer Access Network	17	OFF	192.168.10.0/24	
Cloud Management Network	47	ON	192.168.0.0/21	
External Network	57	OFF		

[Edit Network]

Network Settings: Storage Trunk

The Block Storage Network and the Object Proxy Network must be assigned a VLAN ID, and, if you are using static IP addresses, assigned a CIDR range. installed. If you are using ESXi compute clusters with VMFS storage exclusively, then you do not have to complete the NIC configuration for these two s

Storage Trunk:

Usage:	VLAN ID:	DHCP:	CIDR:
Block Storage Network	67	OFF	192.168.60.0/24
Object Proxy Network	87	OFF	192.168.80.0/24

Figure 8-16 Final network settings

For the management trunk, the FTI reminds you that four networks are required:

- DCM network

- CLM network

- CAN

- EXT network

Helion CloudSystem 9.0 deployment scenarios differ based on whether customers choose to implement VLANs or VxLANs for Provider and Tenant networks and whether CVR or DVR is used:

- Scenario 1: VLAN and CVR

    - Has the same network configuration as CloudSystem 8.1

    - Has Tenant and Provider networks based on VLANs

    - Uses CVR

- Scenario 2: VxLAN and CVR

    - Has Tenant and Provider networks based on VxLANs

    - Uses CVR

- Scenario 3: VxLAN and DVR

    - Has Tenant and Provider networks based on VxLANs

    - Uses DVR

Scenario 1: VLAN and CVR

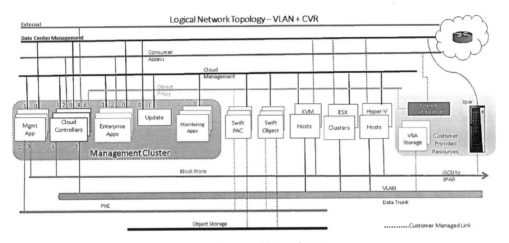

Figure 8-17 VLAN and CVR

As shown in Figure 8-17, this scenario provides a VLAN data trunk to support all Tenant and Provider networks for VM instances. One or more Provider networks are supported from the Neutron Representational State Transfer (REST) APIs without additional setup.

 Note

This scenario is the closest match to what was offered in CloudSystem 8.1.

One change from version 8.1 is the connection of the CLM Network to the Enterprise appliances. This connection is established because the Logstash log management system uses the CLM. Logstash is the mechanism used to collect log files from all appliances and store them on the first management appliance (ma1).

VM instances communicate with the external network via the Neutron router. With the CVR configuration, the Neutron router is scheduled on the CCs, and the external network connectivity is established only on the CCs. In the case of ESX compute, for example, the external network portgroup on the management trunk is connected to the CCs. When network traffic is received from the outside, the Neutron router receives it first, and then it is remapped to the private IPs of the VMs.

The CLM Network is managed by the management appliances. The management appliance provides DHCP and DNS services on this network and assigns IP addresses and hostnames to the appliances on this network. These are different and distinct from hostnames provided in the first-time installer. DNS and DHCP services for the DCM Network should be provided in the data center by the administrator.

Scenario 2: VxLAN and CVR

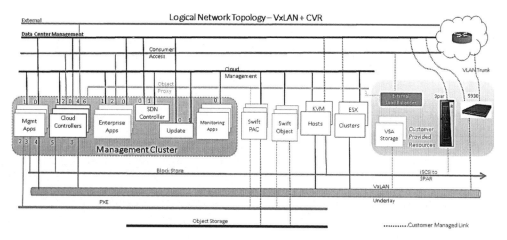

Figure 8-18 VxLAN and CVR

As shown in Figure 8-18, this scenario provides VxLAN-based Provider and Tenant networks with CVR:

- The data trunk is replaced by a VxLAN underlay network, which is a single VLAN.

- A new NIC is created on each management appliance for the VxLAN underlay network, where the DHCP server is running, to lease out IPs to CCs and compute nodes.

- A new SDN controller appliance is automatically deployed to support Neutron VxLAN provider networks.

- The SDN controller is connected to both DCM and CLM networks so that it can reach Neutron services and the physical FlexFabric 5930 ToR switch inside the data center.

- The FlexFabric 5930 hardware switch is needed as an L2 gateway across the cloud VxLAN networks and the legacy VLAN-based networks.

- No iSCSI connection exists for VMware ESX because ESX Cinder block storage only supports FC.

- This scenario supports ESX and KVM compute hosts only.

Scenario 3: VxLAN and DVR

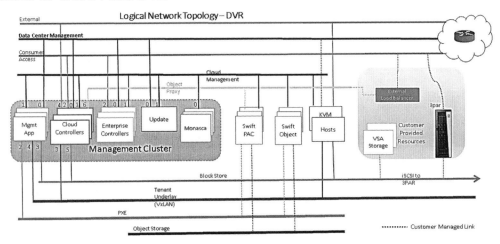

Figure 8-19 VxLAN and DVR

The last supported scenario features VxLANs with DVR technology as shown in Figure 8-19. This is the least-recommended scenario because DVR is not yet mature. There may be issues with the upstream code, and it is not expected that customers would deploy this scenario in a production environment.

One major drawback with this solution is that it does not support Provider networks. Therefore, no FlexFabric 5930 switch or SDN controller is used. DVR is supported only with KVM compute and VxLAN network combination in this release. With DVR, routing services are provided by the compute nodes, and this means that each compute node must be connected to the external network.

Reference installation for ESXi

Figure 8-20 represents a final reference installation in a VMware ESXi environment.

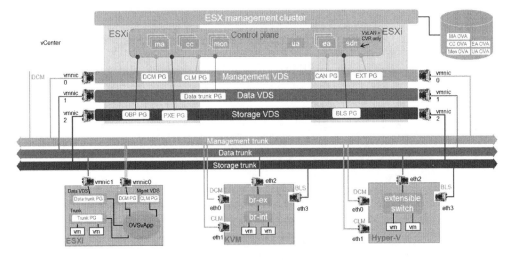

Figure 8-20 Reference installation for ESXi

Reference installation for KVM

Figure 8-21 represents a final reference installation in a KVM environment.

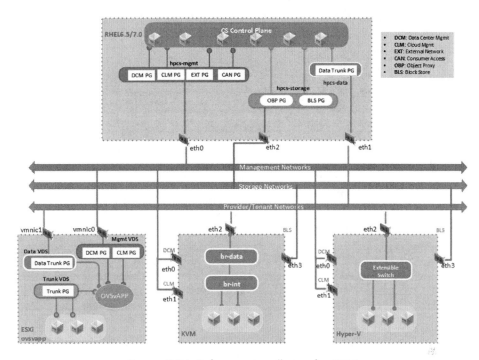

Figure 8-21 Reference installation for KVM

Networks in VMware vCenter after FTI

Figure 8-22 shows the network configuration in VMware vCenter immediately after CloudSystem Foundation is installed.

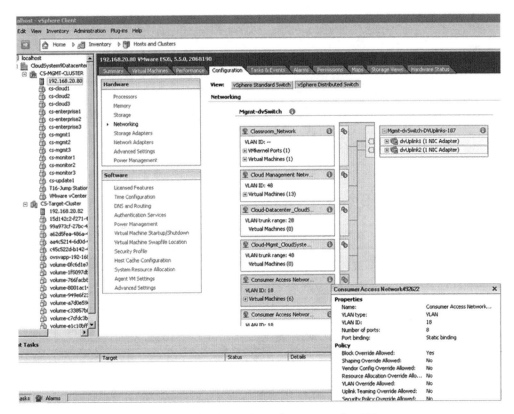

Figure 8-22 Network configuration after FTI

Learning check

1. List the four Helion CloudSystem 9.0 component categories.

_____

_____

_____

_____

2. Which hypervisors are supported for the VxLAN and CVR deployment? (Select two.)

a. ESX

b. KVM

c. Hyper-V

d. Integrity VM

3. Which network is used by providers?

   a. DCM network

   b. CLM network

   c. Consumer access network (CAN)

   d. EXT network

4. Which network provides floating IP addresses?

   a. DCM network

   b. CLM network

   c. Consumer access network (CAN)

   d. EXT

Learning check answers

1. List the four Helion CloudSystem 9.0 component categories.

   - **Management hypervisors and managed resources**
   - **CloudSystem virtual appliances**
   - **CloudSystem user interfaces**
   - **CloudSystem storage**

2. Which hypervisors are supported for the VxLAN and CVR deployment? (Select two.)

   a. ESX

   b. KVM

   c. Hyper-V

   d. Integrity VM

3. Which network is used by providers?

   a. DCM network

   b. CLM network

   c. Consumer access network (CAN)

   d. EXT network

4. Which network provides floating IP addresses?

   a. DCM network

   b. CLM network

   c. Consumer access network (CAN)

   d. EXT

Security

There are certain security concepts to consider when you are working with browsers, certificates, and networks for secure communication and transfer of data among the appliances, networks, and compute nodes in a CloudSystem virtualized environment.

Most security policies and practices used in a traditional environment apply in a virtualized environment. However, in a virtualized environment, these policies might require modifications and additions.

Access & Security in Horizon

Access and security settings are available in Horizon under the **Access & Security** section for each project in the left navigation pane as shown in Figure 8-23.

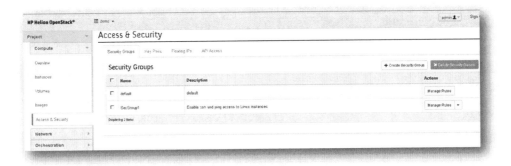

Figure 8-23 Access and security settings

Root certificate for the management hypervisor

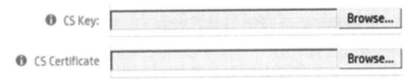

Figure 8-24 Certificate fields in the CloudSystem management appliance installer

A root certificate is the starting point for secure communication in a cloud environment. CloudSystem uses certificates to authenticate and establish trust relationships. The CloudSystem management appliance installer enables a private key and certificate to be specified during installation as shown in

Figure 8-24. One of the most common uses of certificates is when a connection from a web browser to a web server is established. The machine-level authentication is carried out as part of the HTTPS protocol, using SSL. Certificates can also be used to authenticate devices when setting up a communication channel.

Helion CloudSystem supports two methods of applying security certificates to management hypervisors. During installation, you can choose one of these following options:

- Use the customer's existing certificate authority (CA) and import its trusted certificates. During installation, enter the path to the key and certificate from the local CA.

- Use CloudSystem to automatically generate a private key and certificate. During installation, leave the certificate fields blank and CloudSystem will automatically generate a private key and certificate.

Two important considerations about setup and installation include:

- It is important to set the system date and time accurately if you are using a self-signed certificate. You can use the Linux date command to set the system date and time.

- During installation, if you choose to allow CloudSystem to generate the private key and certificate, you cannot decide later to use a local certificate authority.

User authentication

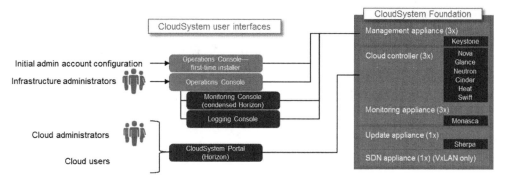

Figure 8-25 User authentication

User authentication can occur locally or by using an enterprise directory. When administering users locally, the Users screen in the Operations Console can be used. An administrator can add new administrator users and modify or remove existing administrator users.

 Note

All users in the Operations Console are administrator users and can perform all tasks.

During the first-time installation, you configure a password for the admin account in the Operations Console and the OpenStack Horizon user portal. These rules apply:

- You cannot change the name of the admin account.

- You cannot change the password of the admin account after the first-time installation.

- If you change the admin account or password in the OpenStack user portal, the change is not propagated to the Operations Console.

The OpenStack Keystone service in the management appliance contains a local directory and this means that local users can log in to the management appliance from the Operations Console and the CloudSystem CLIs. This does not apply to directory service users.

The OpenStack Keystone service in the CC, which hosts the OpenStack user portal, can be configured for directory service authentication using OpenLDAP and Microsoft Active Directory. You can configure directory services for the CC on the Security pane of the Operations Console System Summary screen.

The management appliance (Operations Console) contains only the admin role, whereas the OpenStack user portal (Horizon) contains admin and member roles. The functions of these roles are:

- **Management appliance (Operations Console) admin role**—Consists of infrastructure administrators who can view, create, edit, or remove resources and other admin users managed by the appliance, including management of the appliance itself, through the UI or command line. Infrastructure administrators can create a backup file to recover the appliance. They can also manage information provided by the appliance in the form of activities, notifications, and logs.

- **OpenStack user portal (Horizon) admin role**—Consists of cloud administrators who can view usage and manage instances, volumes, volume types, flavors, images, projects, users, services, and quotas. Cloud administrators can view the Admin and project tabs in the OpenStack user portal.

- **OpenStack user portal (Horizon) member role**—Consists of cloud users who can view and manage resources only in the project to which they are assigned. Cloud users can only view the Project tab in the OpenStack user portal.

 Note

Additional information about users is available in the Manage users section of the *Helion CloudSystem 9.0 Administrator Guide*, available on the Enterprise Information Library.

Access and security for instances—Assessment questions

Assess your existing knowledge of security groups and key pairs. Attempt to answer these questions without looking up the answers.

1. Fill in the blanks to complete each sentence:

   - _____ are virtual firewalls controlling traffic for instances.

   - _____ allow you to use public-key cryptography to encrypt and decrypt login information.

2. For each pair of bold words, circle the correct term:

   - Security groups **are/are not** project-specific and **can/cannot** be shared across projects.

Access and security for instances assessment questions—answers

1. Fill in the blanks to complete each sentence:

   - **Security groups** are virtual firewalls controlling traffic for instances.

   - **Key pairs** allow you to use public-key cryptography to encrypt and decrypt login information.

2. For each pair of bold words, circle the correct term:

   - Security groups **are** project-specific and **cannot** be shared across projects.

Security groups

Figure 8-26 Security groups

Security groups are virtual firewalls that control the traffic for instances. When you launch an instance, you associate one or more security groups with the instance from Security Groups tab of the Horizon portal as shown in Figure 8-26. Administrators create security groups to define a set of IP filter rules that determine how network traffic flows to and from an instance. Cloud users can add additional rules to an existing security group to further define the access options for an instance.

Security groups are project-specific and cannot be shared across projects. After a security group is associated to an instance, the pathway to communicate with the instance is open, but you still need to configure key pairs. The key pair allows you to use Secure Shell (SSH) to connect into the instance.

If a security group is not associated to an instance before it is launched, then you will have very limited access to the instance after it is deployed. You will only be able to access the instance from a virtual network computing (VNC) console.

Key pairs

Figure 8-27 Key pairs

Key pairs allow you to use public-key cryptography to encrypt and decrypt login information. To log in to your instance, you must create a key pair. This can be achieved using the Key Pairs tab in the Horizon portal. The name of the key pair must be specified when you launch the instance, and the private key provided when you connect to the instance.

Key pairs allow you to log into an instance after it is launched without being prompted for a password. Key pairs are only supported in instances that are based on images containing the cloud-init package. You can generate a key pair from the CloudSystem Portal or you can generate a key pair manually from a Linux or Windows system. After the instance is launched, log in using the private key.

Best practices for a secure environment

A partial list of best practices for security that HPE recommends in both physical and virtual environments includes:

- Accounts
  - Limit the number of local accounts in the Operations Console.
  - Integrate the OpenStack user portal (Horizon) with an enterprise directory solution such as Microsoft Active Directory or OpenLDAP.
- Certificates
  - Use certificates signed by a trusted CA if possible.

Note

CloudSystem supports self-signed certificates and certificates used by a CA.

- Updates

  - Ensure that a process is in place to determine if software and firmware updates are available, ensuring that all components in the environment will be updated regularly.

- Overall cloud environment

  - Restrict access to the appliance consoles to authorized users only.

  - If you use an intrusion detection system in the environment, ensure that the solution has visibility into network traffic within the virtual switch.

- Passwords

  - The **admin** password for the Operations Console and the OpenStack user portal is set during first-time installation and cannot be changed from the Operations Console after the first-time installation. It can, however, be changed in the OpenStack user portal on the CC. Ensure that the password follows the password rules described in the *Helion CloudSystem 9.0 Administrator Guide*.

  - The **cloudadmin** password for the CloudSystem appliances is also set during first-time installation. It can be changed in the operating system running on each appliance using the passwd command. If you change the password on one node, you must change the password on the other nodes of an appliance trio (the same appliance type). The passwords on different appliance trios do not have to match. For example, if you change the cloudadmin password on the management appliance trio, you do not have to change the password on the CC trio or the Enterprise appliance trio.

  - Do not change the OpenStack service or internal user account passwords.

  - For local accounts on the management appliance, change the passwords periodically according to the customer's password policies.

Installation and configuration

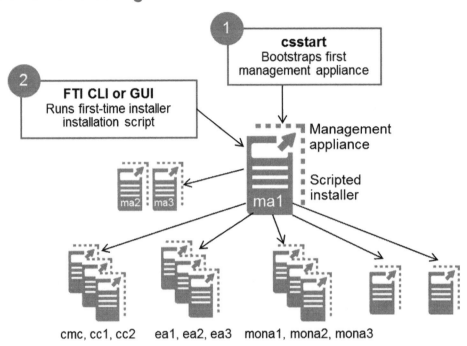

Figure 8-28 CloudSystem 9.0 installation and configuration process

The installation of CloudSystem 9.0 (whether Foundation or Enterprise) is a simple, streamlined process that creates a virtual management platform. This provides IT departments with platform flexibility as well as a fast and infrastructure-efficient installation process. With CloudSystem 9.0, customers will receive an out-of-the-box HA configuration that installs three virtual instances of the management appliances within a cluster of either VMware ESXi or Red Hat KVM hosts. The installation process walks customers through the process, which is summarized in two steps as shown in Figure 8-28:

1. **Step 1**—Run the "csstart" installation tool to bootstrap the initial management appliance on a VM host (which makes up the management platform). This tool is based on a self-guided UI that asks for specifics about a customer environment and starts the first management appliance.

2. **Step 2**—Run the First-Time Installer (FTI), GUI, or CLI that was initiated on the management appliance. This provides a FTI script to set up the remaining instances of the virtual management environment. The scripted procedure then walks the customer through the creation of the remaining virtual management instances, creating the highly available environment that enterprises require for cloud management.

Note

CloudSystem virtual appliances include the following:

- Management appliances—ma1, ma2, ma3

- Cloud controllers—cmc, cc1, cc2

- Enterprise appliances—ea1, ea2, ea3

- Monitoring appliances—mona1, mona2, mona3

- Update appliance—ua

- SDN controller—sdn (VxLAN only)

- OVSvApp appliance—ovsvapp

Figure 8-29 shows the initial FTI screen.

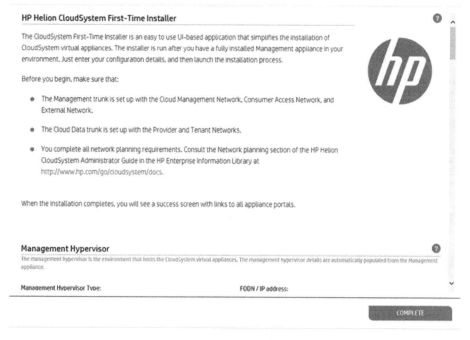

Figure 8-29 Initial FTI screen

Figure 8-30 shows the FTI screen requesting the desired tenant network type (VLAN or VxLAN)

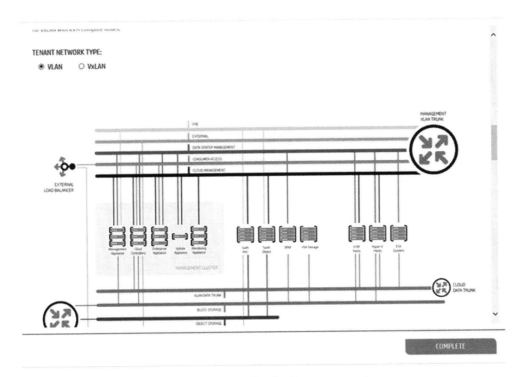

Figure 8-30 FTI screen for the Tenant network type

Installation considerations

CloudSystem supports several different configuration scenarios. The *Helion CloudSystem 9.0 Installation and Configuration Guide* explains how to install the following CloudSystem configurations:

- ESXi or KVM management hypervisor

- CloudSystem installers running in a Windows staging environment

- VLAN cloud data trunk

- CVR

- Enterprise appliance that is included in the initial installation

---

Note

To access the *Helion CloudSystem 9.0 Installation and Configuration Guide* on the Enterprise Information Library, scan this QR code or enter the URL into your browser:

http://www.hp.com/go/CloudSystem/docs

Installation planning worksheet

A CloudSystem Installation Planning Worksheet

Use the worksheet below to gather your CloudSystem environment details before you begin the installation process. Enter your environment details into the fields provided and then use that information as you step through the CloudSystem installers.

Table 3 CloudSystem Installation Planning Worksheet

Installation Preparation	
Images	
Source	release package unpacked and added to your staging environment
Names	
Disk Format	Thin Provision (recommended)
Installation script (csstartgvi.bat)	
Source	release package unpacked and added to your staging environment
Target	
CloudSystem Management Appliance Installer	
Management hypervisor type	
vCenter IP address	
vCenter user name	
vCenter password	
Cluster	
Management appliance image name	
Hostname Example: my.ma.hpiacmgmt.local	
Host IP type	
IP Address *if using static IP address assignment.	
Gateway *if using static IP address assignment.	
Netmask *if using static IP address assignment.	
CS key (optional)	
CS certificate (optional)	
First Time Installer	
NOTE: If you are not sure what information to provide in a field below, refer to Run the First-Time Installer in an ESXi environment (page 18).	
Management hypervisor	Verify the information

Figure 8-31 *Helion CloudSystem 9.0 Installation and Configuration Guide*

The *Helion CloudSystem 9.0 Installation and Configuration Guide* includes an installation planning worksheet, shown in Figure 8-31, that you can use to gather CloudSystem environment details from customers before you begin the installation process. You can fill out the environment details on the worksheet and then use that information as you step through the CloudSystem installers.

Installation and configuration—Learner activity

Use the *Helion CloudSystem 9.0 Installation and Configuration Guide* to answer the following questions.

Installation process

A high-level overview of the CloudSystem installation path is provided in the *Helion CloudSystem 9.0 Installation and Configuration Guide*.

1.   Using the guide, put these steps in the correct order:

Run the Cloud management appliance installer		Run the CloudSystem first-time installer
	1.	
	2.	
Configure ToR switches	3.	Prepare VMware vCenter
	4.	
Finish resource configuration in the Operations Console	5.	

Also mark which installation steps are not covered by this guide and indicate where you can find information pertaining to those steps:

_____

_____

_____

_____

ESXi installation process

A high-level overview of the CloudSystem installation path in an ESXi environment is provided in the *Helion CloudSystem 9.0 Installation and Configuration Guide.*

2. Using the guide, put these steps in the correct order:

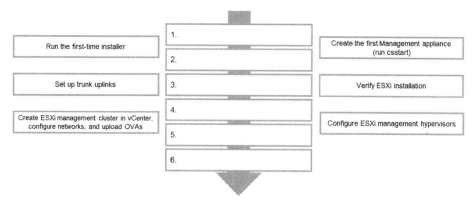

KVM installation process

A high-level overview of the CloudSystem installation path in a KVM environment is provided in the *Helion CloudSystem 9.0 Installation and Configuration Guide.*

3. Using the guide, put these steps in the correct order:

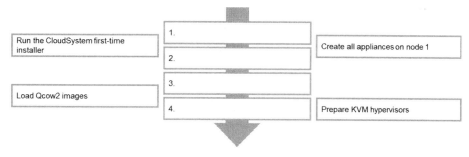

4. Using the guide, fill in these blanks:

If using CVR:

- You can choose _____ or _____ for Provider and Tenant networks.

- Support includes _____, _____, or _____ compute notes.

If using DVR:

- You can choose _____ only (networking option).

- Support includes _____ compute nodes only.

Installation and configuration learner activity—answers

1. Using the *Helion CloudSystem 9.0 Installation and Configuration Guide*, put these steps in the correct order

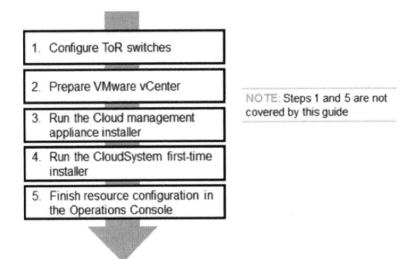

1. Configure ToR switches

2. Prepare VMware vCenter

3. Run the Cloud management appliance installer

4. Run the CloudSystem first-time installer

5. Finish resource configuration in the Operations Console

NOTE: Steps 1 and 5 are not covered by this guide

2. Using the *Helion CloudSystem 9.0 Installation and Configuration Guide*, put these steps in the correct order

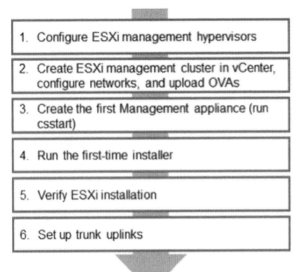

1. Configure ESXi management hypervisors

2. Create ESXi management cluster in vCenter, configure networks, and upload OVAs

3. Create the first Management appliance (run csstart)

4. Run the first-time installer

5. Verify ESXi installation

6. Set up trunk uplinks

3. Using the *Helion CloudSystem 9.0 Installation and Configuration Guide*, put these steps in the correct order

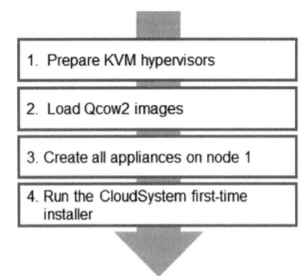

1. Prepare KVM hypervisors

2. Load Qcow2 images

3. Create all appliances on node 1

4. Run the CloudSystem first-time installer

4. Using the *Helion CloudSystem 9.0 Installation and Configuration Guide*, fill in the blanks to complete each sentence

- If using CVR:

    - You can choose **VLAN** or **VxLAN** for Provider and Tenant networks.

    - Support includes **KVM, ESXi**, or **Hyper-V** compute nodes.

- If using DVR:

    - You can choose **VxLAN** only (networking option).

    - Support includes **KVM** compute nodes only.

How to migrate from CloudSystem version 8.x

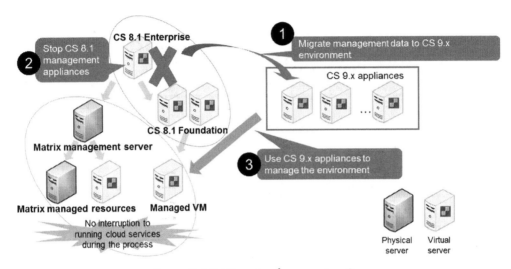

Figure 8-32 Migrating from version 8.x

For existing Helion CloudSystem 8.1 customers, upgrading to version 9.x requires the user to migrate the management plane as illustrated by Figure 8-32. This means they need to bring a new set of Helion CloudSystem 9.x appliances online and migrate the management data from the 8.1 appliances. During the migration process, there is no impact to the running cloud services. At the end of the migration process, the 8.1 appliances are retired, and the new 9.x appliances take over management of the complete Helion CloudSystem environment. The user configuration, service designs, and running services are then managed from the new 9.x appliances.

A summary of this three-step process includes:

1. **Step 1**—Migrate management data to the 9.x environment. This includes OpenStack database, service configuration data, users, networks, compute and volume resources, and (if used) CloudSystem Enterprise with CSA and OO.

2. **Step 2**—Stop the CloudSystem 8.1 management appliances.

3. **Step 3**—Use the 9.x appliances to manage the new CloudSystem environment.

In addition to a migration tool delivered as part of Helion CloudSystem 9.x, Helion Professional Services are available to assist customers. Customers with a Helion CloudSystem 8.0 environment need to first upgrade to 8.1 before upgrading to 9.x.

Migration prerequisites

Before initiating the migration process:

- Prepare the hypervisors for CloudSystem 9.x deployment.

- Ensure the DCM and CAN networks have enough free IP addresses.

- Add a temporary second CLM network to vCenter.

- Update hardware and software as needed:

  - ESX: minimum 5.5 U2 and Enterprise license

  - StoreServ: minimum 3.1.3 MU1

  - KVM: minimum RHEL 6.5

- If using LDAP or Active Directory, add CloudSystem 9.x service users first.

Migration process from 8.1 to 9.x

Figure 8-33 Migration process from 8.1 to 9.x

As shown in Figure 8-33, the migration process from CloudSystem 8.1 to 9.x includes:

1. Installing CloudSystem 9.x

   - Installation of CloudSystem 9.x requires a new hypervisor cluster.

   - The DCM and CAN networks are the same as in version 8.1.

   - VLAN is required.

   - For ESX, OVSvApp is required on compute nodes.

2. Preparing the CloudSystem environment

   - Install the latest CloudSystem 8.1.x update on the old (8.1) management cluster.

   - Upgrade KVM 6.4 and StoreServ firmware.

   - Upgrade ESX licenses to Enterprise Plus.

   - Complete pending management operations.

   - Log off cloud users.

   - Perform system backup.

3. Running csmigrate from the CloudSystem 9.x management appliance 1 (ma1)

   - This action:

     - Automatically exports data and configurations from CloudSystem 8.1 and imports them into CloudSystem 9.x.

     - Installs updated CloudSystem software on management nodes.

- Reconfigures the environment for new management.

- Migrates ESX vNetwork Standard Switches (vSwitch and vSS) to vNetwork Distributed Switches (dvSwitch and vDS).

 Note

CloudSystem requires the use of distributed vSwitches. If the vCenter configuration uses standard vSwitches, then you must migrate the standard vSwitches to distributed vSwitches.

4. Completing the migration

- Manually complete software installation and configuration changes on unreachable compute nodes.

- Resume management and cloud user operations on CloudSystem 9.x.

5. Retiring CloudSystem 8.1

- Shut down CloudSystem 8.1 appliances.

- Retire old management servers.

 Important

If you select VxLAN for the CloudSystem 9.x environment, you will not be able to migrate to future versions of CloudSystem. Migration is not supported in VxLAN configurations.

Learning check

1. If you create a security group in CloudSystem 9.0 and associate it with a compute instance, you do **not** need a key pair to connect to the instance with SSH without having to supply a password.

    ☐ True

    ☐ False

2. What are the two basic steps when installing CloudSystem 9.0 (Foundation or Enterprise)?

    _____

    _____

    _____

    _____

Learning check answers

1. If you create a security group in CloudSystem 9.0 and associate it with a compute instance, you do **not** need a key pair to connect to the instance with SSH without having to supply a password.

    ☐ True

    ☐ **False**

2. What are the two basic steps when installing CloudSystem 9.0 (Foundation or Enterprise)?

    –   **Step 1—Run the "csstart" installation tool to bootstrap the initial management appliance on a VM host (which makes up the management platform)**

    –   **Step 2—Run the Operations Console that was initiated on the management appliance to get a first-time installation script for setup of the remaining instances of the virtual management environment**

Resources

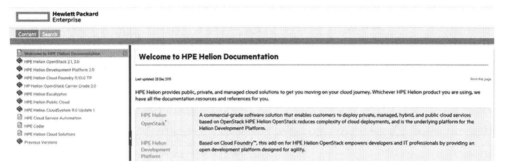

Figure 8-34 HPE Helion documentation

Note

For Helion CloudSystem documentation, scan this QR code or enter the URL into your browser:

http://docs.hpcloud.com/

As shown in Figure 8-34, the HPE Helion documentation website is a source of documentation for Helion CloudSystem 9.0.

The documentation includes:

- Helion CloudSystem 9.0 Installation Guide

- Helion CloudSystem 9.0 Administration Guide

- Helion CloudSystem 9.0 Support Matrix

- Helion CloudSystem 9.0 Release Notes

- Helion CloudSystem 9.0 Command Line Interface Guide

- Helion CloudSystem 9.0 Network Planning Guide

- Helion CloudSystem 9.0 Troubleshooting Guide

Summary

- Helion CloudSystem provides a common, easy-to-use management platform across several environments. Users can design, provision, and manage services from multiple providers. These capabilities produce much higher productivity and greater business agility.

- The Helion CloudSystem portfolio consists of two versions: Helion CloudSystem Foundation and Helion CloudSystem Enterprise. Helion CloudSystem Foundation is the ideal entry point for businesses interested in straightforward, rapid deployment of cloud IaaS solutions. The Helion CloudSystem 9.0 architecture is enterprise-ready and highly available.

- Helion CloudSystem 9.0 networks are organized into three network trunks—management trunk, cloud data trunk, and storage trunk.

- HPE recommends best practices for security in both physical and virtual environments. Factors to consider include accounts, certificates, updates, overall cloud environment, and passwords.

- The installation of CloudSystem 9.0 (whether Foundation or Enterprise) is a simple, streamlined process that creates a virtual management platform. This provides IT departments with platform flexibility as well as a fast and infrastructure-efficient installation process.

9 A Closer Look at CloudSystem Foundation— Management and Administration

WHAT IS IN THIS CHAPTER FOR YOU?

After completing this chapter, you should be able to:

✓ Explain how to manage the CloudSystem Foundation appliances

✓ Describe the process of launching a virtual machine (VM) instance

✓ Identify different ways to monitor resource usage, allocation, and health

Managing the appliances

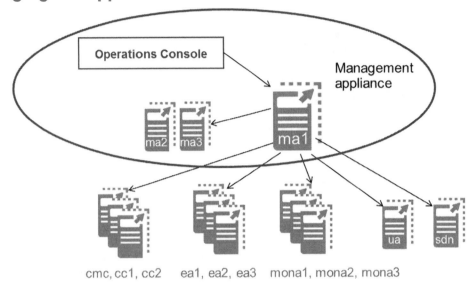

Figure 9-1 Identifying management appliances

The HPE Helion CloudSystem solution identifies the management appliances in the trio by using the internal names assigned on the Cloud Management Network shown in Figure 9-1:

- **ma1** is the first management appliance in the trio. This is the management appliance that was installed by the CloudSystem Management Appliance Installer.

- **ma2** is the second management appliance in the trio.

- **ma3** is the third management appliance in the trio.

You should always access a management appliance through the first management appliance in the trio. To access the first management appliance:

1. Using **cloudadmin** credentials, Secure Shell (SSH) into **ma1**.

2. From **ma1**, SSH into the other management appliances as necessary.

! Important

When performing maintenance on a VM hosting the management appliances, make sure that **ma1** is always the last node shut down and the first node restarted.

To view the management appliance details through the Operations Console, use the **Appliances** screen. From this screen, you can:

- Update CloudSystem appliances

- Install the Enterprise appliance (if you did not choose to install it during the first-time installation)

Managing the cloud controller trio

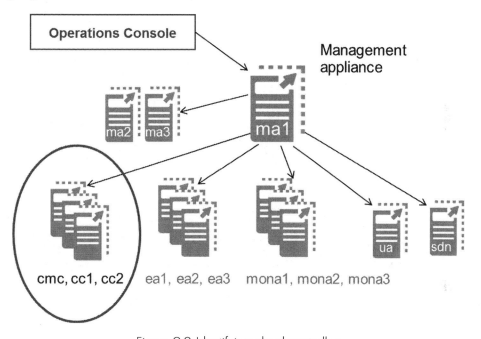

Figure 9-2 Identifying cloud controllers

CloudSystem identifies the cloud controllers in the trio by using the internal names assigned on the Cloud Management Network shown in Figure 9-2:

- **cmc** is the first cloud controller in the trio.

- **cc1** is the second cloud controller in the trio.

- **cc2** is the third cloud controller in the trio.

To access the cloud controllers:

1. Using **cloudadmin** credentials, SSH in to **ma1**.

2. From **ma1**, SSH to the desired cloud controller.

Important

When performing maintenance on a VM hosting the cloud controllers, make sure that cmc is always the last node shut down and the first node restarted.

To view the cloud controller details through the Operations Console, use the **Appliances** screen.

 Note

For a detailed list of the steps to maintain the cloud controller trio, refer to the *Helion CloudSystem 9.0 Administrator Guide* in the Hewlett Packard Enterprise (HPE) Information Library. To access the guide, scan this QR code or enter the URL into your browser.

http://www.hp.com/go/CloudSystem/docs

Listing CloudSystem appliances and their status

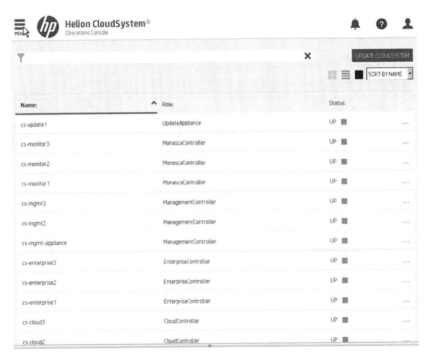

Figure 9-3 Listing CloudSystem appliances and their statuses

To list the appliances in CloudSystem along with their name and status as shown in Figure 9-3, log in to the Operations Console and click **Menu → Appliances**.

Viewing the appliance public networks

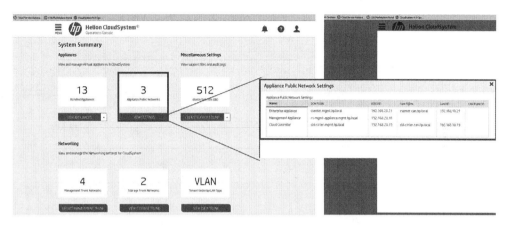

Figure 9-4 Viewing the appliance public networks

The Appliance Public Network Settings are shown in Figure 9-4. To view the appliance public networks and associated settings, log in to the Operations Console and click **Menu → System Summary**. Then click **VIEW SETTINGS** under Appliance Public Networks.

Shutting down and starting up the cloud

You can perform maintenance on a VM that is hosting CloudSystem appliances by using the shutdown action to gracefully stop the guest operating system and release the physical resources.

 Important

CloudSystem appliances are clustered in a trio configuration and must be shut down and then restarted in a very precise order. Make sure that you can identify the first node in each appliance trio because that node is typically the management node and should always be the last node shut down and the first node restarted.

The first appliance typically manages the trio. You can shut down the second and third appliance at any time, but do **not** shut down the first appliance while the other two are still running.

When the first cloud controller appliance (cmc) is shut down, you cannot perform block storage (Cinder) operations such as creating and attaching a volume. However, volumes already created and attached to instances are not affected.

Shutting down a single CloudSystem virtual appliance

Use the following procedure to shut down a single CloudSystem appliance:

 Note

> If the appliance is part of a trio, the remaining appliances in the trio are not affected by this procedure and will continue to run.

1. Use SSH and the cloudadmin credentials to connect to the first management appliance in the trio.

2. From the first management appliance, use SSH to connect to the appliance you want to shut down, and issue the shutdown command.

Example—shutting down the Update appliance

```
ssh cloudadmin@ua1 sudo shutdown -h now
```

Shutting down a trio of CloudSystem virtual appliances

Use this procedure to shut down all three appliances in the trio:

1. Use SSH and the cloudadmin credentials to connect to the first management appliance in the trio.

2. From the first management appliance, use SSH to connect to the **third** appliance in a trio that you want to shut down, and issue the shutdown command.

Example—shutting down mona3

```
ssh cloudadmin@mona3 sudo shutdown -h now
```

3. Repeat Steps 1 and 2 for the **second** appliance in the trio (for example, mona2).

4. Repeat Steps 1 and 2 for the **first** appliance in the trio (for example, mona1).

Shutting down compute nodes

To shut down compute nodes:

1. From the OpenStack user portal (Horizon), shut down all VM instances running on the compute node.

2. For VMware ESXi compute clusters:

    a. Log in to vCenter with the administrator credentials.

    b. Shut down the OVSvApp running on the compute node.

    c. Shut down the compute node.

3. For Linux Kernel-based VM (KVM) compute nodes:

    a. Use SSH and the cloudadmin credentials to connect to the compute node.

    b. Run the shutdown command:

    ```
    sudo shutdown -h now
    ```

    c. Wait for the compute node shutdown process to complete.

4. For Microsoft Hyper-V compute nodes:

    a. Use a remote desktop connection to access the compute node.

    b. From the Settings tab on the bottom-right taskbar, select **Power → Shut down**.

    c. Wait for the compute node shutdown process to complete.

Shutting down the entire cloud

To shut down the entire cloud, the appliances must be shut down in the order shown in Figure 9-5, with step 1 being to use SSH to log in to the first management appliance in the trio, ma1. All other appliances are then shut down in order, finishing with ma1 as shown in step 9 of Figure 9-5.

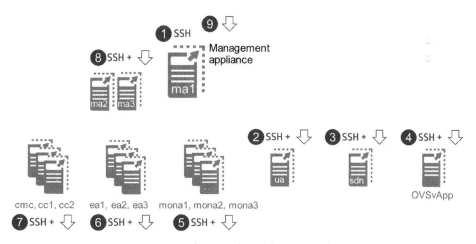

Figure 9-5 Shutting down the entire cloud

The formal procedure to shut down all CloudSystem appliances in the cloud is as follows:

1. Use SSH and the cloudadmin credentials to connect to the first management appliance in the trio.

2. From the first management appliance, use SSH to connect to the Update appliance and run the shutdown command:

```
ssh cloudadmin@ua1

sudo shutdown -h now
```

3. Wait for the appliance shutdown process to complete.

4. Repeat Steps 1 through 3 for these appliances (in order):

    a. SDN controller (if present)

    b. OVSvApp (if present)

    c. Compute nodes

    d. Monitoring appliances (mona3 → mona2 → mona1)

    e. Enterprise appliances (ea3 → ea2 → ea1)

    f. Cloud controllers (cc2 → cc1 → cmc)

    g. Management appliances (ma3 → ma2 → ma1)

Starting up CloudSystem appliances after a shutdown

Starting up the CloudSystem appliances after a shutdown must be done in reverse order:

1. Management appliances (ma1 → ma2 → ma3)

2. Cloud controllers (cmc → cc1 → cc2)

3. Enterprise appliances (ea1 → ea2 → ea3)

4. Monitoring appliances (mona1 → mona2 → mona3)

5. Compute nodes

6. OVSvApp (if present)

7. SDN controller (if present)

8. Update appliance

For additional information on this topic, refer to "Shut down and restart CloudSystem appliances" (Section 9) in the *Helion CloudSystem 9.0 Administrator Guide*. The guide is available in the HPE Information Library.

Backing up, restoring, and recovering CloudSystem appliances

Figure 9-6 Backing up, restoring, and recovering CloudSystem appliances

CloudSystem allows you to save your configuration settings and data to a backup file while the appliances are running. You can also use that backup to restore appliance databases if data loss occurs. Figure 9-6 shows the Backup & Restore control in the HPE Helion CloudSystem Operations Console.

The basic backup process consists of:

1. Setting up the location where the backup file is stored

2. Backing up the CloudSystem appliances

The basic restore and recovery process consists of:

1. Restoring the appliances from a backup file

2. Creating the recovery report

3. Recovering appliances and compute nodes

CloudSystem implements backup and restore as a service using the attis service, which runs on the management appliance. You can use the Operations Console to perform backup, restore, report, and recover actions. You can also execute attis commands from the management appliance.

Use the Operations Console BACKUP & RESTORE screen to:

- Back up, fully or incrementally, the data (databases and files) for the trio of management appliances, cloud controllers, enterprise appliances, and monitoring appliances.

- Restore data for the trio of cloud controllers, enterprise appliances, and monitoring appliances—all from the backup file. Restoring the data for the management appliance trio consists of a self-restore process, which is described in the *Helion CloudSystem 9.0 Administrator Guide*.

- Recover the cloud controller data after the restore.

For additional information, refer to "Backup, restore, and recover CloudSystem appliances" (Section 6) in the *Helion CloudSystem 9.0 Administrator Guide*. For information about the attis CLI, refer to the *Helion CloudSystem 9.0 Command Line Interface Guide*. Both documents are located in the HPE Information Library.

Launching a VM instance demonstration

Watch this 38-minute demonstration of configuring and launching a VM instance in Helion CloudSystem Foundation. The demonstration shows how to use the Helion CloudSystem Operations Console to prepare CloudSystem for VM creation and also shows the monitoring capabilities of the Operations Console. It then shows how to use the Horizon portal to create a VM instance and associated network resources.

 Note

To view this demonstration, scan this QR code or enter the URL into your browser.

https://youtu.be/HU9G3EZ98gA

Write your notes and observations here:

_____

_____

_____

_____

_____

_____

_____

_____

_____

_____

_____

_____

Creating and activating compute nodes—Activity part 1

Using the *Helion CloudSystem 9.0 Administrator Guide*, place the steps in Figure 9-7, outlining the compute node creation and activation process, in the correct sequence:

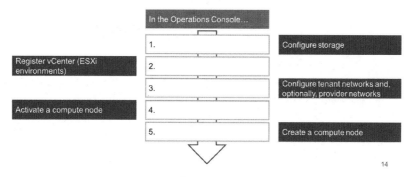

Figure 9-7 Compute node creation and activation process

Creating and activating compute nodes activity part 1—answers

Using the *Helion CloudSystem 9.0 Administrator Guide*, place these steps, outlining the compute node creation and activation process, in the correct sequence:

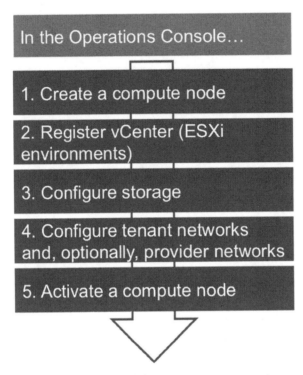

Figure 9-8 Compute node creation and activation process in the correct sequence

Compute nodes manage the resources required to host instances in the cloud. As shown in Figure 9-8, you must first create compute nodes by applying CloudSystem requirements to a cluster or a host. Then, you can activate the compute node, which makes it available to host virtual machine instances.

CloudSystem can simultaneously manage these types of compute nodes:

- ESXi clusters
- Hyper-V compute nodes (clustered or stand-alone)
- KVM compute nodes

When ESXi, Hyper-V, and KVM compute nodes exist in CloudSystem, the topology designs, offerings, and provisioned subscriptions in CloudSystem Enterprise can include ESXi, Hyper-V, KVM, or all three types of compute nodes.

The Compute Nodes screen in the Operations Console displays all available compute clusters and compute nodes along with their resources. You can activate, deactivate, and delete compute nodes from this screen.

 Note

CloudSystem supports ESXi and KVM on compute nodes and as the management hypervisor (the hypervisor software running on the physical server that hosts CloudSystem appliances) Hyper-V is supported on compute nodes.

Creating compute nodes—Activity part 2

The steps for creating compute nodes are different for ESXi, Hyper-V, and KVM.

Using the *Helion CloudSystem 9.0 Administrator Guide*, complete the steps in Figure 9-9.

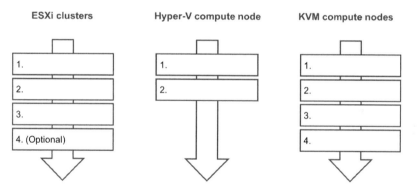

Figure 9-9 Creating compute nodes with ESXi, Hyper-V, and KVM

More information on compute node creation can be found in the *Helion CloudSystem 9.0 Administrator Guide*, under "Compute node creation" (Section 20).

Creating compute nodes activity part 2—answers

The steps for creating compute nodes are different for ESXi, Hyper-V, and KVM.

Using the *Helion CloudSystem 9.0 Administrator Guide*, complete these steps in these flowcharts.

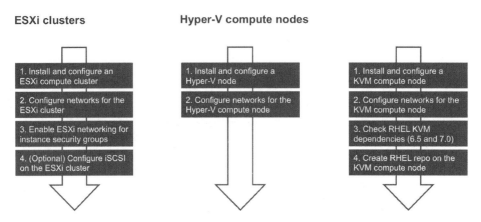

Figure 9-10 Creating compute nodes with ESXi, Hyper-V, and KVM in the correct sequence

Registering VMware vCenter

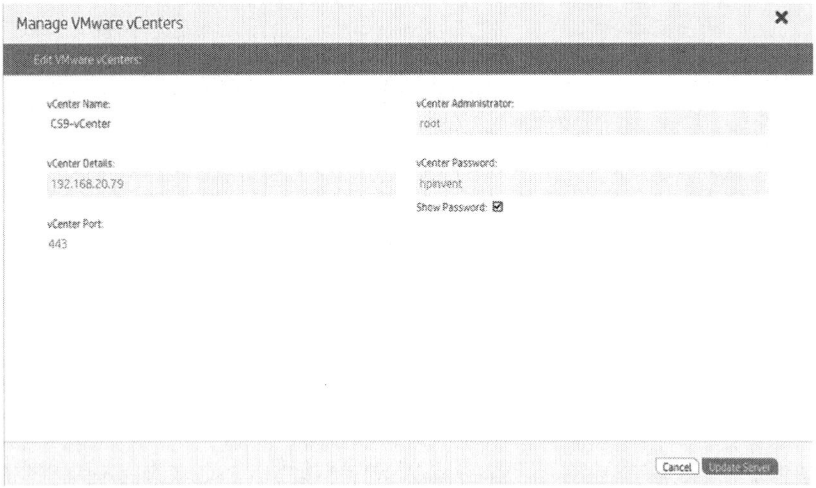

Figure 9-11 Registering VMware vCenter

The Operations Console **Integrated Tools** screen allows you to connect the Operations Console to other data center management software.

In CloudSystem, Manage VMware vCenters is an integrated tool, as shown in Figure 9-7, and is used to establish a connection between VMware vCenter and the CloudSystem management appliance. After VMware vCenter is registered, ESXi clusters can be activated and used as compute nodes. From the Integrated Tools screen, you can also view and edit registered servers and remove the connection between CloudSystem and the VMware vCenter.

For additional information, refer to "Integrated tool connectivity and configuration" (Section 17) in the *Helion CloudSystem 9.0 Administrator Guide*.

Configuring storage—Block storage (Cinder)

CloudSystem 9.0 supports StoreVirtual VSA and expands iSCSI support, in addition to the existing HPE StoreServ and VMware Virtual Machine File System (VMFS) options. Block storage options are summarized in Table 9-1.

Table 9-1 Block storage options

Hypervisor	Image type	Block storage device type	Device is created
ESXi	VMDK	VMware VMFS	Automatically when a vCenter is registered on the Integrated Tools screen
KVM	QCOW2	HPE StoreServ Fibre Channel[1] StoreServ iSCSI StoreVirtual VSA	On the Operations Console **Block Storage Devices** screen
Hyper-V	VHD, VHDX	StoreServ iSCSI StoreVirtual VSA[2]	On the Operations Console **Block Storage Devices** screen

 Note

1. Boot from volume is not supported in KVM-provisioned instances using StoreServ Fibre Channel storage.

2. Boot from volume is not supported in Hyper-V provisioned instances when the Hyper-V compute node is part of a cluster.

Block storage activation

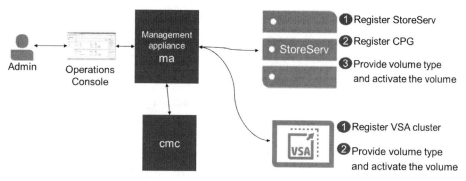

Figure 9-12 Block storage activation

Storage activation is performed in the Operations Console by following the steps shown in Figure 9-12. With StoreServ, the steps include the following:

1. Register the StoreServ storage solution.

2. Register the Common Provisioning Group (CPG).

3. Provide the volume type and activate the volume.

With StoreVirtual VSA, the steps include the following:

1. Register the StoreVirtual VSA cluster.

2. Provide the volume type and activate the volume.

More information on this topic can be found in the *Helion CloudSystem 9.0 Administrator Guide*, under "Storage management" (Section 19).

Configuring storage—Object storage (Swift)

Object storage in Helion CloudSystem Foundation is based on OpenStack Swift technology. CloudSystem allows you to set up a physical object storage solution and connect it to your cloud environment, where a cloud user can store large amounts of unstructured data and retrieve objects stored in publicly accessible physical machines. Object storage has the capacity to scale from a few terabytes (TB) to multiple petabytes (PB) of storage and is designed to scale horizontally, handling large numbers of simultaneous connections.

The recommended object storage model includes a minimum of four nodes:

- **Two Proxy Account Container (PAC) nodes**

  - The Proxy Server ties together the Swift architecture. For each request, it looks up the location of the account, container, or object in the ring and routes the request accordingly. The public API is exposed through the Proxy Server.

  - The Account Server manages container listings. The listings are stored as SQLite database files and replicated across the cluster, similar to how containers are managed. These servers require more I/O speed and less storage space than Object servers.

  - The Container Server manages object listings. It keeps track of the objects in a specific container. The listings are stored as SQLite database files and replicated across the cluster, similar to how objects are managed.

- **Two Object nodes**—The Object server is a simple Binary Large Object (BLOB) storage server that can store, retrieve, and delete objects stored on local devices. Objects are stored as binary files on the file system with metadata stored in the file's extended attributes. Each object is stored using a path derived from the object name's hash and the operation's time stamp. The last write takes precedence over earlier write commands and ensures that the latest object version is served.

The two PAC nodes and two Object nodes are divided evenly into two zones. This means that each zone has one PAC node and one Object node to start. A zone represents redundancy across data centers and can be used to group devices based on physical location or to power or network separations.

More information on this topic can be found in the *Helion CloudSystem 9.0 Administrator Guide*, under "Object storage (OpenStack Swift)" (Section 22).

Configuring networks

CloudSystem is built on Helion OpenStack networking technology. The network administrator creates the underlying network infrastructure before you install the CloudSystem virtual appliances by performing the tasks shown in Table 9-2.

Table 9-2 Configuring networks

Task	User role	User interface
Create pools of VLAN IDs and segmentation ranges that can be assigned to Tenant networks	Infrastructure administrator	Operations Console
Create Provider networks	Infrastructure administrator	Operations Console OpenStack user portal
Create external network Create external network subnet	Infrastructure administrator	OpenStack user portal
Attach Tenant networks to instances	Cloud user	OpenStack user portal
Create routers to connect networks	Cloud user	OpenStack user portal
Manage IP addresses using either dedicated static IP addresses or DHCP	Cloud administrator	OpenStack user portal
Access instances that are on Tenant networks from outside the cloud using floating IP addresses	Cloud user	OpenStack user portal

After CloudSystem is installed, you can create these types of networks that are used to manage instances:

- **Tenant (private) networks** are restricted and can be accessed only by VM instances assigned to the network. Subnets must be defined in the OpenStack user portal before using these networks.

- **Provider networks (optional)** are shared networks in the data center on which users can provision any number of VM instances.

- **External networks** allow you to route VM instances on Tenant networks from the CloudSystem private cloud to the data center, corporate intranet, or the Internet. The external network must be created and subnets must be defined in the OpenStack user portal before using these networks.

More information on this topic can be found in the *Helion CloudSystem 9.0 Administrator Guide*, under "Network configuration" (Section 15).

Activating and managing compute nodes

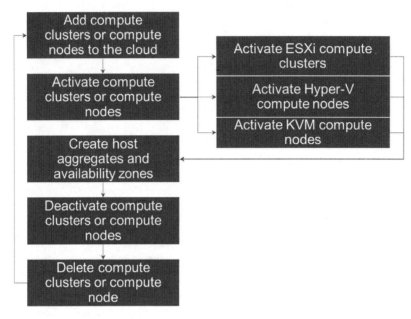

Figure 9-13 Activating and managing compute nodes

From the **Compute Nodes** screen in the Operations Console main menu, you can add, activate, deactivate, and delete compute nodes (compute clusters for ESXi) from the cloud. Figure 9-13 outlines the basic compute node lifecycle. The **Compute Summary** and **Dashboard** in the Operations Console provide additional views of compute node resources.

Within a cloud environment, compute nodes form a core of resources. A compute node provides the ephemeral storage, networking, memory, and processing resources that can be consumed by virtual machine instances. When an instance is created, it is matched to a compute node with available resources. A compute node can host multiple instances until all of its resources are consumed.

Compute resources are always placed in a common resource pool for provisioning. Instances are deployed to a hypervisor based on the image type.

CloudSystem deploys instances to compute nodes based on the optional availability zone you specify when you launch an instance in the OpenStack user portal.

Host aggregates enable the cloud administrator to partition compute deployments into logical groups for load balancing and instance distribution. A host aggregate is a group of hosts with associated metadata. A host can be part of more than one host aggregate.

A common use of host aggregates is to provide information for use with the Nova-scheduler, which deploys instances on specific hosts. For example, you might use a host aggregate to group a set of hosts that share specific flavors or images. You can also use host aggregates to separate different classes of hardware or servers on a separate power source.

A host aggregate is presented to users in the form of an availability zone. When you create a host aggregate, you have the option of providing an availability zone name. If you specify a name, the host aggregate you created is available as an availability zone that can be requested by users when deploying instances.

When users provision resources, they can specify the availability zone in which they want their instance to be deployed. This allows cloud consumers to ensure that their application resources are spread across hosts to achieve high availability in the event of hardware failure.

Detailed instructions for compute node activation and management can be found in the *Helion CloudSystem 9.0 Administrator Guide*, under "Compute node activation and management" (Section 21).

Activating and managing compute nodes — Activity

Using the *Helion CloudSystem 9.0 Administrator Guide*, answer these questions:

1. What does host aggregation accomplish?

    _____

    _____

2. How is a host aggregate exposed to users?

    _____

    _____

3. Why is the status of Hyper-V compute nodes displayed as a question mark in the Operations Console Compute Nodes overview screen?

    _____

    _____

4. When you deactivate an ESXi compute cluster, the disk files remain in the image cache. Why?

    _____

    _____

5. If you delete a compute node and want to bring it back into the cloud, what should you do?

_____

_____

6. What does the Virtual Allocation graph display?

_____

_____

Activating and managing compute nodes activity—answers

1. What does host aggregation accomplish?

   It enables the cloud administrator to partition compute deployments into logical groups for load balancing and instance distribution.

2. How is a host aggregate exposed to the users?

   It is exposed to the user in the form of an availability zone.

3. Why is the status of Hyper-V compute nodes displayed as a question mark in Operations Console Compute Nodes overview screen?

   Because Hyper-V compute nodes are not monitored in this release

4. When you deactivate an ESXi compute cluster, the disk files remain in the image cache. Why?

   Deactivating ESXi compute clusters and Hyper-V and KVM compute nodes does not remove the disk files from the image cache (_base directory). You must remove these files manually to free disk space.

5. If you delete a compute node and want to bring it back into the cloud, what must you do?

   Click Import on the Compute nodes screen in the Operations Console and supply the requested data.

6. What does the Virtual Allocation graph display?

   It displays usage or allocation of virtual CPU, memory, and compute node storage consumed within the last five minutes.

Launching VM instances activity

Using the *Helion CloudSystem 9.0 Administrator Guide*, place these steps, outlining the VM instance launch process, in the correct sequence.

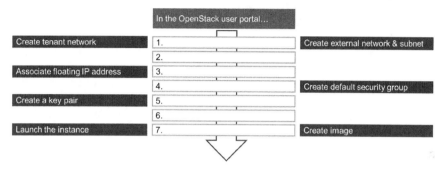

Figure 9-14 The VM instance launch process

Launching VM instances activity—answers

Using the *Helion CloudSystem 9.0 Administrator Guide*, place these steps, outlining the VM instance launch process, in the correct sequence.

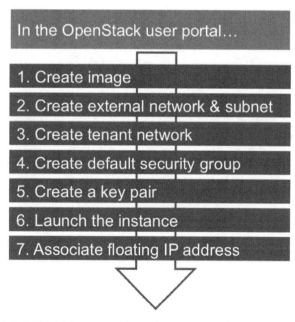

Figure 9-15 The VM instance launch process in the correct sequence

Creating and managing images—Activity

An image contains the operating system for a VM. It defines the file system layout, the operating system version, and other provisioning information. An image can be provisioned to one or more VMs in the cloud. Images that you add (by uploading) are used to boot VM instances in the cloud.

Before VM instances can be provisioned in the cloud, you must create at least one Provider or Tenant network and upload at least one image. There are three ways to upload images using the OpenStack user portal:

● Entering a file server URL

● Selecting a local file

● Creating an image from a snapshot of a currently running instance

Cloud users then use the portal to choose from available images or create their own from existing servers. Users can also create images using the OpenStack application program interface (API) or CLI.

Using the *Helion CloudSystem 9.0 Administrator Guide*, determine what image formats are supported by:

● **ESXi:**_____

● **KVM:**_____

● **Hyper V:**_____

More information on this topic can be found in the *Helion CloudSystem 9.0 Administrator Guide*, under "Image management" (Section 16).

Creating and managing images activity—answers

Determine what image formats are supported by:

● **ESX: Flat and Sparse Virtual Machine Disk format (VMDK) image files with SCSI adapters are supported for VM guest provisioning on VMware ESXi hypervisors**

● **KVM: Quick EMUlator (QEMU) copy-on-write format (QCOW2) formatted image files are supported for VM provisioning on KVM hypervisors**

● **Hyper-V: VHD and VHDX formatted image files are supported for VM provisioning on Hyper-V hypervisors**

Creating the external network and subnet

Create Network

Name

External

Project *

demo

Provider Network Type * ❓

VLAN

Physical Network * ❓

default

Segmentation ID * ❓

58

Admin State *

UP

☑ Shared

☑ External Network

Cancel   Create Network

Description:

Create a new network for any project as you need.

Provider specified network can be created. You can specify a physical network type (like Flat, VLAN, GRE, and VXLAN) and its segmentation_id or physical network name for a new virtual network.

In addition, you can create an external network or a shared network by checking the corresponding checkbox.

Figure 9-16 Creating the external network and subnet

An external network provides external connectivity through a tenant's virtual router to the VM instances connected to private networks. An external network is where the floating (or public) IP pool resides. A cloud user assigns a floating IP address to a VM instance connected to a tenant's private network to make it accessible externally.

 Note

Only one external network is supported in CloudSystem. VMs are not directly attached to the external network. Internal Provider and Tenant networks connect directly to VM instances.

The external network is created in the OpenStack user portal as shown in Figure 9-16, under the **Admin** tab. Select **Networks** → **+ Create Network**.

Creating a Tenant network

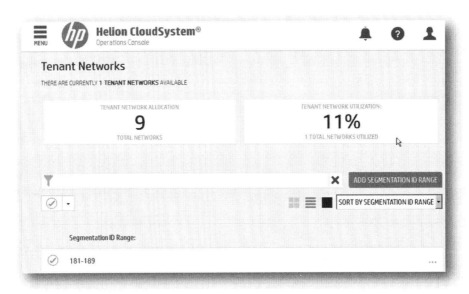

Figure 9-17 Creating a Tenant network

Tenant networks are part of the cloud data trunk and are created from a pool of VLANs, which are configured by using the Operations Console as shown in Figure 9-17. The OpenStack Networking service assigns VLANs from this pool to Tenant networks when they are created by users in the CloudSystem Horizon portal.

To create VLANs for Tenant networks, in the Operations Console Networking section, click **Tenant Networks**.

Users create Tenant networks to associate with their provisioned VM instances. They can then assign Tenant networks to VM instances during VM provisioning. Users create individual Tenant networks using VLANs identified for that purpose.

Creating a default security group

Figure 9-18 Creating a default security group

A security group is a set of security rules that is applied to a VM instance through IP filter rules implemented on the KVM compute nodes that host the VM instance. The rules define networking access to the instance.

All projects have a default security group, which is applied to any instances that have no other defined security group. Unless you change the default rules, the default security group denies all incoming traffic to the instance and allows all outgoing traffic. Tenant administrators and members can create new security groups or edit the default security group of their project to add new rule sets.

To create a new security group, log in to the OpenStack user portal (Horizon), select **Compute** under the **Project** tab, and then select **Access & Security** → **Security Groups** as shown in Figure 9-18.

Creating a key pair

Figure 9-19 Creating a key pair

Key pairs allow you to use public-key cryptography to encrypt and decrypt login information. To log in to your instance, you must create a key pair, specify the name of the key pair when you launch the instance, and provide the private key when you connect to the instance.

Key pairs allow you to log in to an instance after it is launched without being prompted for a password. Key pairs are only supported in instances that are based on images containing the cloud-init package.

You can generate a key pair from the OpenStack Horizon user portal, or you can generate a key pair manually from a Linux or Windows system. To create a key pair using Horizon, use the **Key Pairs** option in the **Access & Security** screen as shown in Figure 9-19.

Launching the instance

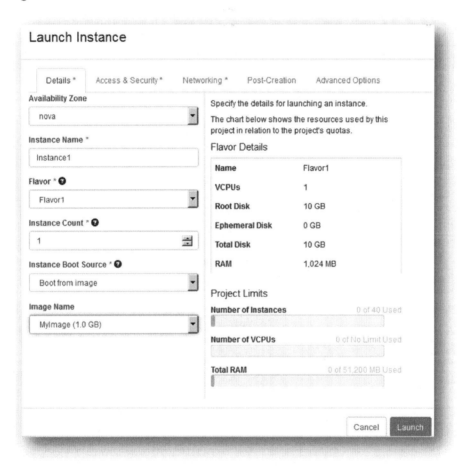

Figure 9-20 Launching the instance

There are two ways to deploy, or launch, instances in the OpenStack user portal:

- From the Project → Instances screen, click **Launch instance** and add instance details in the Launch Instance window shown in Figure 9-20.

- From the Project → Images screen, after creating an image, select the **Launch** button in the Action column for the image. Instance details are captured in the Launch Instance window.

To launch an instance:

- You must have the cloud user privilege at minimum.

- Security groups and security rules must be defined.

- At least one Provider or Tenant network is created.

- At least one image is uploaded.

Associating floating IP addresses

Figure 9-21 Associating floating IP addresses

You can use floating IP addresses to route traffic from the external network to specific virtual machine instances associated with a CloudSystem project.

As shown in Figure 9-21, to allocate and assign floating IP addresses, in the Access & Security screen, click **Floating IPs**.

Monitoring resource usage, allocation, and health

CloudSystem monitors the cloud by watching for problems and usage trends. The information that CloudSystem collects is provided in several ways:

- **Dashboard and Compute summary**—Visual representations of the general health and status of CloudSystem appliances and compute nodes, including resource usage and allocation details.

- **Activity**—Alerts and other notifications about appliance activity and events occurring in the cloud.

- **Logging**—Logs from compute nodes and services running on all CloudSystem appliances.

- **Monitoring**—Detailed data obtained from high-speed metrics processing and querying about the health of appliances and compute nodes (including a streaming alarm engine and notification engine).

- **Support dumps**—Vital diagnostic information from the system, and from OpenStack and CloudSystem services and components, on each appliance.

- **Audit logs**—Security-related actions that are occurring in the cloud.

Dashboard

Figure 9-22 Dashboard

You can select **Dashboard** on the Operations Console main menu to view a summary of system information as shown in Figure 9-22. The charts in the Dashboard provide a visual representation of the general health and status of the CloudSystem virtual appliances and resources in the cloud. From the Dashboard, you can immediately see resources that need attention.

Color-coded graphs provide a quick visual update on appliances, compute nodes, and OpenStack services. You can also see the status of the monitoring service to make sure that the data that is being monitored is current:

- Clicking the center of the **Compute Nodes Summary** graph takes you to the Compute Nodes screen, where you can see the status of various compute nodes, including the state, allocation, and usage of physical and virtual resources.

- Clicking the center of the **Appliance Summary** graph takes you to the Appliances screen, where you can see details about the virtual appliances in the cloud.

Activity Dashboard (OpenStack Kibana)

Figure 9-23 Activity Dashboard

Select **Activity Dashboard** on the Operations Console main menu and then click **Launch Activity Dashboard** to open a new browser window. Then, you can view alerts and other notifications about activities occurring in the cloud environment as shown in Figure 9-23.

 Note

You must log in to Kibana the first time you launch Activity Dashboard, Logging, or Audit Dashboard on the System Summary screen. Enter the Operations Console credentials that were set during first-time installation.

CloudSystem displays activity information using the open-source project Kibana. Kibana is a browser-based analytics and search dashboard. It allows users to search activity data by entering input queries at the top of the page. Related activities are grouped for easier viewing, and progress details of long-running tasks such as compute node activation can be tracked.

Logging Dashboard (OpenStack Kibana and Elasticsearch)

Figure 9-24 Logging Dashboard

From the Logging Dashboard screen in the Operations Console, you can click **Launch Logging Dashboard**, which opens a new browser window shown in Figure 9-24.

CloudSystem collects logs from compute nodes and all services running on all CloudSystem appliances. The logs are displayed in a single user interface, which is launched from the Operations Console.

You can view log information in charts, graphs, tables, histograms, and other forms. Centralized logging helps triage and troubleshoot the distributed cloud deployment from a single location. Users are not required to access the appliances and compute nodes to view the individual log files.

CloudSystem logging uses the open-source projects (Kibana) for data visualization and the Elastic search database for searching and indexing.

Monitoring Dashboard (Monasca)

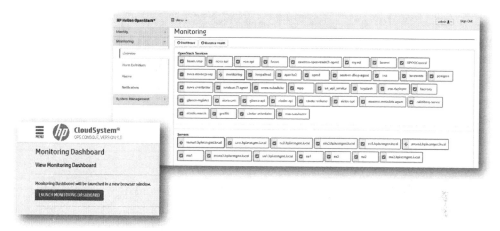

Figure 9-25 Monitoring Dashboard

CloudSystem provides Monitoring-as-a-Service using the open-source project Monasca as shown in Figure 9-25. Monasca is a comprehensive cloud monitoring solution for OpenStack-based clouds. Monasca uses node-based agents to report metrics to a centralized collection point, where alarms are triggered. It enables users to understand the operational effectiveness of the services and underlying infrastructure that make up their cloud and provide actionable details when there is a problem. System status and supporting metrics are constantly monitored, readily available, and trackable, making system management tasks more timely and predictable.

CloudSystem Monitoring-as-a-Service includes:

- A highly performing, scalable, reliable, and fault-tolerant monitoring solution.

- Operational (internal) and customer-facing capabilities. CloudSystem monitoring consolidates and unifies both types of capabilities, which simplifies the number of systems that are required for monitoring.

- Multitenant and authenticated monitoring. Metrics are submitted and authenticated using an access token that is stored and associated with an ID.

From the Monitoring Dashboard screen in the Operations Console, click **Launch Monitoring Dashboard**, which opens a specialized OpenStack Horizon portal running on the management appliance (this is different from the OpenStack user portal running on the cloud controller).

 Important

Monasca does **not** monitor Microsoft Hyper-V compute nodes.

Creating a support dump file—Activity

Using the *Helion CloudSystem 9.0 Administrator Guide*, answer these questions:

1. What are examples of information collected in the support dump file?

   _____

   _____

   _____

   _____

   _____

2. How can you create a support dump file?

   _____

   _____

3. Is the support dump file encrypted?

   _____

   _____

4. When selecting which log files to include in the support dump file, what does HPE recommend?

   _____

   _____

5. Can a support dump file be created from a command line?

   _____

   _____

Creating a support dump file activity—answers

1. What are examples of information collected in the support dump file?

   Configuration files (including first-time installation settings), audit logs, alerts, Hyper-V, and KVM compute node logs (ESXi compute cluster logs are not included), Object storage (scale-out Swift) logs

2. How can you create a support dump file?

   From the System Summary screen in the Operations Console, select Create Support Dump.

3. Is the support dump file encrypted?

   No

4. When selecting which log files to include in the support dump file, what does HPE recommend?

   HPE recommends including all log files, including verbose files.

5. Can a support dump file be created from a command line?

   Yes

Learning check

1. What is the recommended method of connecting to the cloud controller trio?

   a. Through any of the management appliances

   b. Through the first management appliance

   c. Through any of the cloud controller appliances

   d. Through the first cloud controller appliance

2. You need to create compute nodes—after configuring Tenant and Provider networks, what is the next step you need to complete?

   a. Add block storage

   b. Add object storage

   c. Activate compute node

   d. Run csstart

3. To attach Tenant networks to instances, you must log in to the OpenStack user portal as which type of user?

   a. Infrastructure administrator

   b. Infrastructure manager

   c. Cloud administrator

   d. Cloud user

4. Which open-source projects are the Activity Dashboard and Logging Dashboard based on? (Select two.)

   a. Monasca

   b. Kibana

   c. Elasticsearch

   d. Trove

   e. Sahara

Learning check answers

1. What is the recommended method of connecting to the cloud controller trio?

    a. Through any of the management appliances

    b. Through the first management appliance

    c. Through any of the cloud controller appliances

    d. Through the first cloud controller appliance

2. You need to create compute nodes—after configuring Tenant and Provider networks, what is the next step you need to complete?

    a. Add block storage

    b. Add object storage

    c. Activate compute node

    d. Run csstart

3. To attach Tenant networks to instances, you must log in to the OpenStack user portal as which type of user?

    a. Infrastructure administrator

    b. Infrastructure manager

    c. Cloud administrator

    d. Cloud user

4. Which open-source projects are the Activity Dashboard and Logging Dashboard based on? (Select two.)

    a. Monasca

    b. Kibana

    c. Elasticsearch

    d. Trove

    e. Sahara

Summary

- Helion CloudSystem Foundation offers several capabilities for managing the appliances. Using the Operations Console, users can manage the cloud controller trio, access the first management appliance, list CloudSystem appliances and their status, and view the appliance public networks.

- CloudSystem allows users to:

    - Configure networks

    - Manage resources, images, and storage

    - Create and activate compute nodes

    - Launch VM instances

- CloudSystem monitors the cloud by watching for problems and usage trends. The information that CloudSystem collects is provided in several ways:

    - Dashboard and Compute summary

    - Activity

    - Logging

    - Monitoring

    - Support dumps

    - Audit logs

10 A Closer Look at CloudSystem Enterprise

WHAT IS IN THIS CHAPTER FOR YOU?

After completing this chapter, you should be able to:

✓ Describe the CloudSystem Enterprise architecture and components

✓ Provide an overview of the installation, configuration, and management of CloudSystem Enterprise

✓ Recognize how CloudSystem Enterprise integrates with HPE OneView and CloudSystem Matrix

✓ Demonstrate the flow of service provisioning in CloudSystem Enterprise

Review: Helion CloudSystem portfolio

Helion CloudSystem Enterprise
Advanced infrastructure and application services

- Hybrid cloud management with Service Marketplace and designed based on Cloud Service Automation and Operations Orchestration
- Enterprise-class lifecycle management for application services including compliance[3]

HPE Cloud Service Automation and HPE Operations Orchestration— delivered as virtual appliances

Helion CloudSystem Foundation

Matrix Operating Environment

Helion CloudSystem Foundation
Core infrastructure services and PaaS

- Push-button activation of compute and storage nodes
- Out-of-the-box networking services
- Built on Helion OpenStack[1]
- Delivered via virtual appliances
- Helion Development Platform

HPE infrastructure management—HPE OneView[2]

Built on ConvergedSystem or on HPE and third-party infrastructure

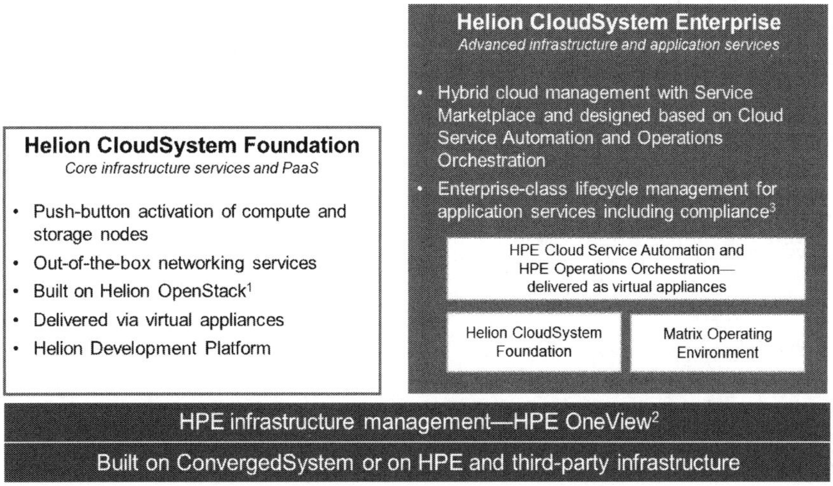

Figure 10-1 Helion CloudSystem portfolio

As previously discussed, the Helion CloudSystem portfolio consists of two versions as shown in Figure 10-1:

- Helion CloudSystem Foundation

- Helion CloudSystem Enterprise

If a customer is just starting their journey to a hybrid infrastructure and on-demand IT, they might need core infrastructure services and Platform-as-a-Service (PaaS). In this case, Helion CloudSystem Foundation is the right place to begin discussions. For a customer who needs a more comprehensive cloud solution, Helion CloudSystem Enterprise may be the optimal choice.

Helion CloudSystem Enterprise is the end-to-end, enterprise-level solution for delivering advanced infrastructure services and optional platform and application services for hybrid cloud environments. Available as a software only or as an integrated combination of storage, servers, networking, and software, customers can build and consume private and hybrid cloud services ideally suited to their business and technology objectives. It is highly integrated, so customers can incorporate cloud services with their existing IT assets.

 Notes

1. Includes Nova, Cinder, Neutron, Keystone, Glance, Horizon, Swift, and Heat

2. Optional

3. Requires Helion Enterprise platform, applications, and analytics add-on suite

Architecture and components of CloudSystem Enterprise 9

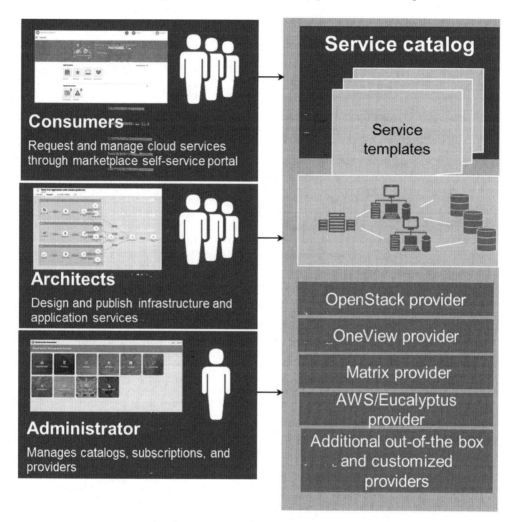

Figure 10-2 CloudSystem Enterprise 9 architecture and components

Helion CloudSystem Enterprise builds upon CloudSystem Foundation. As shown in Figure 10-2, it includes consumer, designer (architect), and administrator portals based on Cloud Service Automation (CSA) and Operations Orchestration (OO). It features:

- **Virtual appliances**, based on VMware ESXi or Red Hat KVM, that are delivered in a three-node high-availability (HA) management cluster with embedded MySQL.

- **Multicloud capabilities** across geographies and data centers using multiple OpenStack pools. With automatic identity management and Keystone integration, CloudSystem Enterprise can manage multiple OpenStack clouds across geographies and data centers. This allows one service design to be provisioned on different clouds, and the portable design includes subscription-time selection of resources, environment, and projects. For example, CloudSystem Enterprise can manage two or three CloudSystem Foundation instances in different locations to support remote offices.

- **Multiple application provisioning services** such as Chef, Puppet, HPE Server Automation (SA), and Docker. With the Topology Designer in CSA 4.5, CloudSystem Enterprise enables users to add application provisioning on top of the infrastructure design.

- **Physical server provisioning** using HPE OneView and ICsp integration.

- **Portable service design** that avoids design sprawl. CloudSystem Enterprise allows deployment of services on different resource pools through late resource binding.

With CloudSystem Enterprise, customers receive:

- CloudSystem Matrix provider, which allows Matrix templates and active services to be used with CSA.

- Virtualization providers such as VMware vCenter and Microsoft Hyper-V.

- Standard public cloud providers such as Amazon Web Services (AWS), which extends to support Helion Eucalyptus private cloud, Microsoft Azure, and OpenStack.

 Note

The out-of-the-box OpenStack provider uses OpenStack application program interfaces (APIs) to support all OpenStack distributions, CloudSystem Foundation, Helion OpenStack, and third-party OpenStack clouds.

- HPE OneView provider, which leverages HPE OneView profiles to create physical server service designs in the CSA service designer portal.

Review: Enterprise-ready HA architecture

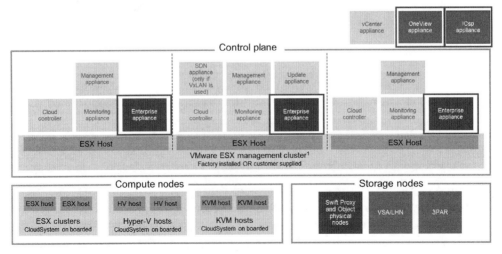

Figure 10-3 Enterprise-ready HA architecture

Figure 10-3 represents the Helion CloudSystem 9.0 architecture and the distribution of CloudSystem appliances and managed resources in a deployed cloud.

The diagram highlights CloudSystem Enterprise appliances (Enterprise appliances, HPE OneView appliance, and Insight Control server provisioning [ICsp] appliance) with a border.

Review: Architecture

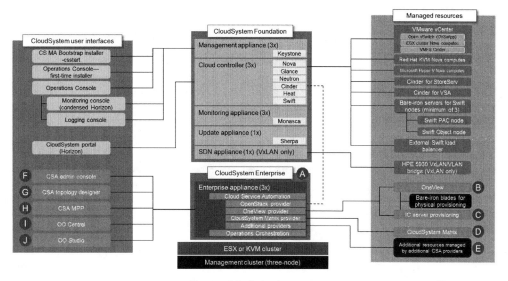

Figure 10-4 Architecture

CloudSystem Enterprise adds a trio of appliances to the CloudSystem Foundation installation. These appliances contain the core functionality of the Enterprise offering, including CSA, the Marketplace Portal, the Topology Designer, and the Sequenced Designer. OO Central is also embedded in the Enterprise appliance.

Figure 10-4 highlights the CloudSystem Enterprise components:

- **CloudSystem Enterprise appliance** (Label A) consists of CSA and OO.

- **OneView** (Label B) manages the converged infrastructure and supports key scenarios such as deploying bare-metal servers, performing ongoing hardware maintenance, and responding to alerts and outages. It is designed for the physical infrastructure needed to support virtualization, cloud computing, big data, and mixed computing environments.

- **Insight Control server provisioning** (Label C) deploys operating systems on ProLiant bare-metal servers; updates drivers, utilities, and firmware; and configures system hardware.

- **CloudSystem Matrix** (Label D) can be configured as an additional provider in CSA.

- **Other CSA providers** (Label E) can be integrated with CSA and include AWS, Microsoft Azure, and Helion OpenStack.

- **CSA Cloud Service Management Console** (Label F) is the administrative portal for the Enterprise appliance. Designs are provisioned as offerings in this console.

- **CSA Topology Designer** (Label G) is an easy-to-use solution for infrastructure provisioning designs.

- **Marketplace Portal** (Label H) displays offerings that can be purchased and applied to a cloud environment as a subscription.

- **CSA Sequenced Designer** (not shown in the diagram) handles more complex application provisioning designs.

- **OO Central** (Label I) provides the ability to run scripted workflows in CSA.

- **OO Studio** (Label J) provides the ability to create and customize new workflows and debug and edit existing workflows. OO Studio is installed separately, using the executable file included in CloudSystem.

When the Enterprise appliance is included as part of the CloudSystem installation, a Helion CloudSystem provider is automatically configured in CSA to support the integration between CloudSystem Enterprise and CloudSystem Foundation. This Helion CloudSystem provider is associated with the CloudSystem Foundation environment and allows all resources in that environment to be managed from CloudSystem Enterprise using the CSA Cloud Service Management Console.

Cloud Service Automation

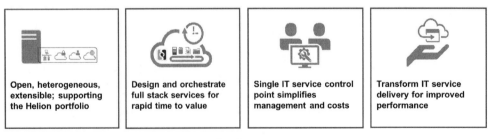

Figure 10-5 Cloud Service Automation

CSA is a cloud lifecycle management solution for heterogeneous cloud environments. Figure 10-5 highlights some of the features. It has three primary feature sets including:

- Highly automated lifecycle management for services

- Enterprise-grade management capabilities

- Flexible, open, and extensible architecture

The platform is highly scalable, both from capacity and capability viewpoints. For example, a customer may initially want to provide Infrastructure-as-a-Service (IaaS) services for development use and also plan to provide Platform-as-a-Service (PaaS) or Software-as-a-Service (SaaS) services to support a mission-critical application in the future. In this scenario, the initial deployment would typically support a few hundred virtual machines (VMs) with a minimum amount of monitoring. Then, in the future, the customer may need to support thousands of VMs with end-to-end service-level management in a secure environment. CSA allows businesses to increase the number of VMs and add more features over time.

Cloud users will access the service catalog through the Marketplace Portal to subscribe to and access services. CSA provides the self-service Marketplace Portal and the service catalog in CloudSystem Enterprise.

 Note

Helion CloudSystem Enterprise 9.0 includes version 4.5 of CSA.

 Note

To view additional information, case studies, and demonstrations or to download a trial version of CSA, scan this QR code or enter the URL into your browser.

http://www8.hp.com/us/en/software-solutions/cloud-service-automation/index.html

CSA concepts

Cloud Service Management Console

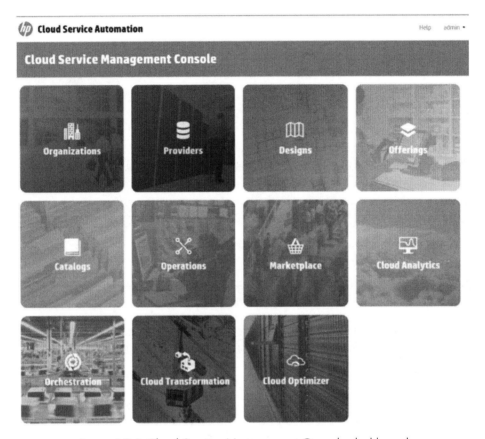

Figure 10-6 Cloud Service Management Console dashboard

Use the Cloud Service Management Console dashboard, shown in Figure 10-6, to navigate to tasks such as configuring organizations and users, creating designs, and managing catalogs. You can always click **CSA** in the title bar of the console to return to the main dashboard.

Depending on your role, you can access specific areas of the Cloud Service Management Console. Click **Help** at the top of the screen for more information about the dashboard and details about managing and configuring cloud services.

Organizations

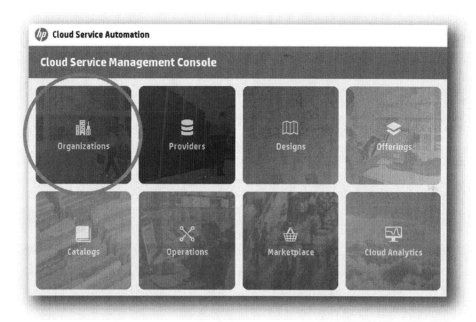

Figure 10-7 Organizations

An organization in CSA determines a member's entry point into the cloud and associates its members with services and resources. An organization typically represents a business entity, such as a company, a business unit, a department, or a group. CSA queries the organization's identity management system to determine the organization's members and groups and uses this information to authenticate and authorize CSA users and their actions. You can manage Organizations by clicking the Organizations tile highlighted in Figure 10-7.

CSA defines one **provider** organization per CSA instance. You can assign provider organization roles to control access to administrative functions. Using the Cloud Service Management Console, members of the provider organization can create one or more **consumer** organizations, manage configured organizations, and manage resources and services (such as designing, offering, and publishing resources and services for consumption). The organizations, resources, and services that can be managed are determined by the roles assigned to the members of the provider organization. For example, the CSA administrator manages all organizations, resources, and services, while the consumer service administrator manages only the organizations.

Members of the consumer organizations, using the Marketplace Portal, subscribe to and consume the resources and services set up by the provider organization. There may be multiple consumer organizations configured; however, each consumer or subscriber sees only the information for the consumer organization of which he or she is a member. CSA uses consumer organizations and catalogs, along with identity management system data, to map service offerings to the appropriate subscribers.

User roles

Specific areas of the Cloud Service Management Console are available to users based on their roles. There are two groups of organization level roles defined in CSA: **provider organization roles** and **consumer organization roles**.

Provider organization roles

Provider organization roles can authorize members to perform specific tasks and access specific parts of the Cloud Service Management Console. These roles are typically configured by the CSA administrator. These roles are:

- **Administrator**—Has access to all functionality in the Cloud Service Management Console

- **Consumer service administrator**—Configures and manages consumer organizations

- **Resource supply manager**—Creates and manages cloud resources, such as resource providers and resource pools

- **Service business manager**—Creates and manages service offerings and service catalogs

- **Service designer**—Designs, implements, and maintains service designs (also referred to as *blueprints*), component palettes, component types, component templates, and resource offerings.

- **Service operations manager**—Views and manages subscriptions and service instances.

Consumer organization roles

Consumer organization roles can authorize access to the Marketplace Portal. These roles are:

- **Consumer organization administrator**—This role can:
  - Create, edit, and delete an organization's catalogs.
  - Manage service offerings in an organization's catalogs.
  - Manage access controls, approval policies, and categories in an organization.

- Manage a user's subscriptions by performing actions on behalf of the original subscriber.

- Use HPE IT Business Analytics to measure and optimize the cost, risk, quality, and value of IT services and processes.

- **Service consumer**—Requests and manages subscriptions offered to the organization through the Marketplace Portal. From the Marketplace Portal, the service consumer can browse catalogs, subscribe to services, view subscriptions, and approve or deny subscription requests. The service consumer cannot log in to the Cloud Service Management Console.

Access control

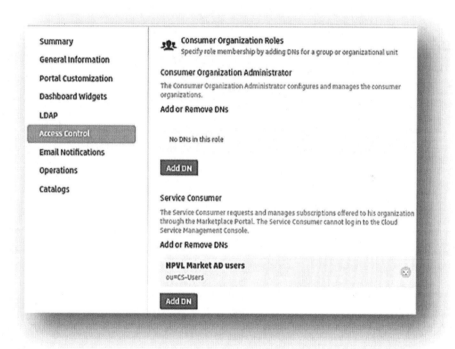

Figure 10-8 Access control

Directory service groups or organization units can be added to a role by associating the organization unit's distinguished name to the desired role as shown in Figure 10-8. Conversely, you can remove groups or units from a role by removing the organizational association from that role. The authenticated users, who are members of a group or an organization unit that is assigned to a role, can perform specific tasks and access specific areas of the Cloud Service Management Console. Users can assign an organization unit's distinguished name to more than one role.

Resource providers

Figure 10-9 Resource providers

Providers are management platforms that offer centralized control over the infrastructure and resources used in a cloud computing environment. For example, the HPE Matrix Operating Environment provider can deploy VMs, while the HPE SiteScope provider monitors applications. The Providers tile shown in Figure 10-9 can be accessed from the Cloud Service Management Console, and it provides access to all Provider functions.

A provider corresponds to the specific instance of an application that CSA can integrate with to help instantiate service designs. For example, to enable service designs that target the Matrix Operating Environment, you must first create a provider (with a Matrix Operating Environment provider type) in the Cloud Service Management Console. When you are creating the provider, you can specify details such as user credentials and the URL for the Matrix Operating Environment service access point.

Components

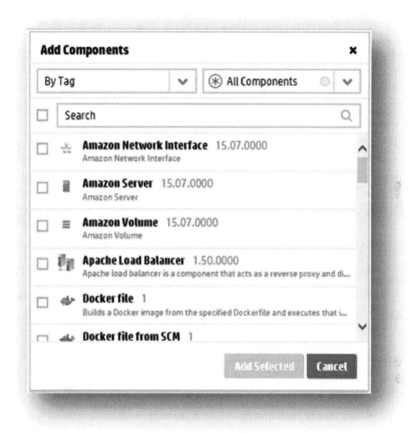

Figure 10-10 Components

Components, shown in Figure 10-10, are elements of service designs that are sequenced or topological. Only topological components are displayed in the Components tab of the Cloud Service Management Console. Sequenced components are not associated with providers or provider types. From the Components tab, you can view the topological components associated with a specific provider instance and manage the topological components.

Service designs and subscriber options

To provide on-demand, automated service delivery, service designers can create, configure, and modify service designs. Service designs include reusable service components and are used as the basis for automating the cloud. Service components and their relationships within a service design define the framework for creating the service.

Service designs also offer options that consumers can select when ordering a service. You can reuse designs for multiple service offerings, and each service offering is customized to meet the needs of different consumer organizations and groups. You can also use service designs shipped with CSA and export or import designs between CSA systems.

Service designers can create topology or sequenced designs:

- **Topology designs** specify components, relationships, and properties. In contrast to sequenced designs, which more explicitly define the provisioning order and sequence of actions, topology designs are declarative and do not include explicit actions or sequencing. The provisioning sequence is inferred by the relationships that exist between components in a topology design.

  Use topology designs for IaaS, PaaS, and SaaS deployments that are enabled via Chef, Puppet, HPE Server Automation, and HPE OO flow-based components.

- **Sequenced designs** specify directed execution of the service component lifecycle and provide mechanisms for controlling resource selection as each component is deployed. When creating sequenced designs, you can specify resource bindings on service components to constrain provider selection. These bindings link the component to one or more resource offerings that provision the component.

  Use sequenced designs for complex services and services that rely on runbook automation, such as integrations with legacy data center systems. Create a sequenced design as a directed component hierarchy to define lifecycle execution. Sequenced designs use components to group multiple automation providers within a single entity, and they permit explicit specification of lifecycle actions.

Subscriber options allow you to expose service design options in the Offerings tile of the Cloud Service Management Console. These are sets of options for a service design. The subscriber options can further be refined by setting pricing for options, hiding options, and setting values for option properties. The subscriber options are then available in the Subscribers tile.

Service offerings

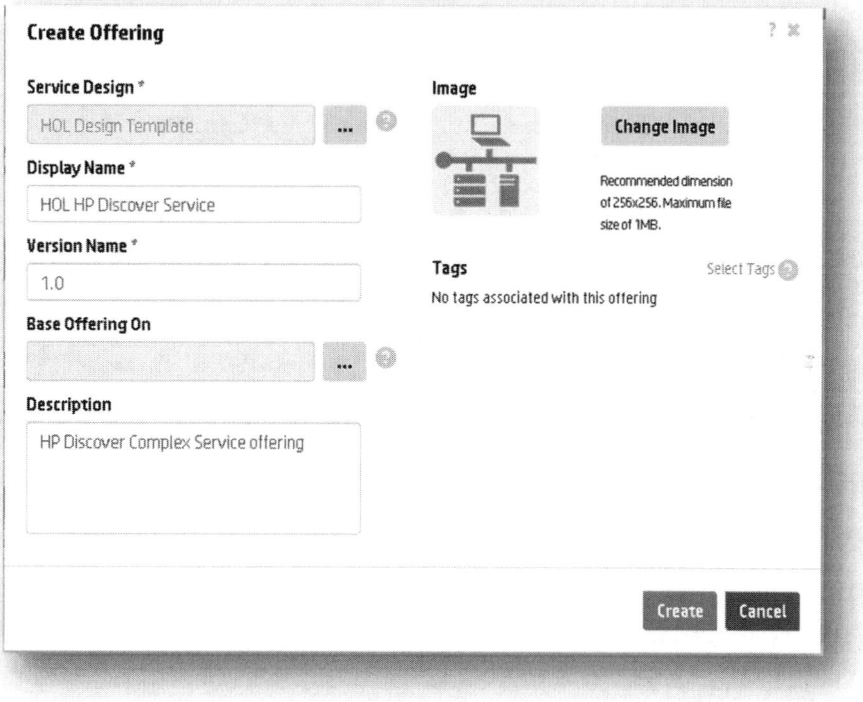

Figure 10-11 Service offerings

Service offerings provide all the information consumers need to select the most appropriate services. Each service offering references a service design, which defines the service options and components of the service. You can tailor service offerings for each consumer group with specifics such as customized terms and conditions, option visibility, and pricing. When you are ready to expose the design to subscribers, you can publish the service offering in a catalog within the Marketplace Portal. The Cloud Service Management Console Create Offering window shown in Figure 10-11 is used to create service offerings, and these can be created from a sequenced or topology design.

Pricing configured on a service offering can be initial, recurring, or option-specific. Pricing details can be hidden from the subscriber in the Marketplace Portal if configured in the service offering. You can attach documents to a standard service offering (for example, service-level agreements, terms, and conditions). Screenshots with images and captions can also be attached in order to provide the user with a visual representation of the offering in the Marketplace Portal.

Service offerings allow you to:

- Customize service offerings for different target groups
- Base customized service offerings on the same service design using different attributes for each group
- Publish a customized service offering in a catalog that is visible to its target group
- Configure these service offering attributes:
  - Offering name, description, image, and tags
  - Option visibility for offerings based on sequenced designs
  - Subscription pricing
  - Attached documents such as service-level agreements or terms and conditions
  - Associated screenshots or other images
  - Multiple versions per service offering
- Link each service offering to its target group by publishing the service offering in a catalog for that group

CSA uses catalogs to constrain the service offerings displayed for each user. CSA manages catalog access through group memberships as configured in the organization's identity management system. CSA does not directly manage the creation or maintenance of individual users or organizational groups. You can specify an identity management system for each CSA organization you create. The groups can be configured to access the organization's user interface. You can also configure catalog access for specific groups within a CSA organization. Each organization's group memberships must exist or be created in its identity management system. Often, existing groups naturally correspond to CSA access control needs. Sometimes you will need to create new groups for specific needs.

Catalogs

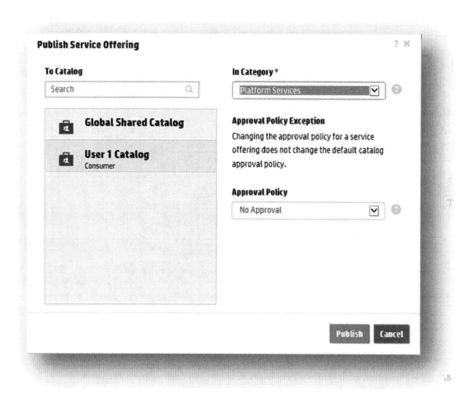

Figure 10-12 Catalogs

Catalogs map service offerings to specific groups within a consumer organization. Publishing a service offering in a catalog makes the offering visible in the Marketplace Portal to groups associated with the catalog. You can configure each catalog as visible to specific groups within the catalog's associated organization, and you can also specify the default approval process and the available approval policies for each catalog. In addition, you can publish a service offering in multiple catalogs to make it visible to more than one set of consumers.

You can configure the automatically created, default catalog (Global Shared Catalog) shown in Figure 10-12, or you can manually create a new catalog and associate it with an organization. Multiple catalogs can be associated with the same organization, and any changes made to the Global Shared Catalog will be visible in every organization's Marketplace Portal.

Service instances

Figure 10-13 Operations

Use the Operations tile of the Cloud Service Management Console shown in Figure 10-13 to view and manage subscriptions and service instances for all consumer organizations. A subscriber who is using a service catalog in the Marketplace Portal creates a *subscription request* to ask for the delivery of cloud services. After a subscription request is approved, a service instance is created.

CSA constructs service instance artifacts during service deployment and updates service instances during service management. Service instances encapsulate all details of the deployed service and its components (for example, provisioned IP details for a network segment component). CSA bases service instances on the service design configured for the service offering and on consumer demand.

Shopping for cloud services in Marketplace Portal

Figure 10-14 Marketplace Portal

CSA delivers cloud services through an innovative, enterprise-ready Marketplace Portal. As the home page of the Marketplace Portal, the dashboard shown in Figure 10-14 allows users to shop for service offerings and manage catalogs, service offerings, requests, and subscriptions. Users can scroll up and down the rows of tiles to select and order service offerings by category.

In the dashboard banner, users can click **Start Shopping** to browse and order services using certain categories, keyword search, or quick links. Users can browse for services and subscriptions using the global search feature.

The Sidebar Menu offers quick and direct navigation to and from any view in the Marketplace Portal.

When users are logged in as a tenant administrator, they can access the Administration tiles in order to:

- Manage a user's subscriptions in their organization on behalf of the original subscriber

- View, create, edit, and delete catalogs in their organization

- View, create, delete, publish, and unpublish services offerings in their organization's catalogs

- Manage access controls, approval policies, and categories in their organization

- Launch HPE IT Business Analytics, which automatically gathers metrics from CSA to build key performance indicators

Why orchestrate?

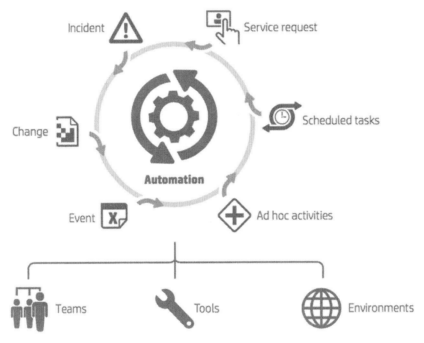

Figure 10-15 Why orchestrate?

In many companies, the issues illustrated by Figure 10-15 can result in poor service quality, delayed time-to-market, and high operating costs. These include the following:

- **Incidents**—Floods of alerts, unnecessary escalations

- **Changes and releases**—Too many manual errors, lack of audit trails

- **Process management**—Lack of processes for complex tasks (for example, disaster recovery)

- **Virtualization**—Inconsistent management of physical and virtual assets

Orchestration is the coordination of automated tasks across teams, tools, and environments. Orchestration enables several types of automation:

- Element automation of either networks, servers, or storage, spanning tasks from provisioning and change management to compliance enforcement and reporting

- Runbook automation of common and repeatable IT processes across all infrastructure tiers, IT groups, and systems

- Integrated automation of applications, servers, networks, storage, and common processes across the data center

- Automation of business services, with continuous control of each phase of the service lifecycle, across the data center and client endpoints, from automated operations to monitoring and ticketing

Operations Orchestration

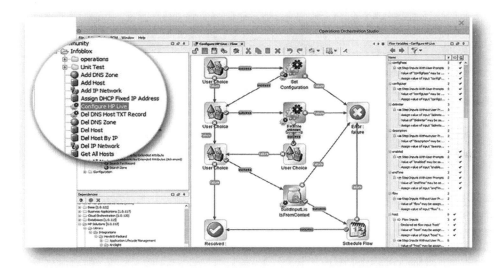

Figure 10-16 Operations Orchestration

OO, shown in Figure 10-16, is the next-generation IT process automation solution that is supported across the entire IT environment from traditional data center to cloud. It is the orchestration and policy engine in CloudSystem Enterprise.

Functional architecture and components

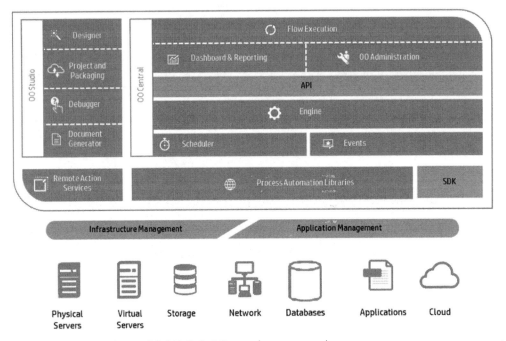

Figure 10-17 OO 10.x architecture and components

As illustrated by Figure 10-17, OO 10.x is composed of four main functional components:

- OO Studio
- OO Central
- OO Remote Action Services (RAS)
- OO Content (Process Automation Libraries)

Together, the components of OO enable you to manage various services and devices across the organization and their lifecycle.

OO Studio

OO Studio is a desktop-based application that is used by flow authors to create OO flows. Studio enables the author to design, debug, and package flows. It provides automation via code capabilities, such as integration to source control management software, project separation, and collaboration via multiauthoring.

OO Studio includes:

- **Designer**—OO Studio provides a drag-and-wire graphical designer to formulate flows out of various operations and subflows.

- **Project and packaging**—OO Studio lets users break down a set of authored content into granular projects. Each project groups together contents that are similar; for example, according to functionality, development owner, geographic location, and release time lines.

- **Debugger**—OO Studio provides a debugger to test the designed flows. The debugger reflects the behavior of the flow in the OO Central environment.

- **Document generator**—OO Studio lets users generate documents for every flow or group of workflows. These autogenerated documents include information about the flow, including its graphical presentation.

OO Central

OO Central is the runtime environment of OO. It is used for running flows, monitoring the various runs, and generating reports. Administrators, users, and integrators can access OO Central's web-based UI and a set of APIs. This environment is available as a web archive (WAR) file to be included within an application server or as a stand-alone installation.

It includes:

- **Flow execution**—OO Central provides execution capabilities for the deployed flows. The execution is performed from a web-based UI and from Representational State Transfer (REST) APIs. The execution capabilities include browsing the flow library, launching an execution, and tracking an execution until its end.

- **Dashboard and reporting**—OO Central provides reports about the various executions. These include running flows, finished flows, successful and failed flows, and so on. In addition, the rich set of RESTful APIs allow users to gather execution information and compound a more advanced set of dashboard and reports.

- **OO administration**—OO Central allows administrators to administrate OO from system and operational perspectives. It covers areas such as setting permissions, defining system components, configuring LDAP, and more.

- **API**—OO Central provides a complete set of RESTful APIs for every capability. In fact, every capability of the web UI is implemented on top of public RESTful APIs, enabling users to implement their own web UI and combine OO capabilities within web applications.

- **Engine**—The engine works behind the scenes of OO Central. This is the back-end component that processes the entire flow execution. The engine manages the step execution, persistency, and manual interaction with users.

● **Scheduler**—OO Central includes an out-of-the-box scheduler, which enables the OO administrator to define various recurrence patterns for flow executions, along with tracking and managing them.

OO RAS

The OO RAS enables execution in remote data centers and networks. A Remote Action Service is an instance of a service that enables a flow to run on machines, using commands, or against APIs that would be otherwise inaccessible to HP OO flows.

For example, you could use a RAS to accomplish the following:

● Restart a server in a different domain from the one in which OO Central is installed

● Carry out an operation on a Windows server when Central was installed on a Linux server

RAS periodically accesses OO Central in order to execute operations, so users need to open ports for inbound communication only in OO Central. Moreover, to achieve high availability of RAS, users can simply add another RAS and point it to OO Central.

In addition, these services support a grouping mechanism, which enables you to correlate between a step in the flow and the type of remote services that can execute this step. Therefore, the binding between the flow steps and the services is dynamic.

OO Content

OO provides a rich set of out-of-the-box operations and flows that enable users to author complex flows, orchestrating various services. OO Content is delivered as a set of granular content packs that can be downloaded, deployed, and managed individually. These are the *process automation libraries*.

In addition, OO provides wizards for generating additional content over other services such as Web Service Wizard. OO provides Java and .NET software development kits (SDKs) to enable developing custom content and operations. By using OO Content, users can build a rich set of process automation libraries.

Integration of OO in Helion CloudSystem Enterprise

Name:	Role:	Status:	
cs-update1	UpdateAppliance	UP	...
cs-monitor3	MonascaController	UP	...
cs-monitor2	MonascaController	UP	...
cs-monitor1	MonascaController	UP	...
cs-mgmt3	ManagementController	UP	...
cs-mgmt2	ManagementController	UP	...
cs-mgmt1	ManagementController	UP	...
cs-enterprise3	EnterpriseController	UP	...
cs-enterprise2	EnterpriseController	UP	...
cs-enterprise1	EnterpriseController	UP	...
cs-cloud3	CloudController	UP	...
cs-cloud2	CloudController	UP	...
cs-cloud1	CloudController	UP	...

Figure 10-18 Integration of OO in Helion CloudSystem Enterprise

OO Central is included as part of Helion CloudSystem Enterprise. It contains a set of default workflows that allows users to manage administrative tasks associated with the private cloud. OO Central is automatically installed as part of the CloudSystem Enterprise appliance trio highlighted in Figure 10-18. CloudSystem supports full OO functionality, but only the workflows in the predefined bundle are available for use. Customers can optionally install OO Studio to edit existing workflows or create new workflows. After workflows are edited in OO Studio, users can load them back to OO Central and use them to perform administrative tasks.

Accessing OO in Helion CloudSystem Enterprise

Figure 10-19 Accessing OO

Access OO Central using a supported browser at this address:

```
https://<Enterprise_CAN_VIP>:9090/oo
```

You will reach a login prompt as shown in Figure 10-19. Use administrator as the username. The OO Central password is defined during the first-time installation.

 Note

To view additional information, case studies, and demonstrations or to download a trial version of OO, scan this QR code or enter the URL into your browser.

http://www8.hp.com/us/en/software-solutions/operations-orchestration-it-process-automation/index.html

Learning check

1. Helion CloudSystem Enterprise supports _____ provider organization(s) and _____ customer organization(s) per CSA instance.

2. Which CSA designer should be used for IaaS, PaaS, and SaaS deployments that are enabled via Chef, Puppet, HPE Server Automation, or OO?

    a. Topology designer

    b. Sequenced designer

3. _____ _____ _____ is the default CSA catalog.

4. Which Cloud Service Management Console tile should be used to view and manage subscriptions and service instances for all consumer organizations?

    a. Organizations

    b. Catalogs

    c. Operations

    d. Offerings

5. Which OO component is optional in CloudSystem Enterprise and must be installed manually to edit existing workflows or to create new workflows?

    a. OO Central

    b. OO Studio

    c. OO RSA

    d. OO Content

Learning check answers

1. Helion CloudSystem Enterprise supports **one** provider organization and **multiple** customer organizations per CSA instance.

2. Which CSA designer should be used for IaaS, PaaS, and SaaS deployments that are enabled via Chef, Puppet, HPE Server Automation, or OO?

    a. **Topology designer**

    b. Sequenced designer

3. **Global Shared Catalog** is the default CSA catalog.

4. Which Cloud Service Management Console tile should be used to view and manage subscriptions and service instances for all consumer organizations?

    a. Organizations

    b. Catalogs

    c. Operations

    d. Offerings

5. Which OO component is optional in CloudSystem Enterprise and must be installed manually to edit existing workflows or to create new workflows?

    a. OO Central

    b. OO Studio

    c. OO RSA

    d. OO Content

Installation, configuration, and management

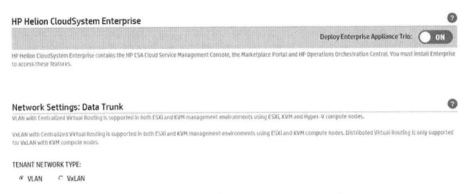

Figure 10-20 CloudSystem First-Time Installer

During the first-time installation, CloudSystem Enterprise can be installed by sliding the **Deploy Enterprise Appliance Trio** button to **ON** in the First-Time Installer as shown in Figure 10-20. If you do not plan to use CSA or OO, you can disable installation of the Enterprise appliance in the First-Time Installer by selecting **Off** for **Deploy Enterprise Appliance Trio**.

If you need to install CloudSystem Enterprise after first-time installation, you must:

1. **Install the Enterprise appliance driver**—Download it from the HPE Software Depot and install it on the Management appliance (ma1) in the boot directory for each Enterprise appliance.

2. **Install the Enterprise appliances**—Click **Install Enterprise** from the Operations Console **Appliances** screen. This option is shown only if Enterprise was not installed during the first-time installation.

Detailed steps are available in the *Helion CloudSystem 9.0 Administrator Guide*, under "Installing the Enterprise appliance after First-Time Installation."

Managing the Enterprise appliance trio

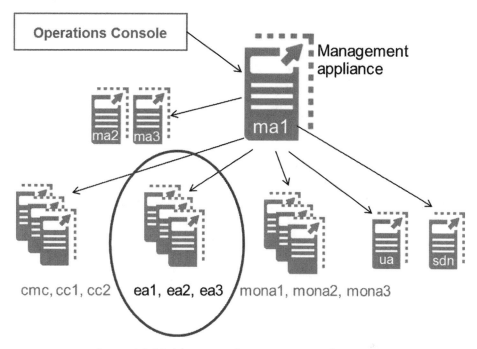

Figure 10-21 Managing the Enterprise appliance trio

As illustrated by Figure 10-21, CloudSystem identifies the Enterprise appliances in the trio by the internal name assigned on the Cloud Management Network:

- **ea1**—The first Enterprise appliance in the trio

- **ea2**—The second Enterprise appliance in the trio

- **ea3**—The third Enterprise appliance in the trio

You should always access an Enterprise appliance through the first Management appliance in the trio as follows:

1. Using the cloudadmin credentials, Secure Shell (SSH) into ma1.

2. From ma1, SSH to the Enterprise appliance you want to access.

 Note

When performing maintenance on a VM hosting the Enterprise appliances, make sure that ea1 is always the last node shut down and the first node restarted.

CloudSystem Enterprise is a separate virtual appliance that runs CSA. However, all appliance management tasks are performed through the CloudSystem Operations Console and all OpenStack functionality included in CloudSystem Foundation is also included with CloudSystem Enterprise.

Changing default passwords

If you deploy the Enterprise appliance from the Appliances screen instead of during first-time installation, the password for the cloudadmin user on the Enterprise appliance is set to a default value of cloudadmin. It is not set to the password you specified for cloudadmin in the first-time installer user interface.

After the Enterprise appliance trio is installed, you must execute cURL commands to set a new password for the cloudadmin account. Detailed steps are provided in the *Helion CloudSystem 9.0 Administrator Guide*, under "Changing the Enterprise appliance password when Enterprise is deployed after First-Time Installation."

Figure 10-22 shows the cURL command used to set a new password for the cloudadmin account (execute on each ea).

```
curl -X PUT http://192.168.0.20:6666/rest/pavmms/v1.0/ea1
-H "X-Auth-Token: <token-ID>" -H "Content-Type: application/json"
-d '{"sys_creds": {"username": "cloudadmin", "password": "cloudadmin", "newpassword": "<new-password>"}}'
```

Figure 10-22 CURL command

The default credentials for the Cloud Service Management Console and the Marketplace Portal are shown in Table 10-1.

Table 10-1 Cloud Service Management Console and Marketplace Portal default credentials

Cloud Service Management Console	Marketplace Portal
Username: admin	Username: consumer
Password: cloud	Password: cloud

Changing these default passwords is described in the *Helion CloudSystem 9.0 Administrator Guide*, under "Logging in and changing the default CSA and Marketplace Portal password."

Integrations

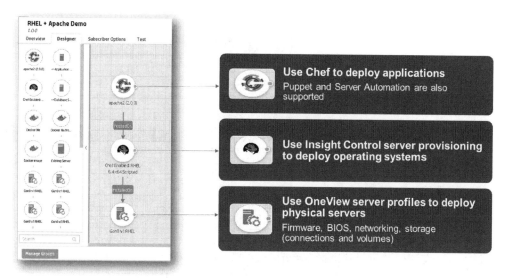

Figure 10-23 Integration with HPE OneView

In CloudSystem Enterprise, users can provision physical servers using HPE OneView profiles. They can also use the CSA Topology Designer to lay out a design that looks similar to Figure 10-23.

After you select the HPE OneView profile in Topology Designer, you can add Insight Control server provisioning to deploy the required operating system. Furthermore, you can also use existing tools such as Chef or Puppet to add application deployment.

Note

The rights to use HPE OneView are **not** granted by the CloudSystem Enterprise software licenses. The CloudSystem software and HPE OneView are often sold and delivered together. However, HPE OneView and CloudSystem are distinct software products and are licensed independently under their respective license agreements.

OneView provisioning management in CSA—Activity

Watch the seven-minute demonstration showing how to enable your customers to order SaaS on physical servers from the CSA Marketplace Portal.

Note

To access this demonstration, scan this QR code or enter the URL into your browser.

https://vrp.glb.itcs.hpe.com/SDP/Content/ContentDetails.aspx?ID=4645

After watching the demonstration, answer these questions:

1. What does CSA use as its process engine to execute the workflows?

   _____

2. In HPE OneView, what allows IT experts to rapidly define or customize all server configuration settings (firmware, BIOS, storage, and networking) in one place?

   _____

3. What does CSA use as a template to deploy new physical servers?

   _____

4. What label should be used in HPE OneView to make servers available to CSA for deployment?

   _____

5. What must you do before creating and publishing the service design in CSA?

   _____

OneView provisioning management in CSA activity—answers

1. What does CSA use as its process engine to execute the workflows?

   - **OO**

2. In HPE OneView, what allows IT experts to rapidly define or customize all server configuration settings (firmware, BIOS, storage, and networking) in one place?

   - **Server profiles**

3. What does CSA use as a template to deploy new physical servers?

   - **Unassigned server profiles**

4. What label should be used in HPE OneView to make servers available to CSA for deployment?

   - **The hpcsaUNASSIGNED label**

5. What must you do before creating and publishing the service design in CSA?

- **Import OneView server profiles, ICsp OS build plans, and Chef cookbooks**

Integration with CloudSystem Matrix

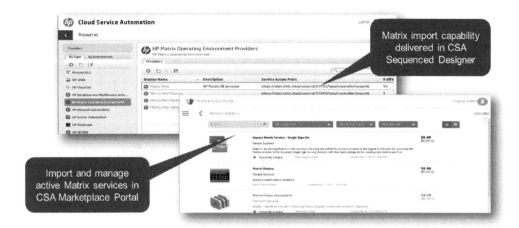

Figure 10-24 Integration with CloudSystem Matrix

CloudSystem Matrix is delivered as part of CloudSystem Enterprise. If a customer already has Matrix templates created when they upgrade to CloudSystem Enterprise, they can have users request Matrix services through the Marketplace Portal as shown in Figure 10-24. With this capability, users only need to access one self-service portal to consume cloud services along with other services provided by Helion OpenStack or other providers.

Demonstration

Watch this two-minute demonstration explaining how to create a service offering in CSA using the Matrix Operating Environment templates and then subscribe to that service from the CSA self-service portal.

 Note

To access this demonstration, scan this QR code or enter the URL into your browser.

https://vrp.glb.itcs.hpe.com/SDP/Content/ContentDetails.aspx?ID=3449

Using CloudSystem Enterprise

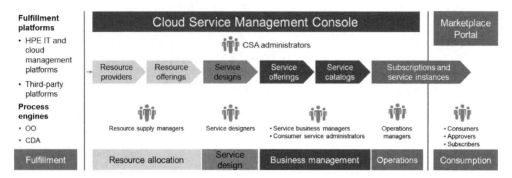

Figure 10-25 Flow of service provisioning in CSA

CSA is a unique platform that orchestrates the deployment of compute and infrastructure resources and complex multitier application architectures. It integrates and leverages the strengths of several HPE data center management and automation products, adding resource management, service offering design, and a customer portal to create a comprehensive service automation solution. The flow of service provisioning in CSA is illustrated in Figure 10-25.

The CSA subscription, service design, and resource utilization capabilities address three key challenges:

● The CSA Marketplace Portal provides a customer interface for requesting new cloud services and for monitoring and managing existing services with subscription pricing to meet business requirements.

● The CSA graphical service design and content portability tools simplify developing, leveraging, and sharing an array of service offerings that can be tailored to customer needs.

● The CSA lifecycle framework and resource utilization features ease the complexity of mapping the cloud fulfillment infrastructure into reusable, automated resource offerings for on-time and on-budget delivery.

CSA addresses these challenges from a task- and role-based perspective. Customers request new cloud services and manage their existing services with the CSA Marketplace Portal. Tasks that can be performed include:

● Request subscriptions for services offered to user groups within CSA.

● View status of pending requests, pending approvals, and approved subscriptions, including detailed component information.

● Request service modifications and other actions for cloud services.

● Cancel subscriptions.

Marketplace Portal users include these groups:

- **Consumers**—Use cloud services.

- **Subscribers**—Request cloud services for themselves or on behalf of others.

- **Approvers**—Grant requests from others for services when an approval is required.

The high-level steps for provisioning through CSA are:

1. Accessing the CSA console (Cloud Service Management Console).

2. Creating, configuring, and publishing service designs using the Topology Designer or the Sequenced Designer.

3. Creating and publishing service offerings to the Marketplace Portal.

4. Deploying services from published service offerings.

5. Managing subscriptions.

CSA demonstration

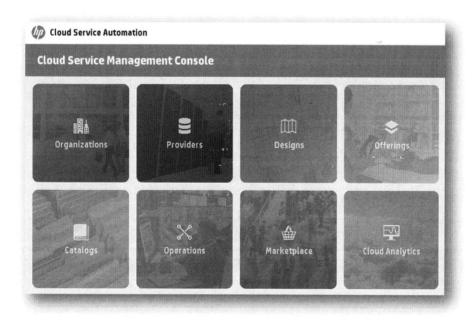

Figure 10-26 Cloud Service Management Console dashboard

Watch this 32-minute demonstration of these tasks in CloudSystem Enterprise:

1. Accessing the Cloud Service Management Console

2. Creating, configuring, and publishing service designs

3. Creating and publishing service offerings to the Marketplace Portal

4. Deploying services from published service offerings

5. Managing subscriptions

 Note

To view this demonstration, scan this QR code or enter the URL into your browser.

https://youtu.be/79-70tJKB54

Write your notes and observations here:

_____

_____

_____

_____

_____

_____

_____

Step 1: Accessing the Cloud Service Management Console

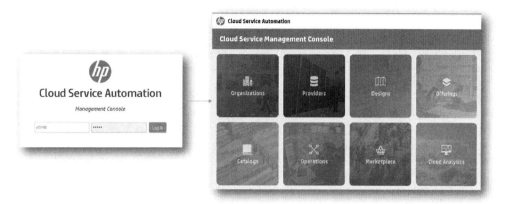

Figure 10-27 Accessing the Cloud Service Management Console

You can access the Cloud Service Management Console in two ways:

- Navigate to: https://[Enterprise_CAN_VIP address]:8444/csa

- Use the Menu launcher in Operations Console, navigate to **CloudSystem Enterprise**, and click **Start CloudSystem Enterprise Admin Console**

Then, log in with the default username of *admin* and the default password of *cloud*.

The console provides several tiles for managing different aspects of the product. To provide a brief recap, these tiles include:

- **Organizations**—Manage tenant organizations in a multitenancy configuration. An organization represents a department, a business unit, or a customer.

- **Providers**—Manage resource providers utilized by CSA to provision application and infrastructure services published in the service catalog. Helion CloudSystem is configured as the default resource provider upon installation of CloudSystem Enterprise. Additional resource providers for previous generations of CloudSystem, VMware, AWS, Microsoft Azure, and many more can be defined in the Providers tile. CloudSystem Enterprise with CSA is an enabler for hybrid cloud services. It provides a single platform to manage and broker cloud usage and services across multiple deployment models (private, public, and managed clouds).

- **Designs**—Provide access to Sequenced Designer and Topology Designer, which are two utilities used by cloud architects to create application and infrastructure service design templates (also known as blueprints). Topology Designer allows users to create service templates for cloud providers based on OpenStack, such as Helion CloudSystem. Sequenced Designer is used to create service templates for all other resource providers. Additional configuration steps are required before cloud architect can use the Sequenced Designer.

- **Offerings**—Allow users to work with a service design template to create one or more service offerings that the cloud architect can publish in the service catalog. Cloud users can then subscribe to those service offerings.

- **Catalog**—Enables management of the global service catalog (shared among all organizations) and service catalogs specific to organizations.

- **Operations**—Provide access to monitoring of in-flight operations such as provisioning of services.

- **Marketplace Portal**—Provides access to the default consumer Marketplace Portal to browse the service catalog and subscribe to a published service offering.

- **Cloud Analytics**—Enable direct access to the HPE IT Business Analytics reports and the dashboard for cloud architect and business managers. This provides insight into how the cloud-based service offerings are consumed.

- **Orchestration**—Provides direct access to OO Central. OO orchestrates deployment and lifecycle actions of infrastructure and application services submitted by cloud users.

Step 2.1: Creating service designs

The goal of CSA is to provide fast, easy, and reliable provisioning of cloud services to customers. Subscribers use the Marketplace Portal to browse through the available offerings, select available service options, and place orders.

The "recipes" for creating service offerings are defined using CSA service designs. Service designs describe subscription options, provider resources, and processes that realize the requests. Service designs encapsulate manual IT processes to deploy, modify, and retire a variety of services in an automatic and repeatable manner.

Topology model and Topology Designer

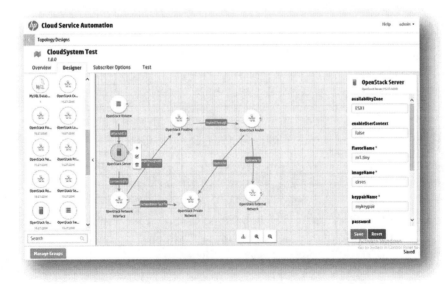

Figure 10-28 Topology Designer

The OO Topology Designer, shown in Figure 10-28, enables rapid development of cloud services based on Cloud OS. A Topology Developer (architect) creates services by specifying high-level components, options, and relationships. The implementation logic is automatically provided by CSA.

Before a Topology design can be created, CSA queries a Cloud OS provider, and the design references the resources, properties, and relationships of that particular instance. In turn, Cloud OS interacts with one or more OpenStack instances.

Because Cloud OS is constantly adding new functionality, each provider may furnish a version-specific description of objects and capabilities.

Sequenced model and Sequenced Designer

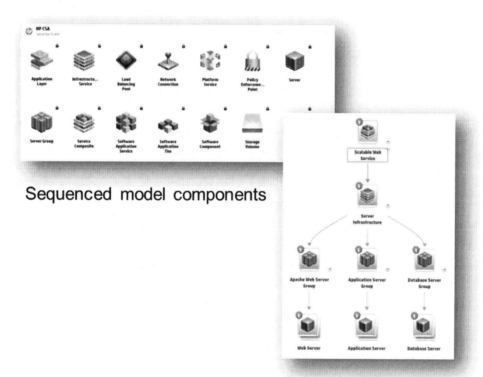

Sequenced model components

Sequenced component design

Figure 10-29 Sequenced model and Sequenced Designer

Sequenced Designer, shown in Figure 10-29, is used to create complex application service designs. The Sequenced model is used when developers want total control of the components and actions used in the lifecycle of a service. This model works well when automating and aligning an IT process with a customer's business process. HPE provides a broad range of sample content for the Sequenced model.

The Sequenced model allows designers to define a hierarchy of components that describe the elements of a service and the steps for deploying, modifying, and retiring that service. The hierarchy describes components, such as servers, network elements, and applications. Designers can use predefined components from the CSA palette, or they can modify or create their own.

 Note

The Topology Designer does not support the Matrix Operating Environment. The Sequenced Designer does support the Matrix Operating Environment. In order to create service designs using the Matrix Operating Environment, you have to use Sequenced Designer.

Step 2.2: Configuring service designs

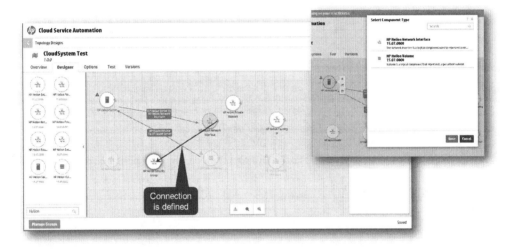

Figure 10-30 Configuring service designs

In the Cloud Service Management Console, users can import, create, and configure components to add to the service design. To do so, use the **Designer** tab, drag elements from the left component panel, and drop them into the design area (the canvas) as shown in Figure 10-30. Hold the mouse over the warning icons to see what connections need to be made. Drag the connection handle from one component to another. Click a component in the design to display its pop-up property sheet. Automatic attribute verification displays an alert if any required properties are not set for the components.

For additional information on creating and configuring service designs, use the *Cloud Service Management Console Help* guide and look for the "Topology Designs" and "Sequenced Designs" chapters, under the "Components" section.

 Note

To access the help guide, scan this QR code or enter the URL into your browser.

http://h10032.www1.hp.com/ctg/Manual/c04220319

Step 2.3: Publishing service designs

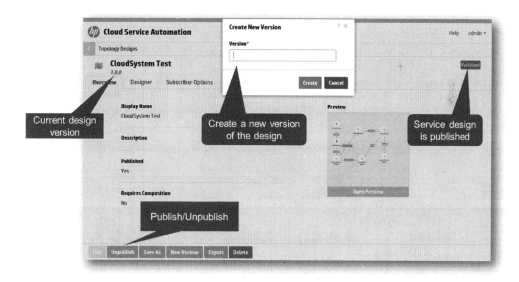

Figure 10-31 Publishing service designs

When cloud architects complete a service design, they can publish it to make it available for use as shown in Figure 10-31. Then, cloud architects can create one or more service offerings based on that service design. This makes the service design available for inclusion in an offering.

Step 3: Creating and publishing service offerings to Marketplace Portal

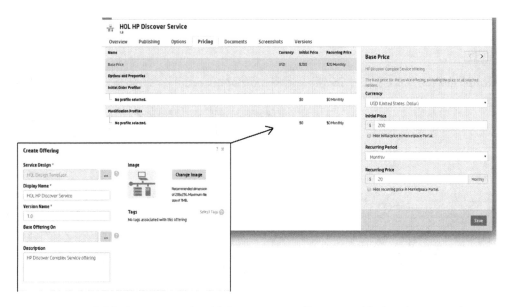

Figure 10-32 Creating and publishing service offerings to Marketplace Portal

Service offerings are listings in the Marketplace Portal that may contain custom features. These features can be viewed and edited in the subscription, and pricing is based on published designs.

Figure 10-32 shows the creation of a service called HOL HP Discover Service, and the Pricing screen on the right side is where the price of the service can be set. After creating the service offering, you must publish it in the service catalog for cloud users to subscribe to and deploy.

For additional information on creating and publishing service offerings, use the *Cloud Service Management Console Help* guide and look for the "Create Offerings" and "Publishing Tab" chapters.

Step 4.1: Deploying services from published service offerings—Exploring the Marketplace Portal

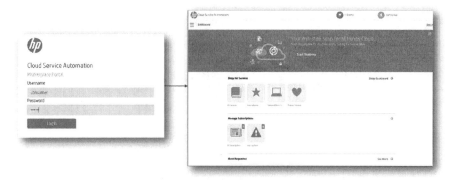

Figure 10-33 Deploying services

In order to deploy services from published service offerings, follow these steps (Figure 10-33):

- From the **Menu** launcher in Operations Console, navigate to **CloudSystem Enterprise** and click **Start CloudSystem Enterprise Marketplace Portal.** Alternatively, you can use this URL: https://[Enterprise CAN VIP consumer cloud address]:8089/org/CSA_CONSUMER

- Then, log in with the default username of *consumer* and the default password of *cloud.*

- From the Dashboard, users can check requests, subscriptions, and existing services. Users can also browse the catalog to see the available services and subscribe to them.

 Important

After a service is deployed from the Marketplace Portal, it must be managed through the CSA console. Do not modify the service outside of the CSA console. If you modify the service using native APIs, for example, the CSA console users will not see the change. The CSA console cannot register modifications to services that happen outside of the console.

Step 4.2: Deploying services from published service offerings—Ordering a service from the catalog

Figure 10-34 Ordering a service

A cloud user (consumer) can order a service from the catalog by following the steps shown in Figure 10-34:

- From the Sidebar menu, use the **Browse Catalog** option to see published service offerings.

- Select the service offering of interest.

- Specify options such as the initial number of instances to deploy.

- Click **Checkout**, provide a name for the service, and **Submit** request. The cloud user will receive a notification that your request was submitted.

This set of actions triggers a service offering request that is sent to the cloud administrator. A cloud administrator can perform the following tasks:

- Use the Cloud Service Management Console (the Operations tile) to see the incoming request.

- Use the OpenStack user portal (Horizon) to check the underlying OpenStack platform and determine which VMs were created.

- Use the OO console to monitor the OO flows that run in the background to deploy or undeploy the services.

Step 5: Managing subscriptions

Figure 10-35 Managing subscriptions

A cloud user (consumer) can use the Marketplace Portal to manage subscriptions by following the steps shown in Figure 10-35:

- In the Marketplace Portal, use the Sidebar menu to select **My Services**.

- Then, to manage subscriptions (for example, scaling out, canceling, or deleting the service), use the **Manage Subscription** option.

Learning check

1. What two HPE solutions provide the foundation for CloudSystem Enterprise?

   _____

   _____

   _____

   _____

2. Fill in the missing CloudSystem Enterprise appliances:

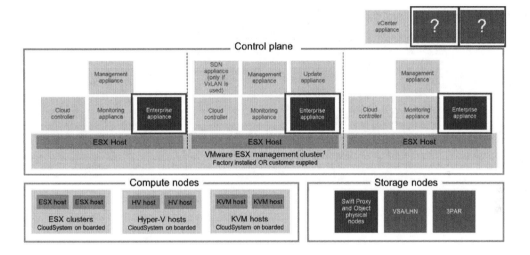

3. Which console is used to perform administrative functions in CSA?

   a. Operations Console

   b. OpenStack Horizon console

   c. Cloud Service Management Console

   d. Service design console

4. What are the default CloudSystem Enterprise credentials?

Cloud Service Management Console	Marketplace Portal
Username:	Username:
Password:	Password:

5. If a customer already has Matrix templates created when they upgrade to CloudSystem Enterprise, they can have their users request Matrix services through the Marketplace Portal.

☐ True

☐ False

6. List the five main tasks that can be performed using CloudSystem Enterprise.

_____

_____

_____

_____

7. Only the Sequenced Designer supports the Matrix Operating Environment.

☐ True

☐ False

8. Which Cloud Service Management Console menu option (tile) provides access to the Topology Designer and the Sequenced Designer?

a. Orchestration

b. Providers

c. Designs

d. Offerings

e. Operations

Learning check answers

1. What two HPE solutions provide the foundation for CloudSystem Enterprise?

- **CSA**

- **OO**

2. Fill in the missing CloudSystem Enterprise appliances:

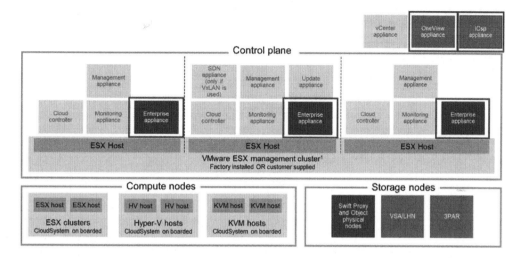

3. Which console is used to perform administrative functions in CSA?

    a. Operations Console

    b. OpenStack Horizon console

    c. **Cloud Service Management Console**

    d. Service design console

4. What are the default CloudSystem Enterprise credentials?

Cloud Service Management Console	Marketplace Portal
Username: admin	Username: consumer
Password: cloud	Password: cloud

5. If a customer already has Matrix templates created when they upgrade to CloudSystem Enterprise, they can have their users request Matrix services through the Marketplace Portal.

    ☐ **True**

    ☐ False

6. List the five main tasks that can be performed using CloudSystem Enterprise.

    • **Accessing the Cloud Service Management Console**

    • **Creating, configuring, and publishing service designs**

    • **Creating and publishing service offerings to the Marketplace Portal**

- **Deploying services from published service offerings**
- **Managing subscriptions**

7. Only the Sequenced Designer supports the Matrix Operating Environment.

   ☐ **True**

   ☐ False

8. Which Cloud Service Management Console menu option (tile) provides access to the Topology Designer and the Sequenced Designer?

   a. Orchestration

   b. Providers

   c. Designs

   d. Offerings

   e. Operations

Resources

Helion CloudSystem documentation can be found on the HP Helion Documentation website as shown in Figure 10-36.

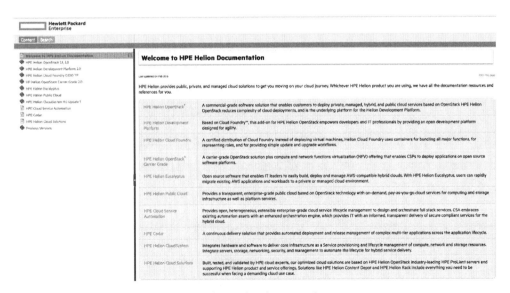

Figure 10-36 Helion CloudSystem documentation

Note

To access Helion CloudSystem documentation, scan this QR code or enter the URL into your browser.

http://docs.hpcloud.com/

You can also access CSA and OO resources on the HPE Live Network by scanning these QR codes or entering the URLs into your browser.

CSA resources

https://hpln.hp.com/group/cloud-service-automation

OO resources

https://hpln.hp.com/group/operations-orchestration

Summary

- Helion CloudSystem Enterprise builds upon CloudSystem Foundation. It includes service marketplace and designers based on CSA and OO. It features:

    - Virtual appliances

    - Multicloud capabilities

    - Multiple application provisioning services

- Physical server provisioning

- Portable service design

- During the first-time installation, CloudSystem Enterprise can installed. CloudSystem Enterprise is a separate virtual appliance that runs CSA. However, all appliance management tasks are performed through the CloudSystem Operations Console and all OpenStack functionality included in CloudSystem is available to Enterprise.
  In CloudSystem Enterprise, users can:

  - Provision physical servers using OneView profiles

  - Use the CSA Topology Designer to lay out a design

  - Add Insight Control server provisioning to deploy the required operating system

  - Use existing tools such as Chef or Puppet to add application deployment

- CSA is a unique platform that orchestrates the deployment of compute and infrastructure resources and complex multitier application architectures. It integrates and leverages the strengths of several HPE data center management and automation products, adding resource management, service offering design, and a customer portal to create a comprehensive service automation solution.

11 Customer Engagement Tools

WHAT IS IN THIS CHAPTER FOR YOU?

After completing this chapter, you should be able to:

✓ Discuss what buyers consider when making IT purchasing decisions.

✓ Describe Hewlett Packard Enterprise (HPE) maturity models and assessment tools, including the Converged Infrastructure Capability Model.

✓ Outline HPE tools that can assist in your customer engagements

Before proceeding with this section, assess your existing experience with customer engagement by answering the following questions.

Assessment activity

1. Are you comfortable having the right conversation with your customers? Do you know what matters the most to them?

   _____

   _____

   _____

2. Have you had an experience where you or someone you know did not connect with the customer or did not make the sale even with a superior solution? Do you know why?

   _____

   _____

   _____

   _____

3. How do you determine today where your customers are on their journey to hybrid infrastructure?

_____

_____

_____

_____

Customer buyer roles

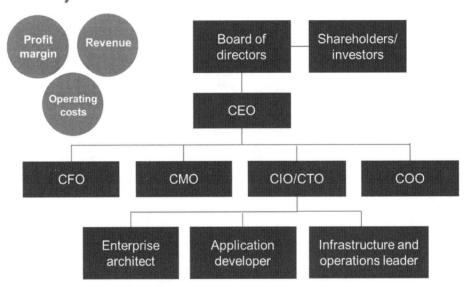

Figure 11-1 Customer buyer roles

The majority of today's business leaders recognize the role technology plays in their success. As illustrated by Figure 11-1, an increasing number of executives are directly involved in IT purchasing decisions. The shift from information technology to business technology and the associated complexity of how companies acquire and consume technology present both an opportunity and a challenge.

As they adopt buyer roles, executives choose which vendors to invest time and resources in and which ones to bypass. According to the Forrester Research report _Executive Buyer Insight Study: Defining the Gap Between Buyers and Sellers_, executives choose vendors who:

- Understand the company and its nuances
- Focus on driving business results

- Exhibit the attributes of a strategic supplier that is aligned with the company's goals and collaborates to meet those goals

- Offer a clear path to solving business challenges

 Note

For more information about executive roles scan this QR code or enter the URL into your browser.

https://en.wikipedia.org/wiki/List_of_corporate_titles

Maturity models and assessment services

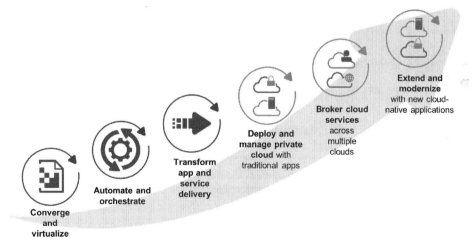

Figure 11-2 The journey to a hybrid infrastructure

Customers who want to embark on the journey to a hybrid infrastructure and on-demand IT, illustrated in Figure 11-2, have different starting points, destinations, budgets, and time frames. HPE and various channel partners offer a range of tools and services to help customers assess their journey. These tools and services evaluate the customer's IT infrastructure, business processes, people readiness, and so forth.

You might have experience with assessment tools, services, and methodologies from the channel partners. Or you might be familiar with assessment tools from industry analysts and advisors, such as Gartner. Gartner's Five-Stage Demand-Driven Maturity Model provides a methodology for assessing a customer's IT infrastructure, including advice for these five stages: react, anticipate, integrate, collaborate, and orchestrate.

Gartner also developed another tool, the ITScore, which evaluates IT organizations in terms of how IT services are provided and how the enterprise is affected.

 Note

For more information about ITScore, scan this QR code or enter the URL into your browser.

http://www.gartner.com/technology/research/methodologies/it-score.jsp

---

The HPE DevOps maturity model is another assessment tool option, delivered as part of the HPE Business Service Management approach. This model describes five levels of maturity (initial, managed, defined, measured, and optimized). For each level, three dimensions are examined—process maturity, automation maturity, and collaboration maturity.

 Note

For more information about this maturity model, scan this QR code or enter the URL into your browser:

http://community.hpe.com/t5/Business-Service-Management/DevOps-and-OpsDev-How-Maturity-Model-Works/ba-p/6042901#.VlQ22_7ruUk

---

Converged Infrastructure Capability Model

The HPE Converged Infrastructure Capability Model (CI-CM) is delivered by Enterprise Group presales architects or Technology Services Consulting resources. It provides a high-level assessment of the current state of a customer's IT infrastructure, including suggested next steps and recommended projects to help drive toward a converged data center. It also helps define an action-oriented, customer-specific road map. A CI-CM evaluation also compares a customer's IT environment to others in their industry and region. CI-CM should be used when customers are interested in moving toward a converged infrastructure and want an external assessment of where they are in the journey.

The next-generation data center moves away from traditional silos and utilizes converged infrastructure across servers, storage, and networking. Leading analysts agree that this new data center improves IT across three characteristics:

- Operational efficiency

- Quality of service (QoS)

- Speed of IT project implementation

Assessing the journey forward

A successful strategy for developing a next-generation data center requires both tactical and strategic dimensions. Tactically, companies need to break down the IT silos by standardizing, modularizing, and virtualizing their IT environment. They should also focus on automating error-prone manual processes, along with simplifying and tightening management control. Strategically, they need to change how IT interacts with the business and realign people, processes, and technology to the new vision.

First, CI-CM is used to establish the current state of the IT infrastructure for each business. Then, the model is used to define the target state for each metric.

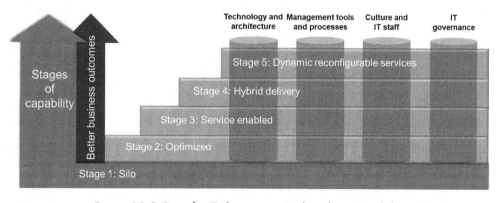

Figure 11-3 Specific IT domain areas that drive capability

CI-CM domains

CI-CM measures a company's maturity in four domains as shown in Figure 11-3:

- **Technology and architecture**—This domain focuses on hardware as well as data center facilities, including how the company's infrastructure is utilized and how well IT services can withstand failures. The goal here is for hardware, software, and network resources to be shared, virtualized, and progressively automated. This means moving from dedicated, high-cost IT silos to low-cost, highly available, pooled IT assets.

- **Management tools and processes**—This domain focuses on deploying resources to plan, manage, and deliver IT services more efficiently. The aim is to move toward more standardized, integrated processes using tools that align IT more closely with business goals.

- **Culture and IT staff**—Moving toward a business-centric IT environment calls for the IT organization structure, roles, and functions to be redefined. This area, though challenging, is crucial to achieve and sustain a more adaptive IT environment.

- **Demand, supply, and IT governance**—This involves outlining the organizational structures and processes in order to better align the enterprise IT environment with overall business strategies and objectives. Here, roles in the business strategy development process are defined, and the mechanisms for executing the business strategy with adherence to the IT strategy are established. In essence, this aligns IT supply with business demands.

CI-CM stages

The infrastructure's current state and targeted state are rated against five stages of capability in each of the domains:

- **Stage 1: Silo**—This domain is primarily a dedicated, project-based technology. A business unit specifies an application, which specifies the infrastructure. Then, the application and infrastructure are moved into the data center, and IT managers run the operations. In this stage of evolution, the technology is built around the project itself. This, in turn, limits the sharing assets and the consistency of infrastructure, operating systems, and processes.

  The results of this type of project-based architecture is a one-application-to-one-server environment—in essence, a compartmentalized approach. This approach tends to lead to old legacy systems that are not upgraded because systems "belong" to the business unit and after depreciation, they are perceived to be "free." As a result, data centers can slowly collect old equipment and applications that are high maintenance, often with a very inefficient cooling architecture.

- **Stage 2: Optimized**—This stage consists of rigid standardization of all IT architecture elements. It also requires standardization of management tools and processes. From an operational governance perspective, there is a significant shift in management within a standardized IT environment. For example, in Stage 1, a business unit decides what a new system does and how it does it. In Stage 2, a business unit stills decides what a new system does, but IT decides how it will be done—application, infrastructure, and so on. This approach enables IT to have

  - Consistency of technology and operating system images

  - Standardized applications and databases

  - Standardized operational practices often independent of technology type

- Reduced maintenance labor

- Higher reliability

- **Stage 3: Service enabled**—In this stage, IT moves to a consolidated, virtualized, and shared infrastructure. All aspects of IT are rationalized including architectures, processes, and tools. All technology aspects are managed by IT and are not under business unit control. This enables IT to share the assets across applications without business unit intervention. In this stage, there tends to be pervasive use of virtualization; a standards-based, shared infrastructure; disaster tolerance; and consolidated, environmentally efficient data centers. The objectives include increased asset utilization, greater ability to handle periods of peak business demand, faster deployment of new applications, improved service levels, maximized reliability, and operational efficiency.

- **Stage 4: Hybrid delivery**—Infrastructure and other enterprise architecture artifacts are offered as a service, on demand with tiered service levels. These offerings are supported by service-centric, integrated IT processes. The service levels are not measured on the technology, but on the service as a whole. This means that each layer of the IT environment is offered as a service, starting with the infrastructure services that provide server and storage resources. Tangible benefits include faster time to market for practically any IT service.

- **Stage 5: Dynamic reconfigurable services**—IT provides all services from a pooled, shared, automated source of supply, which can be inside or outside the enterprise IT environment (for instance, cloud-based). At this stage, automated management is enabled and infrastructure is reallocated based on business process needs. Manual labor is eliminated from most provisioning and resource utilization improves. The IT infrastructure is virtualized, resilient, orchestrated, optimized, modular, and fully converged.

Current and future states versus industry average

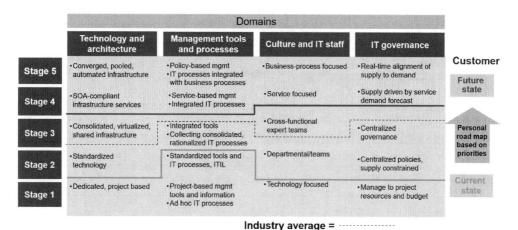

Figure 11-4 Current and future states versus industry average

After a company's IT capability has been assessed, the next step is to identify which metrics are most valuable. Determining which metrics to use when measuring progress during a specific time frame helps create a road map toward a converged infrastructure. An example CI-CM is shown in Figure 11-4 and illustrates the current state, desired future state, and industry average for a particular business.

It is important to underline that there is no "right" answer with respect to a targeted state of capability. It is not simply a technological discussion, either. The model provides a framework for a guided discussion about a consistent approach. Then, a plan to accomplish that approach can be developed.

Each business is unique, and each approach to IT is different. Each stage of capability and the IT organization should be aligned with the overall business strategy. These business objectives ultimately guide the final decision regarding IT capability.

Which stage describes most customers today?

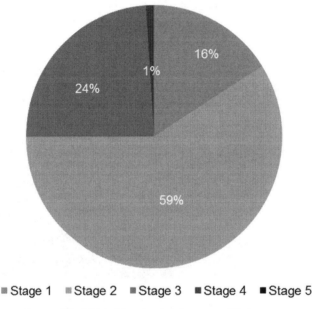

■ Stage 1 ■ Stage 2 ■ Stage 3 ■ Stage 4 ■ Stage 5

Figure 11-5 Worldwide distribution—All domains

Figure 11-5 shows the distribution of the five stages of capability based on the CI-CM database. The CI-CM database contains more than 1500 customer profiles across various industries and geographies. On average, more than 50% of customer profiles in the database fall into Stage 2. The number of organizations that meet the classifications at the higher maturity levels decreases significantly. Very few organizations meet the requirements for Stage 5.

Interestingly, there are not many discrepancies between the stages of maturity across the domains. This corroborates the need for organizations to consider all domains along the maturity curve.

The key takeaway is that many organizations have yet to move forward on their journey to hybrid infrastructure and on-demand IT. This fact presents many opportunities for HPE and its channel partners.

 Note

> This is not an HPE proprietary model. The data points are based on industry best practices and standards. All the data points have been gathered by third-party market research firms.

HPE business value tools

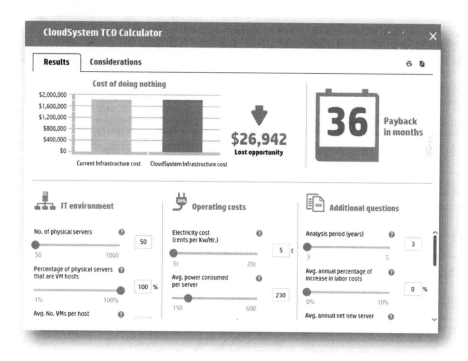

Figure 11-6 HPE business value tools

After determining the stage of a customer's IT, you can begin to articulate the financial benefit of an HPE solution. HPE has worked extensively with partners to develop total cost of ownership (TCO) and return on investment (ROI) calculators to provide useful information for C-level executives and other decision makers. The CloudSystem TCO Calculator is shown in Figure 11-6.

Note

To access these calculators, scan this QR code or enter the URL into your browser.

http://www8.hp.com/us/en/business-solutions/tco-calculators.html

What is the difference between TCO and ROI?

TCO measures the economics of IT assets over their useful service life. Many decision-makers fail to look beyond the up-front investment. Before customers decide on a new IT platform aligned with their business strategy, they need to assess all direct and indirect costs involved over the entire life cycle.

The TCO program at HPE supports organizations in identifying potential savings, including shifts from capital expenditures to operational expenditures, that help free up budget for IT innovation aligned with business objectives.

ROI is a numerical representation (expressed as a percentage) of the earning power of a company's assets. ROI is calculated as the ratio of a company's net income to its average equity. ROI can be calculated for the entire company or a specific project. This ratio can be compared to the company's cost of capital to determine if a company or a project is financially viable.

 Note

For more information on how to reduce TCO, scan this QR code or enter the URL into your browser.

http://www8.hp.com/us/en/business-solutions/tco-reduce.html

Note

For more information on the HPE Storage Quick ROI calculator, as an example, scan this QR code or enter the URL into your browser.

https://roianalyst.alinean.com/ent_02/AutoLogin.do?d=
11840239389855150

Learning check

1. When executives assume a buyer role, what characteristics do they look for in a vendor?

   _____

   _____

   _____

   _____

2. After a company rationalizes all aspects of IT, including architectures, processes, and tools, which CI-CM stage has the organization achieved?

   a. Stage 1

   b. Stage 2

   c. Stage 3

   d. Stage 4

   e. Stage 5

Learning check answers

1. When executives assume a buyer role, what characteristics do they look for in a vendor?

   Executives look for vendors who:

   – **Understand the company's pressures, challenges, responsibilities, and requirements**

   – **Exhibit attributes of a strategic partner aligned with the company's goals**

   – **Offer clear solutions to achieve business outcomes**

2. After a company rationalizes all aspects of IT, including architectures, processes, and tools, which CI-CM stage has the organization achieved?

    a. Stage 1

    b. Stage 2

    c. Stage 3

    d. Stage 4

    e. Stage 5

HPE tools

HPE has an array of tools that can help you in your engagement with customers.

HPE selectors and configurators

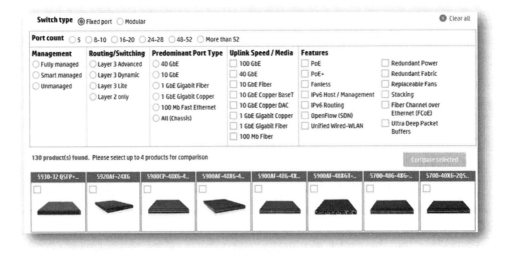

Figure 11-7 HPE selectors and configurators

Similar to sizers, HPE selectors and configurators are designed to help you select the correct product for each customer and configure it appropriately. These selectors and configurators, one of which is shown in Figure 11-7, are accessible online.

Sales Builder for Windows

Sales Builder for Windows (SBW) is the premier HPE configuration and quotation tool for the HPE sales force and channel partners. SBW is a downloadable tool designed to support the complete HPE product portfolio of servers and related service and support products.

The SBW Configurator displays the system diagram and modifications for HPE clusters, servers, and storage running HPE-UX, Windows, Linux, or mixed environments. You can use this tool to configure and customize technical solutions for new systems, upgrades, and add-ons. Components of the Configurator include:

- **Whiteboard**—The center of SBW that shows technical solutions and forms the unit of quoting and storing.

- **Configurator worksheet**—Main configuration tool that enables you to configure complete technical solutions.

- **System diagram**—A graphical view of your configuration that can be used for modifications.

Other components of SBW include:

- **Quoter**—Prepares a budgetary quotation for the customer that shows numbers, descriptions, and prices of all the products in the solution.

- **Price book**—Data files containing the latest product descriptions and prices that are updated every two weeks.

- **Knowledge base**—Data file containing the rules and product modeling used by SBW to check configurations.

SBW exports schematic diagrams in Microsoft Visio format. SBW is accessible via the HPE Unison Partner Portal.

Additional HPE selectors and configurators includes:

- **Networking Switch Selector**—This web-based tool helps you select the correct HPE networking product based on specific requirements, such as switch type, port count, management capabilities, routing and switching capabilities, predominant port type, and others.

 Note

To access the Networking Switch Selector tool, scan this QR code or enter the URL into your browser:

http://h17007.www1.hpe.com/us/en/networking/products/switches/ switch-selector.aspx#.VyNPpKQrKUk

---

- **Networking Online Configurator**—This configurator enables you to create price quotations for HPE networking products quickly and easily using your web browser. You can save quotation files to your hard drive or export them in several formats including Microsoft Excel.

 Note

To access the Networking Online Configurator, scan this QR code or enter the URL into your browser.

http://h17007.www1.hp.com/th/en/products/configurator/index.aspx

---

- **One Config Simple (OCS)**—This guided, self-service tool replaces several older configuration tools. OCS helps you provide customers with initial configurations in three to five minutes. You can then send the configuration on for special pricing or configuration help, or use it for your existing ordering processes.

 Note

To access the OCS, scan this QR code or enter the URL into your browser:

http://h22174.www2.hp.com/SimplifiedConfig/Index

Solution Demonstration Portal

Figure 11-8 Solution Demonstration Portal

The Solution Demonstration Portal (SDP) shown in Figure 11-8, formerly known as the *Virtual Resource Portal*, provides a central location for all demonstrations, webinars, and supporting collateral that showcase how HPE technologies lead, innovate, and transform enterprise businesses. Live and prerecorded demonstrations feature HPE hardware, software, services, and partnerships in an exciting multimedia format, illustrating how HPE can help solve business and IT problems.

 Note

To access the SDP, scan this QR code or enter the URL into your browser.

https://vrp.glb.itcs.hpe.com/SDP/default.aspx

More HPE tools and resources

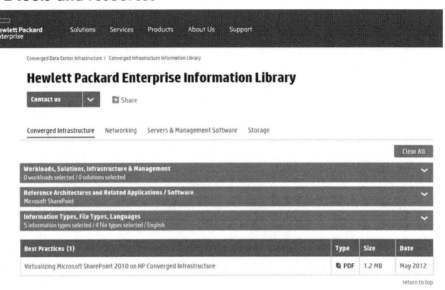

Figure 11-9 HPE Information Library

The Enterprise Information Library, shown in Figure 11-9, is a search-based tool used to access user guides, technical documentation, and other collateral for a variety of products and solutions.

 Note

To access the Enterprise Information Library, scan this QR code or enter the URL into your browser.

http://www.hp.com/go/cloudsystem/docs

The Single Point of Connectivity Knowledge (SPOCK) portal provides detailed information about supported HPE storage product configurations.

You need an HPE Passport account to enter the SPOCK website.

 Note

To access SPOCK, scan this QR code or enter the URL into your browser:

https://h20272.www2.hp.com/spock/

HPE Live Network is a value-added service available to customers who purchase products enabled through Live Network. Through Live Network, customers can access the most current content add-ons and extensions for their HPE IT Performance Suite products.

 Note

To access the Live Network, scan this QR code or enter the URL into your browser:

https://hpln.hp.com/

Learning check

1. Sales Builder for Windows can only be used to configure and customize technical solutions for upgrades and add-ons.

   ☐ True

   ☐ False

2. The Simplified Configuration Environment uses a guided solution wizard to build a solution based on the applications needed.

   ☐ True

   ☐ False

Learning check answers

1. Sales Builder for Windows can only be used to configure and customize technical solutions for upgrades and add-ons.

   ☐ True

   ☐ **False**

2. The Simplified Configuration Environment uses a guided solution wizard to build a solution based on the applications needed.

   ☐ **True**

   ☐ False

Summary

An increasing number of executives are directly involved in IT purchasing decisions, and it is important to understand how each customer acquires and consumes IT—including pressures, challenges, responsibilities, and requirements.

Business executives choose IT vendors who:

- Understand the company and its nuances,

- Focus on driving business results.

- Exhibit the attributes of a strategic supplier that is aligned with the company's goals and collaborates to meet those goals.

- Offer a clear path to solving business challenges.

HPE offers a wide range of maturity models, tools, and assessment services to guide customers along their journey to a hybrid infrastructure and on-demand IT.

CI-CM measures a company's maturity in four domains (technology and architecture, management tools and processes, culture and IT staff, and IT governance) across five stages of capability (silo, optimized, service enabled, hybrid delivery, and dynamic reconfigurable services).

12 Competitive Strengths of Converged Infrastructure and Cloud Solutions

WHAT IS IN THIS CHAPTER FOR YOU?

After completing this chapter, you should be able to:

- ✓ Review the factors that are causing enterprises to move toward a hybrid infrastructure and on-demand IT

- ✓ Discuss how HPE Helion OpenStack solutions fit into the competitive cloud landscape for private and hybrid clouds

- ✓ Summarize the strengths of Helion solutions

Review: Industry shifts and transformation to a hybrid infrastructure and on-demand IT

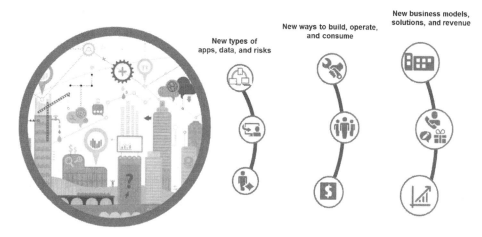

New types of apps, data, and risks

New ways to build, operate, and consume

New business models, solutions, and revenue

Figure 12-1 The idea economy

As illustrated by Figure 12-1, the idea economy presents an opportunity and a challenge for most enterprises. On the one hand, cloud, mobile, big data, and analytics give businesses the tools to accelerate time to value. This increased speed allows organizations to combine applications and data to create dramatically new experiences, even new markets.

On the other hand, most organizations were built with rigid IT infrastructures that are costly to maintain. This rigidity makes it difficult, if not impossible, to implement new ideas quickly.

Creating and delivering new business models, solutions, and experiences requires harnessing new types of applications, data, and risks. It also requires implementing new ways to build, operate, and consume technology. This new way of doing business no longer just supports the company—it becomes the core of the company.

Review: Faster application development enables accelerated innovation

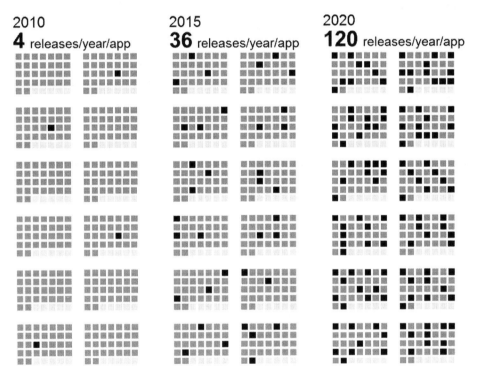

Figure 12-2 Average application release cycle

From 2010 to 2015, much changed from an application development perspective. In 2010, the average application release cycle was four releases per year, per application. In 2015, this number went up to 36 releases per year, per application. It is projected that by 2020, there will be 120 releases per year, per application (30 times more releases than in 2010) as shown in Figure 12-2.

Considerations from *Better Outcomes, Faster Results: Continuous Delivery and the Race for Better Business Performance*, the Forrester Thought Leader Paper commissioned by Hewlett Packard Enterprise (HPE), help summarize this trend:

- Agility is paramount.

- Even when delivering at cadences of less than a week, 20% of the organizations want to go even faster.

- Developers need flexibility.

- Organizations expect to deploy 50% to 70% of the code to cloud environments by 2015.

- Companies want open, flexible architectures for application portability and lock-in prevention.

Review: Long time to value is costly

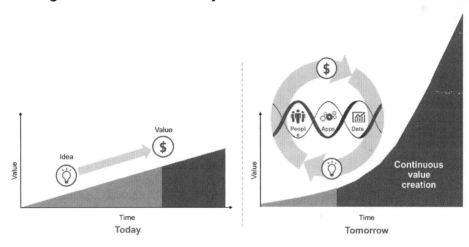

Figure 12-3 Time to value

As shown in Figure 12-3, success today is defined by how quickly an enterprise can turn ideas into value—how quickly a business can experiment, learn, test, tune, and make things better. Speed is a key differentiator in all industries.

Uber did not invent a new technology. Instead, the company took advantage of the explosion of smartphones and mobile applications to design a compelling customer experience, ultimately creating a new way of doing business.

This example is not only about Uber executing a good idea. It is also about the taxicab industry's inability to act quickly to transform its business models to compete. Examples such as Uber serve as a warning. Every Fortune 1000 company is at risk of missing a market opportunity, failing to secure its enterprise, and being disrupted by a new idea or business model.

Timelines for IT projects and new applications used to be planned over years and months. Today, these projects take weeks or days. Increasingly, it is shrinking to hours. Now is the time for a company to ask the following:

- How quickly can the company capitalize on a new idea?

- How rapidly can the company seize a new opportunity?

- How fast can the company respond to a new competitor that threatens the business?

The good news for established companies is that the same technologies making it easy for new companies to get started are also enabling enterprises to adapt quickly to changing business models and achieve faster time to value.

Review: Thriving in the idea economy requires a new style of business

Thriving in the idea economy requires enterprises to adopt a new style of business. This new style:

- **Is experience and outcome driven**—Rapidly compose new services from any source to meet the evolving needs of customers and citizens.

- **Proactively manages risks**—Remain safe and compliant in a world of rapidly changing threat landscape.

- **Is contextually aware and predictive**—Harness 100% of data to generate real-time instant insights for continuous improvement, innovation, and learning.

- **Is hyper-connected to customers, employees, and the ecosystem**—Deliver experiences that enable employees and engage customers in a persistent, personalized way.

In the idea economy, applications and information are the products.

Review: IT must become a value creator that bridges the old and the new

Figure 12-4 IT must shift focus

To respond to the disruptions created by the idea economy, IT must transform from a cost center to a value creator as shown in Figure 12-4. In order to evolve, IT must shift focus:

- From efficiently hosting workloads and services to continuously creating and delivering new services

- From simply providing hardened systems and networks to proactively managing and mitigating risks

- From just storing and managing data to providing real-time insight and understanding

- From using software to automate business systems to differentiating products and services

Customers need to make IT environments more efficient, productive, and secure as they transition to the new style of business. They need to enable their organizations to act rapidly on ideas by creating, consuming, and reconfiguring new solutions, experiences, and business models.

One of the first steps in achieving this kind of agility is to break down the old infrastructure silos that make enterprises resistant to new ideas internally and vulnerable to new ideas externally. Designing compelling new experiences and services does not work if the infrastructure cannot support them.

The right compute platform can make a significant impact on business outcomes and performance. Examples include storage that "thinks" as much as it stores; networking that moves information faster and more securely than ever before; and orchestration and management software that provides predictive capabilities.

Each company is on a unique journey to the cloud, custom-made for the way it consumes and allocates resources, transforms to the changing landscape, implements financial models, and achieves desired outcomes.

Review: Why hybrid infrastructure and on-demand IT?

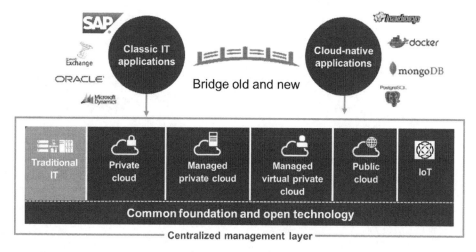

Figure 12-5 Bridge old and new

Most customers do not have the luxury of abandoning their existing investment and jumping straight into new technologies. They need to continue using their traditional IT and simultaneously move to new computing platforms as illustrated by Figure 12-5, bridging the old technologies and applications with new ones.

The public cloud provides convenience, and the private cloud gives control. A combination of public and private cloud can provide the right destination for the right applications at the right cost.

The key considerations for achieving a hybrid infrastructure and on-demand IT include the following:

- Security

- Availability

- Resilience

- Compliance

- Data sovereignty

- Performance

- Openness

- Cost

Review: The journey to hybrid infrastructure and on-demand IT

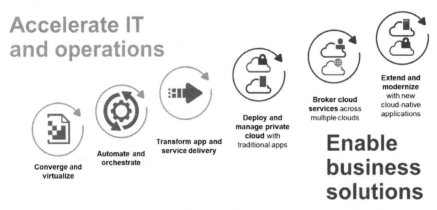

Figure 12-6 The journey to hybrid infrastructure and on-demand IT

Chief information officers (CIOs) around the world are being challenged to add more value to the business by rapidly developing and deploying new solutions—solutions that enable new products, new services, new business models, and new customer experiences, all to help compete. That is the ultimate goal and business outcome of the transformation to hybrid infrastructure and on-demand IT as shown in Figure 12-6.

The journey is not easy, and it is not linear. Customers embarking on this journey start in different places, have different destinations, and different timeframes. The journey requires that customers accelerate transformation of their current IT environment and operations to achieve maximum agility and performance. Businesses need to modernize, converge, and virtualize their infrastructure. They need to build a bridge between the current environment and the future—for applications, workloads, and data. Then, businesses need to automate and orchestrate resources for continuous delivery. These transformations create an infrastructure that enables business solutions CIOs are looking for.

Most IT organizations are looking to become an internal service provider that can deploy and manage private clouds, broker applications, and services regardless of where they are hosted, and extend and modernize the environment by empowering the developers to build, manage, deliver, and improve cloud-native applications.

In other words, customers are implementing a next-generation infrastructure environment, extending it to open and hybrid cloud, and transforming applications and workloads from old to new. Customers are building the foundation for the new style of business to create and deliver new value instantly and continuously.

Review: What the journey requires

For hybrid infrastructure and on-demand IT to be effective, purposeful innovation is required in these areas:

- **Converged technologies**—HPE offers a complete portfolio of modern technologies based on common architectures across servers, storage, and networking. Designed for convergence, these technologies can integrate into shared pools of resources and be managed though one platform.

- **Converged management**—Converged management is quickly becoming the central nervous system of the data center. HPE offers several management solutions for simple data center management that is aligned with applications.

  OneView is one example of converged management that brings the best of the consumer world to the enterprise. It is the industry's first intuitive IT management platform offering a modern and an integrated workspace. IT teams can collaborate to capture processes, configurations, and best practices using a software-defined approach. By eliminating manual operations, OneView empowers IT teams to deploy resources at the push of a button, reduce cost and errors of lifecycle management, and readily adapt to the enterprise as it evolves.

- **ConvergedSystem**—For customers looking to gain the value of convergence fast, HPE offers a complete ConvergedSystem portfolio. These workload-optimized, engineered systems are built for the way people work. They provide a new level of seamlessness from virtualization to cloud and from physical to virtual.

  Each system shortens the procurement lifecycle with a set of core platforms that are easy to configure, quote, order, and deploy. Customers can go from ordering to implementing operations in 15 to 30 days. Each system includes simplified management at the systems level, not the component level, leveraging and extending OneView capabilities. These integrated systems improve standardizing and automation capabilities without losing application performance.

- **Converged expertise**—Converged Infrastructure consulting, support, education, and financial services work with a customer's IT teams, processes, facilities, and infrastructure on transitioning to hybrid. These resources help create a smooth transition from a traditional, siloed, product-focused IT infrastructure to an app-centric, converged infrastructure.

This transition also requires partners with capabilities that go well beyond product-specific consulting and implementation. Today, the HPE partner community can implement complex solutions based on customer requirements. For example, HPE channel partners designated as Platinum partners are considered Converged Infrastructure Specialists.

Competitive cloud landscape for private and hybrid clouds

This section provides an overview of third-party products that compete with Helion products. It reviews the strengths and weaknesses of specific vendors and provides HPE differentiators.

Table 12-1 shows a high-level overview of competitors to HPE Helion solutions and what products they offer in each area.

Table 12-1 Competitors to Helion and their comparable products (as of 26 June 2015)

Competitor	Helion OpenStack	Helion CloudSystem	Helion Development Platform
redhat	Red Hat Enterprise Linux OpenStack Platform	Red Hat Cloud Infrastructure	OpenShift
MIRANTIS	Mirantis OpenStack	N/A	Cloud Foundry
CANONICAL	The Canonical Distribution of Ubuntu OpenStack	N/A	N/A
Pivotal	N/A	N/A	Pivotal Cloud Foundry
CISCO	UCS Integrated Infrastructure for Red Hat Enterprise Linux OpenStack Platform	CIAC + VCE/UCS/FlexPod, Intercloud	N/A
vmware	vCloud Suite/Integrated OpenStack	vRealize/VCE	Pivotal Cloud Foundry
Microsoft	System Center/Azure	Hyper-V/System Center/CPS	Azure
IBM	Cloud Manager with OpenStack, SoftLayer	SmartCloud Entry on PureFlex System	Bluemix
Do it yourself	Download the OpenStack code and go from there	Use existing servers with management, virtualization, and storage	Download the Cloud Foundry code and go from there

Note

In the Cisco row, VCE is listed as a competitor to Helion CloudSystem. VCE is a combined offering from VMware (virtualization), Cisco (hardware), and EMC (storage). Together, they provide a full offering to compete against Helion CloudSystem.

Some customers prefer NetApp or did not want to pay the high cost of VMware and EMC, so Cisco branched off and created FlexPod. FlexPod combines an offering from Cisco (hardware), NetApp (storage), and either VMware or Microsoft (virtualization).

 Note

Some companies do not have an offering that matches specifically to an HPE solution. Table 12-1 indicates this with "N/A." This information reflects positioning as of 26 June 2015.

Top competitors by product

These specific vendors are in direct competition to HPE for the products listed:

- Helion OpenStack
  - Red Hat
  - Mirantis
  - Canonical
- Helion CloudSystem
  - Cisco
  - VMware
- Helion Development Platform
  - Red Hat
  - Pivotal
  - VMware

The next few pages look at each competitor and analyzes its claims and strengths compared to HPE differentiators.

Competitive analysis: Red Hat

Red Hat promotes itself as the "open" solution, but in reality, its solutions are a path to lock-in with Red Hat products. Its solutions are a combination of third-party solutions that only support Red Hat Enterprise Linux (RHEL) and OpenShift. To demonstrate an advantage over Red Hat, focus on complete cloud offerings from HPE including private, hosted, and platform-as a-service (PaaS) solutions. This allows customers to get a complete platform from one company. It provides true choice for customers with hardware, software, and development flexibility without vendor-based lock-in.

Table 12-2 lists the claims and strengths of Red Hat, along with HPE advantages and differentiators.

Table 12-2 Claims and strengths of Red Hat

Competitor claims	Competitor strengths	HPE differentiators
The leader in Linux operating system, Linux virtualization, and open-source software, so OpenStack technology is natural	Large Linux operating system customer base and strong association with open-source software	Offers private, hosted, and PaaS cloud platforms, all from a single company
Has a broad portfolio with all the tools and capabilities customers need	Has broad portfolio with OpenStack IaaS, OpenShift PaaS, and integration with RHEL	Commercial-grade features, experience running large-scale OpenStack solutions, and professional services to help customers deploy
A truly "open" implementation	Open-source implementation from a trusted open-source software company	More hardware, software, and development choices, providing flexibility without vendor-based lock in

Competitive analysis: Mirantis

Mirantis promotes itself as a "pure" OpenStack offering. In reality, it is small scale and forces customers into custom deployments. Mirantis lacks the full range of cloud options (PaaS and managed) and the full portfolio (it just started with DevOps). To demonstrate value over Mirantis, point out that:

- HPE offers commercial-grade solutions with choice and flexibility. It does not require a custom installation for each deployment.

- HPE has the global presence and experience needed to deploy OpenStack projects at scale.

- Enterprises trust HPE for their cloud deployments.

Table 12-3 lists the claims and strengths of Mirantis, along with HPE advantages and differentiators.

Table 12-3 Claims and strengths of Mirantis

Competitor claims	Competitor strengths	HPE differentiators
Strong services offering focused on OpenStack technology Technologies based on OpenStack, without additional code or vendor-specific improvements, and customized for each customer Early leader in OpenStack deployments and integration	One of the founding members of the OpenStack Foundation Strong cloud professional services, supporting implementations Partnership with Ericsson for service provider accounts Excellent training and certification program for OpenStack technology Fuel installer can install third-party distributions Easy distributions (download and deployment)	Commercial-grade solutions with choice and flexibility, without requiring a custom installation for each deployment Global presence and experience needed to deploy OpenStack at scale Mirantis lacks range of cloud options (PaaS and managed) and full portfolio (just started into DevOps) Trusted by enterprises for cloud deployments

Competitive analysis: Canonical

Canonical says that it offers a straightforward implementation of cloud. In reality, it offers a bare-bones, do-it-yourself OpenStack product that lacks the features and support enterprises need. Canonical implementations are custom projects with support that depends on Canonical. The best way to demonstrate value against Canonical is to point out that HPE is a leader in OpenStack projects and IT services. HPE has deep expertise with service providers and major enterprises. HPE also has a better portfolio to build and integrate an OpenStack cloud.

Table 12-4 lists the claims and strengths of Canonical, along with HPE advantages and differentiators.

Table 12-4 Claims and strengths of Canonical

Competitor claims	Competitor strengths	HPE differentiators
Deploying OpenStack projects with Ubuntu is the best way to ensure a straightforward implementation Has all the software infrastructure for customers to build and deploy their own OpenStack cloud Enterprise-grade professional services to help build and deploy, even a turnkey build-and-run option with BootStack service	Early adopter of OpenStack technology, leading to some first-mover advantages User-friendly, free, and easy to use Strong brand association with Linux customers and early OpenStack adopters	Offers complete cloud solutions including private, hosted, and PaaS platforms, all with integrated management (Canonical has limited offerings) A leader in OpenStack technology and IT services with deep expertise with service providers and major enterprises (Canonical has limited enterprise experience)

Competitive analysis: VMware

As VMware customers move to cloud, VMware claims to be a cloud vendor. In reality, its solutions are essentially virtualization solutions. It is still fundamentally a virtual infrastructure company, with cloud as an extension. VMware also lacks the experience running cloud systems and services, enterprise professional services, and data centers. It also lacks application platform and development capabilities.

Table 12-5 lists the claims and strengths of VMware, along with HPE advantages and differentiators.

Table 12-5 Claims and strengths of VMware

Competitor claims	Competitor strengths	HPE differentiators
Future based on the software-defined data center, centered on VMware products that build the infrastructure for "one cloud, any application, any device"	Established leader in on-premise virtualization with large and loyal customer base	Cloud is at the core of HPE offerings (for VMware, virtual infrastructure is at the core, with cloud as an extension)
VMware leading cloud arena with solutions from private to public, including strong cloud management and OpenStack capabilities	Growing portfolio of cloud, expanding private cloud, cloud management/automation, and vCloud Air solutions	Unlike HPE, VMware lacks experience running cloud systems/services, enterprise professional services, or data centers
Leverage existing VMware install base to help transition to the cloud	Partnerships with major players across cloud landscape, including HPE, Dell, Cisco, EMC, Pivotal, NetApp, and VCE	VMware lacks core application platform and development capability

Competitive analysis: Cisco

Cisco promotes its hardware and automation software but fundamentally lacks core cloud capabilities and experience. Cisco has limited experience running any large-scale clouds, and OpenStack support is basic.

To demonstrate value over Cisco, focus on the cloud platform from HPE, including private, hosted, and PaaS solutions with integrated management. HPE also provides full-stack automation across servers, databases, middleware, and applications. Cisco does not have the capability or the integration to do so.

Table 12-6 lists the claims and strengths of Cisco, along with HPE advantages and differentiators.

Table 12-6 Claims and strengths of Cisco

Competitor claims	Competitor strengths	HPE differentiators
Growing portfolio of cloud products and services (UCS servers, Nexus virtual networking, Intercloud)	Growing portfolio of cloud products and services (UCS servers, Nexus virtual networking, Intercloud)	Offers cloud solutions including private, hosted, and PaaS platforms, all with integrated management (Cisco solutions are fragmented islands of management)
Private cloud solutions and converged infrastructure management	Private cloud solutions and converged infrastructure management	Cisco has limited experience running any large-scale clouds and OpenStack technology support is basic at best
Strong partnerships with VMware/EMC (VCE) and NetApp (FlexPod)	Strong partnerships with VMware/EMC (VCE) and NetApp (FlexPod)	HPE provides full-stack automation across servers, databases, middleware, and applications (Cisco lacks both capability and integration)

Competitive analysis: IBM

IBM promotes itself as a strong, trusted advisor for cloud solutions. In reality, it offers a patchwork mix of cloud solutions, none of which are enterprise ready. IBM's public cloud is based on a completely proprietary system, whereas HPE is committed to OpenStack. Lastly, HPE has strong integration across private, hosted, and hybrid clouds, and IBM has no integration across hybrid deployments.

Table 12-7 lists the claims and strengths of IBM, along with HPE advantages and differentiators.

Table 12-7 Claims and strengths of IBM

Competitor claims	Competitor strengths	HPE differentiators
Strong, trusted, traditional enterprise that vendor companies rely on	Strong, trusted, traditional enterprise that vendor companies rely on	Committed to the OpenStack project (IBM's public cloud runs on a completely proprietary solution)
Has a complete cloud platform, including private, hosted, public, and PaaS solutions	Has a complete cloud platform, including private, hosted, public, and PaaS solutions	Strong integration across private, hosted, and public clouds (IBM has no integration across hybrid deployments)
Large-scale, experienced professional services organization to help customers deploy	Large-scale, experienced professional services organization to help customers deploy	Strong integration across private, hosted, and hybrid clouds (IBM has no integration across hybrid deployments)

Competitive analysis: Microsoft

Microsoft is focused on the cloud of the future, but its offering fundamentally locks customers into its proprietary platform. Microsoft is fundamentally a proprietary software company with little open-source and enterprise cloud experience. Its cloud software offering is fragmented and poorly integrated. To demonstrate HPE value over Microsoft, point out that:

- HPE provides customers with an open-source cloud, including choices for hardware, software, and development platforms.

- HPE focuses on cloud with open-source software.

- HPE uses an integrated platform across its cloud solutions.

Table 12-8 lists the claims and strengths of Microsoft, along with HPE advantages and differentiators.

Table 12-8 Claims and strengths of Microsoft

Competitor claims	Competitor strengths	HPE differentiators
Experience and end-to-end cloud solutions including private, hosted, public, and PaaS solutions	A growing, complete cloud platform including private, hosted, public, and PaaS solutions from a single company	Provides customers with an open-source cloud with choices for hardware, software, and development (Microsoft's platform is a proprietary lock-in solution)
Strongest cloud solution with a single platform to go from on-premise to hybrid to public cloud	Strong enterprise presence with Windows Server	HPE vision focuses on open-source software at the core of its future (Microsoft is still focused on locking in customers to legacy Microsoft software)
Cloud solutions that embrace open-source software and non-Microsoft products and services	Strong focus on growing Microsoft Azure and cloud offerings	Microsoft claims an integrated cloud experience, but the actual integration of legacy and cloud is poor

Competitive analysis: Customers DIY

You might face organizations that want to attempt a do-it-yourself (DIY) approach to implementing OpenStack technology. This route can be complex, time-consuming, and disastrous. Counter the associated arguments for DIY advantages by stating that:

- HPE has contributed to the OpenStack project and made significant investments to bring Helion OpenStack to market.

- The cost of reinventing the HPE contribution to and integration with OpenStack is too big.

- The organization should focus its internal time and resources on developing customizable solutions that create competitive advantages, instead of integrating OpenStack and making it work.

- Unlike the full professional services offering from HPE, DIY offers no support or structured guidance, leaving room for errors, frustration, and lost time.

Table 12-9 lists the claims and strengths of customers who want to build open-source solutions themselves, along with HPE advantages and differentiators.

Table 12-9 Claims and strengths of customers who want to build open-source solutions themselves

Customer claims	Customer strengths	HPE differentiators
They have the infrastructure and development staff to build and maintain an open-source private cloud	No upfront licensing costs	Offers cloud solutions including private, hosted, and PaaS platforms, all from one company
They can use OpenStack technology to create building blocks for a private, in-house cloud, which avoids software licensing costs	Ability to create fully customized solution with full control	Commercial-grade features, experience running large-scale OpenStack solutions, and professional services to help customers deploy (DIY is completely unsupported)
They can have more control that is highly customizable	No vendor lock in	

Competitive landscape summary

Figure 12-7 Competitive landscape

In general, competitors to Helion, shown in Figure 12-7, claim that they:

- Have complete cloud solutions when they only have parts of the solution stack

- Have "open" solutions but instead are full of vendor-specific services and software, which results in vendor lock-in for customers

- Are leaders in open-source, enterprise, or virtualization services, but none have the cloud platform leadership of HPE

In response to these claims, HPE:

- Offers a complete cloud platform, including private, hosted, and PaaS solutions, all from a single company

- Provides true choices for customers across hypervisors, hardware, software, and application development

- Runs one of the world's largest clouds, built on OpenStack technology, providing unique experience and knowledge

Strengths of Helion solutions

Business and IT executives are looking at emerging technologies such as cloud computing as a new platform for creating and delivering business value. According to a recent Gartner survey of more than 2000 CIOs, the top business priorities for CIOs are to increase technology's potential to drive enterprise growth, deliver operational results, reduce the overall cost structure, and improve efficiency. When used effectively, cloud capabilities offer numerous opportunities to meet these objectives.

 Note

To read the Gartner survey, scan this QR code or enter the URL into your browser.

http://www.gartner.com/newsroom/id/2304615

Cloud technologies enable enterprises to create and deliver significant business value. Cloud computing offers three main business benefits:

- Faster innovation and accelerated business growth

- Enhanced business agility

- Improved efficiency and lowered costs

These benefits directly affect financial and operating key performance indicators (KPIs) that companies care most about, such as:

- Revenue growth

- Earnings

- Market share

- Time to market

- Cost of acquiring new customers

- Return on equity

- Cash flow

How does cloud enable business transformation?

The cloud speeds innovation by removing barriers to creating, building, and testing applications and services anywhere and anytime. Companies can quickly develop new products or services, such as a new financial service or promotional program. They can instantly access resources needed to shorten time to market and drive new revenue streams.

Cloud capabilities enhance business agility, enabling companies to deploy and redeploy IT resources dynamically and quickly where they are needed most. This approach allows businesses to respond efficiently to an emerging opportunity, a competitive threat, or changing customer demands.

Through resource sharing and automation, cloud solutions enable IT organizations to increase asset utilization, streamline workflow, and improve employee productivity. Employees and partners can collaborate and share information using a common platform. This improves data, document, and process collaboration.

Data and document collaboration increase efficiency by sharing documents in a cloud repository that everyone can access immediately. No data replication is needed.

Process collaboration between business partners allows them to perform their work in the same cloud or software-as-a-service (SaaS) solution. For example, engineering design, supply chain coordination, and development and testing applications with codes stored in the cloud. Cloud can also integrate business and enterprise applications that companies use, even if they use different applications, versions, or processes.

HPE in the cloud marketplace

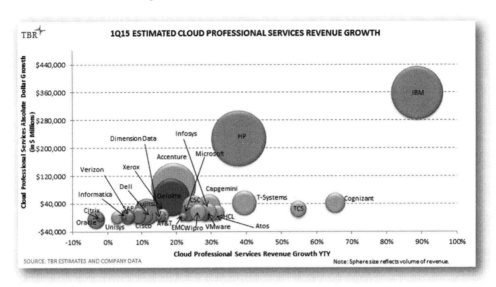

Figure 12-8 Estimated cloud professional services revenue growth

Figure 12-8 shows the results of a study conducted by Technology Business Research (TBR). It shows estimated cloud professional services growth on the *X*- and *Y*-axes, with the size of the circles reflecting the volume of revenue. HP is clearly viewed by TBR to be poised for significant growth.

In a recent study, Forrester evaluated private cloud solutions from IT vendors across 61 criteria including "current offering," "strategy," and "market presence." HPE achieved the highest scores for eight out of 15 categories.

The study also found that 33% of hardware decision-makers have already adopted cloud, and 55% plan to prioritize building internal private clouds, mostly by layering software solutions on their existing infrastructure. According to Forrester's analysis, top solution trends in the private cloud market include:

- Vendors who are focused on IaaS with application templates in addition to infrastructure templates, a trend Forrester calls IaaS+

- Vendors who are developing tools to help both development and operations (DevOps) professionals

- IT service management functions that are expanding

- Most current solutions have relatively basic hybrid-cloud-enabling features

Forrester concluded that HPE "leads the pack" because it offers "a clean and navigable interface that wraps substantial breadth and depth of capabilities into the fewest number of interfaces. Unlike other vendors in this space, HPE adds functionality into a single interface as a rule, without making the overall experience less intuitive."

Note

To read the Forrester Wave Report, scan this QR code or enter the URL into your browser.

http://on.hp.com/forresterwavereport

Helion cloud stories

Figure 12-9 Helion cloud stories

HPE enables the selection of the best method for service delivery, whether that is using traditional IT or private cloud architectures. HPE can also manage, outsource, and finance cloud implementations.

Note

Stories describing how Helion helps businesses move to the cloud are listed on the Helion Cloud Stories website, shown in Figure 12-9. To access the website, scan this QR code or enter the URL into your browser.

http://hp-cloudstories.com/Overview

Helion cloud solution value propositions

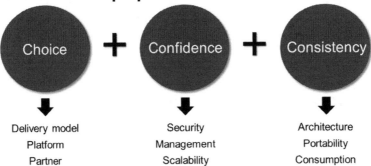

Figure 12-10 Helion cloud solution value propositions

The HPE approach to cloud solution design aims to eliminate the fragmented, siloed environment that can grow out of the traditional approach, where different management and security tools are required for each of the deployment models (traditional IT along with private and managed cloud). Helion is integrated by design for business and governments. It delivers three fundamental value propositions shown in Figure 12-10:

- **Choice**—Helion is an open, standards-based approach. It also brings traditional IT into the mix. This is an important point because HPE understands that for many customers, traditional IT will be a critical part of their operating environment for many years. Helion supports multiple hypervisors, operating systems, development environments, and third-party infrastructure.

  HPE has a long history of working with a wide spectrum of business partners, from independent software vendors to service providers to distributers to value-added resellers to system integrators to outsourcing providers. Helion embraces this extensible partner ecosystem.

- **Confidence**—Management and security are major concerns for many customers, and Helion instills confidence by delivering management and security solutions that extend across information, applications, and infrastructure.

- **Consistency**—Helion is built on a single common architecture across all delivery models, which enables portability across models. Helion CloudSystem is integrated by design with one consistent strategy, providing a single consumption experience.

HPE unique differentiators

HPE cloud computing solutions offer several key characteristics:

- Support for traditional IT, private cloud, and hybrid cloud environments

- The ability to include a broad range of business applications, all leading hypervisors, and multiple operating systems

- Unified delivery and control of services across both cloud and traditional IT environments

- Automated lifecycle management of applications and infrastructure

- End-to-end security

- Scalability to meet unpredictable business demands

Hybrid world

HPE research shows that as organizations move toward the cloud, most will adopt a hybrid form of cloud computing, leveraging the best of private, managed, and public cloud, in addition to traditional IT.

A recent HPE study revealed that:

- 75% of IT executives plan to pursue a hybrid delivery model

- 72% said that portability of workloads between cloud models is important to their cloud implementations

- 65% are concerned with vendor lock-in

Adopting a hybrid approach means looking at the portfolio of applications in the organization, segmenting them according to their respective service-level agreement (SLA) requirements, and then aligning them to the appropriate deployment models. This allows businesses to achieve the most optimal use of internal and external resources.

In this new hybrid world, the role of the CIO and the IT organization expands from the traditional builder of services to a builder and broker, creating a seamless experience for users, independent of service sources.

Open and heterogeneous

Helion is based on open standards and supports multivendor, heterogeneous software and hardware. Helion CloudSystem is the choice for organizations of any size that want to deploy a cloud environment with the most open architecture and a rich set of cloud enablement software. Support for heterogeneous resource pools and multiple hypervisors provides customers with choice today and the flexibility of deployment options going forward. Helion CloudSystem provides the fastest road to the cloud.

Compute capabilities at the core of cloud services can be built on a wide range of server types and vendors. The combination of server types and vendors within a customer environment is important to understand because it can impact which software products to recommend.

The Helion portfolio of cloud products and services creates an open ecosystem with a common management structure, and it integrates easily into existing infrastructures through a wide range of delivery models, providing an open, secure, and agile environment.

The HPE cloud architecture is built with partners in mind, allowing HPE to leverage an already thriving partner ecosystem. Close-to-trunk releases of the Helion OpenStack Community supply customer environments with the latest capabilities of OpenStack technology. The Helion OpenStack Community does not mix proprietary and third-party tools.

 Note

Helion OpenStack Community can be downloaded for free, under an open-source license with an optional paid support contract. Scan this QR code or enter the URL into your browser.

http://www8.hp.com/us/en/cloud/hphelion-openstack.html

Additional advantages of Helion

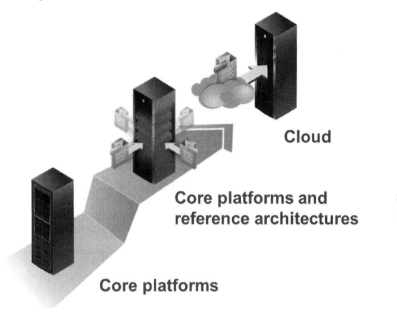

Figure 12-11 Additional advantages of Helion

Additional advantages of Helion include the following:

- **Easy upgrade paths**—HPE protects customer investments in BladeSystem with conversion services to Helion CloudSystem as depicted in Figure 12-11. Most competitors in the cloud market base their systems on hardware and software that are not compatible with their cloud offerings. As a result, customers cannot leverage their existing equipment using a migration to third-party cloud systems. Easy upgrade paths simplify the evolution from CloudSystem Matrix to Helion CloudSystem Foundation and Helion CloudSystem Enterprise. Open-source standards leave the path to future upgrades open and accessible. Customers can also integrate CloudSystem Matrix into CloudSystem Enterprise.

- **The industry's first flat SAN architecture**—BladeSystem with StoreServ arrays improves one-way storage network latency by removing the SAN switches. Storage traffic traverses multiple layers of the SAN fabric infrastructure, which can cause latency, performance, and complexity issues to arise. With Virtual Connect direct-attach Fibre Channel for StoreServ systems, customers can cut the cost of the SAN fabric in half. With 86% fewer components, complexity is significantly reduced compared to similar configurations from other vendors. By flattening a complex, multitier SAN infrastructure into a single tier, HPE customers enjoy 55% lower latency and provisioning that is more than two times faster.

- **Efficiency and agility at cloud scale**—Efficiency and agility at the scale needed in many cloud environments is built in with Virtual Connect direct-attach Fibre Channel storage. With a single StoreServ P10000 storage system, customers can attach 48 BladeSystem c7000 enclosures, filled with 768 ProLiant server blades that can support more than 20,000 virtual machines. This massive scale does not require a dedicated infrastructure.

Learning check

1. Why are customers moving to a hybrid infrastructure and on-demand IT?

_____

_____

_____

_____

_____

2. Which companies have a product in direct competition with the Helion Development Platform? (Select three.)

   a. Red Hat

   b. Pivotal

   c. Canonical

   d. Cisco

   e. VMware

   f. Citrix Systems

3. How would you respond to a customer that wanted to try a DIY approach to implementing an open-source private cloud?

_____

_____

_____

_____

_____

4. Complete a few statements regarding the Helion business advantages by filling in the blanks:

   – Speed _____ and accelerate business growth

   – Enhance business _____

   – Increase _____ and lower _____

Learning check answers

1. Why are customers moving to a hybrid infrastructure and on-demand IT?

   - **It offers security, availability, resiliency, compliance, data sovereignty, performance, and openness**

   - **It provides a common foundation and centralized management layer**

   - **It allows customers to use their existing IT during the transition to cloud**

   - **A combination of public and private cloud can provide the right destination for the right applications at the right cost**

2. Which companies have a product in direct competition with the Helion Development Platform? (Select three.)

   a. **Red Hat**

   b. **Pivotal**

   c. Canonical

   d. Cisco

   e. **VMware**

   f. Citrix Systems

3. How would you respond to a customer that wanted to try a DIY approach to implementing an open-source private cloud?

   - **HPE offers complete cloud solutions including private, hosted, and PaaS platforms**

   - **HPE has experience running large-scale OpenStack solutions**

   - **HPE has made significant contributions to the OpenStack project, and the cost of reinventing the HPE integration with OpenStack is too big**

   - **Professional services from HPE can guide customers in developing customizable solutions that create competitive advantages**

   - **DIY offers no support or structured guidance, leaving a lot of room for errors, frustration, and lost time**

4. Complete a few statements regarding the Helion business advantages by filling in the blanks:

   - Speed **innovation** and accelerate business growth.

   - Enhance business **agility.**

   - Increase **efficiency** and lower **costs.**

Summary

Most businesses need to continue using existing IT infrastructures, making it important to bridge old technologies with new ones. Key considerations for achieving a hybrid infrastructure and on-demand IT include:

- Security
- Availability
- Resilience
- Compliance
- Data sovereignty
- Performance
- Openness
- Cost

In general, competitors to Helion claim that they are leaders in open-source, enterprise, or virtualization services. They claim to have complete cloud solutions. In reality, they can typically only offer parts of the solution stack, providing vendor-specific services and software.

HPE offers a truly complete cloud platform including private, hosted, and PaaS solutions. HPE customers have choices across hypervisors, hardware, software, and application development. Additionally, HPE runs one of the world's largest clouds, built on OpenStack technology, providing unique experience and knowledge.

Business and IT executives are looking at emerging technologies such as cloud computing as a new platform for creating and delivering business value. Helion solutions provide business outcomes including:

- Faster innovation and accelerated business growth
- Enhanced business agility
- Improved efficiency and lowered costs
- Enhanced agility and flexibility
- Testing and development in a private cloud
- High-performance computing for research

13 Planning and Designing Customer Solutions

WHAT IS IN THIS CHAPTER FOR YOU?

After completing this chapter, you should be able to:

✓ Specify the four steps of the solution proposal preparation process

✓ Describe the predeployment and installation planning process considerations

✓ Explain how to access Hewlett Packard Enterprise (HPE) support

Assessment activity

A critical step in planning and designing data center and cloud solutions is to prepare your solution proposal, including the technical components, and then deliver it to the customer. Before continuing, consider the following questions.

1. What experience do you have using HPE tools? Which ones have you found most helpful and why?

   _____

   _____

   _____

   _____

2. When you work on a proposal, do you focus primarily on the technical aspects of the solution? Do you address the business value of the proposed solution in your proposal? If so, how?

   _____

   _____

   _____

   _____

3. Based on your experience, describe some of the activities that have been most effective in moving proposals to the next stage with the customer. Is there an approach you have tried that has been unsuccessful? Describe some of the actions you take to achieve customer buy-in.

_____

_____

_____

_____

_____

Preparing a solution proposal

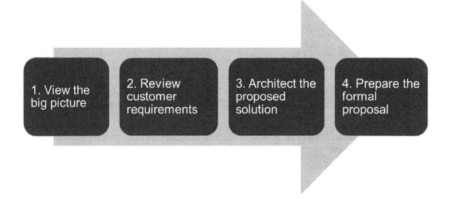

Figure 13-1 Preparing a solution

As illustrated in Figure 13-1, preparing a solution to meet customer requirements requires a systematic four-step approach:

1. View the big picture.

2. Review customer requirements.

3. Architect the proposed solution.

4. Prepare the formal proposal.

Step 1: View the big picture

You can gain valuable insight on each customer's situation in several ways:

-   Gain perspective during meetings with their representatives and business leaders

-   Review the request for information (RFI) or request for proposal (RFP) documents

- Analyze the financial statements and other documents provided by the customer or gathered from the customer's website

- Use the Converged Infrastructure Capability Model (CI-CM) or a similar assessment tool

However, building the best solution for your customer requires specific details. Make sure you have important information by using this checklist:

- **Scalability**—Is the customer global? Does the customer have multiple data centers?

- **Heterogeneity**—Does the customer prefer third-party or rack-mounted hardware?

- **Time to implement**—Does the customer require IT services or solutions to be up and running in a short time?

- **Brand-new or existing infrastructure**—Is the customer willing to purchase new hardware, or do they have existing infrastructure that they want to reuse?

- **Workload characteristics**—What are the characteristics of the customer's typical workload? Understanding the nature of the workloads early in the process can reveal the components that should be present in the solution.

- Specific considerations for cloud customers:

    - **Build only on private cloud or use hybrid resources**—Does the customer need to use both private and public cloud services?

    - **Key cloud characteristics**—Does the customer need Infrastructure-as-a-service (IaaS), Platform-as-a-Service (PaaS), Software-as-a-Service (SaaS), or everything-as-a-service (XaaS)? Does the company need to extend its private cloud to use public cloud resources? What is the customer's ideal situation for monitoring, billing, and so forth?

Step 2: Review customer requirements

Reviewing the customer's current IT environment in order to verify that it can support the solution you are architecting is a crucial step in preparing the proposal. It can be useful to contextualize your considerations according to the four primary domains of CI-CM: technology and architecture; management tools and processes; culture and IT staff; and demand, supply, and IT governance. Channel partners might use their company's assessment model or the Gartner Five-Stage Demand-Driven Maturity Model. Depending on which assessment model you use, or whether you apply a combination of approaches, the content, terminology, and applicability of the questions will vary. However, all customer assessments should analyze the primary categories and questions covered by CI-CM.

Technology and architecture

This category encompasses technical infrastructure, including hardware and data center facilities as well as applications. Some of the areas and questions you should consider are:

- **CPU utilization**—What is the average utilization across all servers in the data center? How much does it range?

- **Storage**—What is the average utilization of disk storage? What is the preferred disk and SAN technology? What are the requirements for performance, data protection, and more?

- **Standards and policies**—How consistently and at what level are standards applied across technology stacks, including data centers, networking, storage, operating systems, software, and more?

- **Energy costs**—Are there practices in place to control energy costs? Are there strategies for using cost savings to benefit the business or to improve IT?

- **Availability**—What level of data availability is currently achieved? What are the specifics regarding hours of downtime per year, duration of downtime, acceptable performance degradation in a failure situation, protection against failures (server, disk subsystem, network, and power), and 24×7 availability?

- **Disaster recovery**—What are the recovery-time and recovery-point objectives for business-critical applications?

- **Networking**—Is there consistent low-latency storage and data networking? What types of networking are available (LAN, WAN, and SAN)? Are there automated, shared networks that enhance agility?

- **Service-oriented infrastructure**—How much of the infrastructure is delivered as a service?

- **Provisioning**—When a request is made, how long does it take to provision infrastructure resources?

- **Performance**—What are the specifications for number of concurrent users, benchmark results, response times, or any other relevant metrics?

- **Backup**—What is the time limit, frequency, and type of backups (offline, online, local, or across the network)?

- **Applications**—How many users do applications need to accommodate? What IT resources do applications require? Are there applications that can be consolidated?

- **Application change management**—How long does it take to add or change (patch and upgrade) an application?

Management tools and processes

This category involves the resources that plan, manage, and deliver IT services. To help you move the customer toward standardized and integrated processes that align IT with their business goals, consider these questions:

- **Standardization**—What is the level (or desired level) of standardization regarding IT processes and management tools?

- **Compliance**—Is there compliance with IT processes? If so, what is the percentage of this compliance?

- **Remote management**—What level of remote management is currently in place?

- **Automation**—What is the extent of automation of infrastructure management processes?

- **Service-level agreements (SLAs)**—Are SLAs already defined? At what level of the infrastructure are they managed? Are they adopted throughout the data center and in service delivery?

 Note

SLAs provide a common understanding about services, priorities, responsibilities, guarantees, warranties, and penalties. Several factors need to be determined and agreed upon with the customer, including expectations for performance levels and availability, legal compliance requirements, system and data security, and data retention policies. The key consideration for SLAs is whether the solution design prioritizes the solution or the price. For example, the SLA might be modified by minimizing it to meet financial requirements or expanding it to provide an even more robust solution.

- **IT security and risk**—To what extent do existing policies, processes, and procedures manage IT security and risk? How well have security and risk management procedures been adopted?

Culture and IT staff

This category involves the definition of the IT department's structure, roles, and functions—all of which are important to a readily adaptable IT environment. Consider these questions:

- **IT staff organization**—How is the IT infrastructure staff organized (centralized or decentralized)? What is the reporting structure?

- **Information retention**—How is information retention managed across structured and unstructured data to support business decisions and compliance?

- **Environmental and green practices, carbon footprint**—Does the company have policies about carbon footprint or other environmentally responsible practices?

- **Skills and expertise**—Are the right people with the right skills in place for the current and future business needs?

- **Training**—Is there clear and consistent training on IT policies and practices?

- **Change risk and management**—How does the company manage and plan for the risk associated with change? How are new applications and application changes handled to ensure they will function correctly, perform to requirements, and have minimal security risk?

- **Knowledge**—Is the IT staff proficient at implementing the company's IT processes and policies, particularly regarding end-to-end service delivery?

- **Culture**—What is the IT organization's culture concerning IT innovation and adoption of new technologies?

Demand, supply, and IT governance

This category involves how organizational structures and processes can enable and improve alignment between the customer's IT environment and their business strategies and objectives. Consider these questions:

- **Funding**—How is funding for the IT infrastructure handled?

- **Policies**—What role do common policies, such as those for networking and security, have in operations management decisions?

- **IT asset management**—How are IT assets managed so that the company can increase its return on investment (ROI) for IT and meet compliance requirements? How are physical IT assets managed and moved through their lifecycle?

- **Governance**—Are enterprise architecture policies and principles standardized and documented?

- **Security management**—Are security management policies and standards formalized in support of the IT delivery model?

- **Time to infrastructure**—How long does it typically take to deploy and integrate infrastructure for new projects? How long does it take to make changes to the existing systems?

- **Business intelligence**—What is the business intelligence architecture? Does it present a single, consistent view of the company?

- **Hybrid delivery and management**—Does the company implement hybrid delivery, such as SaaS? How are these services managed in regard to costs, governance, and business priorities?

- **IT operations evaluation**—How will projects improve IT operations? What are potential obstacles to implementation?

Step 3: Architect the proposed solution

During this stage, you should focus on various aspects of the solution, including addressing architectural and transitional issues such as functional and technical design, organizational design, technology governance, and change management. HPE sizing and configuration tools can help you design specific solutions. You should also consider licensing options and include them in your proposal. Be sure to check the product QuickSpecs for specific licensing information.

In your proposal, you should include how the proposed solution will provide business value for the customer. You can architect the solution to meet the customer's complete needs (business, technical, and financial) by:

- Developing a logical architecture to host the solution, including:
  - Network layout
  - Server requirements
  - Application services
  - Storage space requirements
- Sizing the solution and building the configuration
- Incorporating licensing options
- Outlining how to integrate your solution into the customer's IT infrastructure

Step 4: Prepare the formal proposal

After you architect the technical aspects of the solution and consider how services and licensing will be integrated, you can start preparing the proposal you will present to your customer. It is not enough to present the technical aspects of the solution. Remember that you want to present the value of the solution in business terms that matter to the customer.

The creation of timely, accurate, proposal-ready content requires a rigorous development and maintenance process. From a content perspective, high-quality documents ideally include:

- Key benefits and differentiators
- Latest HPE marketing messages and value propositions
- Customer and analysis quotes
- High-level technical information
- Proof points

The order in which you present information might be dictated by points of focus in the RFI or RFP. Depending on the solution you are proposing and the resources required, you might also include support information from channel partners, program managers, special interest groups, and others.

Proposal Web

Figure 13-2 Proposal Web

Proposal Web is a dedicated portal where you can create customer proposals with time-saving, ready-to-use content. You can use Proposal Web to assist with some or all of your customer solution proposal.

As shown in Figure 13-2, Proposal Web is a solution for HPE sales, presales and partners, and features content such as:

- Boilerplates
- Cover letters
- Main selling points
- Case studies

Other features of Proposal Web include the following:

- Wizards
- Executive summary generator
- Online quote editor
- RFP builder
- Multiple language portals and content

 Note

Proposal Web is a restricted website requiring registration. There is a three-minute demonstration of Proposal Web on the homepage. To visit the website, scan this QR code or enter the URL into your browser.

https://proposalweb.ext.hpe.com

Proposal Web activity

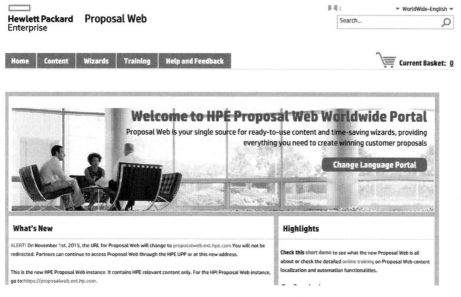

Figure 13-3 Proposal Web

Follow these steps to use Proposal Web (Figure 13-3):

1. Access the Proposal Web homepage.

 Note

To visit the Proposal Web homepage, scan this QR code or enter the URL into your browser.

https://proposalweb.ext.hpe.com

---

2. Optionally view the **Proposal Web demo**.

3. Select a portal—use **WorldWide-English**.

4. Click the **Content** tab, download the **1_Welcome to Proposal Web** and **2_HP proposal Web FAQs** documents and use them to answer these questions:

   - What two benefits does the Online Quote Editor provide?

     _____

   - What types of proposal wizards are available?

     _____

   - How is Proposal Web available to channel partners?

     _____

   - Is there content available in languages other than English?

     _____

5. Click the **Wizards** tab and launch the **CloudSystem Wizard**. Follow these steps:

   - Under **Group/Elements**, choose Cover Letter, HPE Strategy and General Information, ProLiant Servers, ESSN Software, Executive Summary, and HPE Cloud Services.

   - Under **Documents**, expand each element category and include documents of your interest.

   - Under **Re-Arrange Documents**, make changes to the order in which the selected documents appear based on your preference.

   - Under **User Data Entry**, fill out your information and the customer's information.

   - Under **Metadata**, enter the name of your proposal and a description.

   - Under **Download**, download your proposal and take a look at the included information.

6. Click the **Content** page and see what CloudSystem-related collateral you can find. Use the **Search** capabilities or expand the **Enterprise Group** in the left pane and then select CloudSystem.

   - How many documents did you find?

     _____

   - What are the names of some of the documents?

     _____

7. Expand **Cloud (Helion)** in the left pane and see what is available. What optimized cloud solutions are listed?

8. Log out of Proposal Web and close your browser.

Proposal Web activity—answers

1. Access the Proposal Web homepage at: **https://proposalweb.ext.hpe.com**

   Note that this website requires logon credentials.

2. Optionally view the **Proposal Web demo**. This demo is available from the Proposal Web home page.

3. Select a portal. For this activity, use **WorldWide-English**.

4. Click the **Content** tab, download the **1_Welcome to Proposal Web** and **2_proposal Web FAQs** documents, and use them to answer these questions:

   a. What two benefits does the Online Quote Editor provide?

   It integrates quotes from HPE pricing tools such as Watson and Sales Builder for Windows into proposal documents and combines quotes with boilerplates from the Proposal Web Library

   b. What types of proposal wizards are available?

   Pan-HPE wizards and wizards specific to HPE business groups

   c. How is Proposal Web available to channel partners?

   Partners with valid Passport credentials can access Proposal Web directly at: https:// proposalweb.ext.hpe.com

   d. Is there content available in languages other than English?

   Proposal Web supports 20+ business, country, and region portals with content in several languages

5. Click the **Wizards** tab and launch the **CloudSystem Wizard**. Provide these answers:

   a. Group/Elements Selection: **Cover Letter**, **Strategy and General Information**, **ProLiant Servers**, **ESSN Software**, **Executive Summary**, and **Cloud Services**

   b. Documents: Expand each element category and include documents of your interest

   c.   Re-Arrange Documents: Make any changes to the order in which the selected documents appear

   d.   User Data Entry: Fill out your information and customer information

   e.   Metadata: Enter the name of your proposal and a description

   f.   Download: Download your proposal and take a look at the included information

6. Click the **Content** page and see what CloudSystem-related collateral you can find. Use the **Search** capabilities or expand the **Enterprise Group** in the left pane and then select CloudSystem.

- How many documents did you find? **Eleven (subject to change)**

- What are some of the documents available? **Executive Summary, Helion OpenStack, Helion OpenStack Community, Matrix Operating Environment for HP-UX, and others**

7. Expand **Cloud (Helion)** in the left pane and see what's available there. What optimized cloud solutions are listed? **Helion High Performance Computing (HPC) (subject to change)**

8. Log out of Proposal Web and close your browser.

Writing a scope of work

A scope of work is a project overview that you prepare for the proposal. It is crucial for ensuring a mutual understanding between you and your customer. The scope of work should provide a summary of the plan you create for the solution, including the following:

- Overall time frame

- Completion milestones for each aspect of the solution

- Resources required

  - Channel partners, HPE sales representatives, HPE services, and any other parties involved in delivering the solution

  - Executive support for the project (name, position, and more)

 Note

    A scope of work should not be confused with a statement of work, which is a final project overview that is prepared for billing purposes.

Learning check

1. What are the steps you need to take to develop a proposal for your customer?

_____

_____

_____

_____

2. Explain the difference between scope of work and statement of work:

   ● A scope of work is

   _____

   _____

   ● A statement of work is

   _____

   _____

Learning check—answers

1. What are the steps you need to take to develop a proposal for your customer?
   - **View the big picture**
   - **Review customer requirements**
   - **Architect the proposed solution**
   - **Prepare the formal proposal**

2. Explain the difference between scope of work and statement of work
   - **A scope of work is an overview you prepare for the proposal—it is crucial for ensuring a mutual understanding between you and your customer**
   - **A statement of work is a final project overview that is prepared for billing purposes**

Predeployment and installation planning

There are several licensing options and considerations available for specific IT implementations. It is important to consider the options, including Helion CloudSystem 9.0 licensing, hybrid cloud licensing, preinstallation preparation of the environment, and additional factors. Discuss these considerations with your customer to make sure they understand how it will affect the solution strategy.

CloudSystem licensing

This section covers the licensing options available for Helion CloudSystem 9.0 and how they affect your planning.

Two licensing options

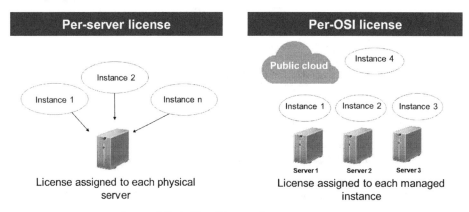

Figure 13-4 CloudSystem licensing options

CloudSystem software licensing is based on two options shown in Figure 13-4:

- **Per-server license**—A license is purchased for each managed physical server. This type of license allows an unlimited number of virtual machines to be deployed and managed on a limited set of licensed servers. This licensing model aligns with how customers buy most of their other data center infrastructure software. Per-server licenses are connected to the physical server that the software runs on for the life of that server. Licenses are not transferable to new hardware, meaning that the license expires when the server is retired.

  - Customers who elect to license their on-premise infrastructure on a per-server basis may still purchase per-OSI licenses to manage public cloud instances.

- **Per-OSI license**—A license is purchased for each managed operating system instance. This type of license allows a limited number of virtual machines (VMs) to be deployed and managed across an unlimited number of physical servers in a private, hybrid, or public cloud infrastructure. The license "counts" the number of VM instances that you run. The licenses do not expire, which means that they can be transferred from one physical server to another as a workload migrates between servers or clouds in a hybrid cloud environment.

- Per-OSI licensing is required for managing public cloud workloads because servers running public clouds are not controlled by the customer (and thus cannot have server licenses assigned to them). Per-OSI licensing provides customers with rights to new software versions as long as they have an active support agreement. This means that customers can avoid needing to repurchase licenses over time.

Basic licensing scenarios (on-premises cloud)

For basic licensing scenarios, applicable to on-premises cloud deployments, consider these scenarios:

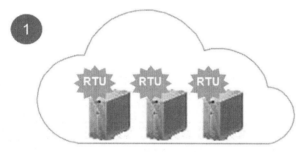

Three server licenses assigned to three managed servers

Figure 13-5 Scenario 1

- **Scenario 1,** shown in Figure 13-5, represents a customer with a single, private, on-premises cloud. This customer can use per-server licensing and needs three server licenses to be assigned to three managed physical servers. These three managed and licensed servers can run an unlimited number of instances.

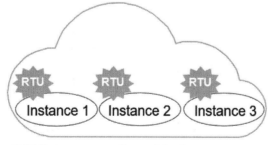

Three OSI licenses assigned to three managed instances

Figure 13-6 Scenario 2

- **Scenario 2**, shown in Figure 13-6, represents a customer that also has a single, private, on-premises cloud. This customer can choose per-OSI licensing and needs three OSI licenses to be assigned to three managed instances. These instances can run on any physical server.

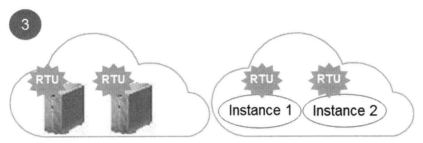

Two server licenses assigned to one cloud and two OSI licenses assigned to another

Figure 13-7 Scenario 3

- **Scenario 3**, shown in Figure 13-7, represents a customer with two private, on-premises clouds. This customer can use the per-server license for one cloud and the per-OSI license for the other cloud. Each cloud then consumes licenses that correspond to its licensing option. This means the per-server licensed cloud consumes server licenses for each physical server, and the per-OSI licensed cloud consumes OSI licenses for each instance.

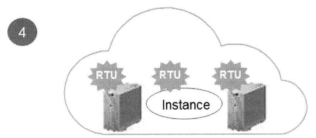

Mix of server and OSI licenses running in a single cloud

Figure 13-8 Scenario 4

- **Scenario 4**, shown in Figure 13-8, represents another customer with a single, private, on-premises cloud. The mix of per-server and per-OSI licenses within a single cloud is not permitted. This customer has to choose one licensing option.

Hybrid cloud licensing scenarios (on-premises cloud)

For licensing options in a hybrid cloud environment, consider Scenario 5 and Scenario 6.

 Note

It is permitted to mix license types across clouds in a hybrid cloud environment.

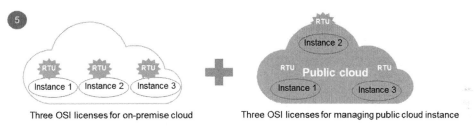

Figure 13-9 Scenario 5

- **Scenario 5**, shown in Figure 13-9, represents a customer using the per-OSI licensing model for the private, on-premises cloud and applying the same model to managing public cloud instances. This example consumes a total of six OSI instances.

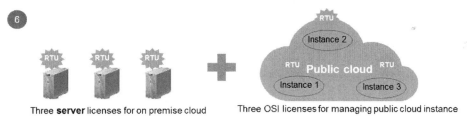

Figure 13-10 Scenario 6

- In **scenario 6**, shown in Figure 13-10, the customer must use the per-OSI license option for instances running in the public cloud (hence, consuming three OSI licenses there). However, the per-server licensing option can be used for the private, on-premises cloud, which consumes three server licenses in this example.

CloudSystem Enterprise per-server licensing

Figure 13-11 CloudSystem Enterprise per-server licensing

Per-server licensing in CloudSystem Enterprise enables a special licensing right for managing public clouds. Each CloudSystem Enterprise per-server license also gives customers the right to manage one instance running in the public cloud, as illustrated by Figure 13-11.

Per-OSI license consumption with Matrix OE

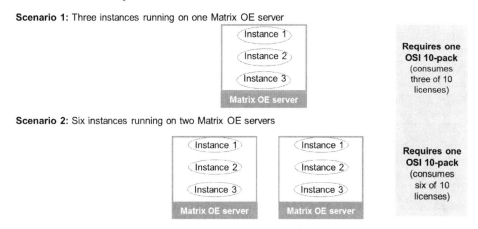

Figure 13-12 Per-OSI license consumption with Matrix OE

The scenarios shown in Figure 13-12 cover customers with per-OSI rights to Matrix Operating Environment (OE), such as the OSI 10-pack. Thus, the Matrix OE environment is not tied to physical servers. Under this licensing model, count your total instances (Matrix OE instances and those managed by CSA) to determine the number of licenses consumed (you are not anchored to a single server deployment).

 Note

These scenarios apply to CloudSystem Enterprise only.

Per-server license consumption with Matrix OE

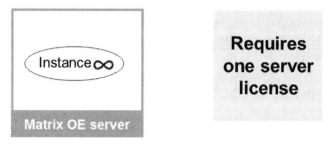

Figure 13-13 Per-server license consumption with Matrix OE

When a customer decides to use the per-server licensing model with Matrix OE, each physical server that is hosting Matrix OE consumes one server license as shown in Figure 13-13. In this case, you count all servers running Matrix OE to determine the number of consumed per-server licenses.

Note

This scenario applies to CloudSystem Enterprise only.

Choosing between per-server or per-OSI license models

To help customers choose between the per-server or per-OSI license models, consider these customer environment characteristics:

- Choose the **per-server** license model if:

  - The customer is taking initial steps into the cloud (or extending their data center capabilities into the cloud).

  - VM densities per server are already high or are expected to increase.

  - Physical server licenses are easier to track than instance licenses.

  - Server refresh rate is low.

  - The customer has a desire to deploy Matrix OE with a traditional per-server license.

- Choose the **per-OSI** license model if:

  - The customer is frequently moving instances between private and public clouds.

  - VM densities per server are low and are expected to stay low.

  - The customer is able to effectively track instances on different servers.

  - Server refresh rate is high.

  - The customer has a desire to deploy Matrix OE on multiple servers.

Note

There are multiple factors that influence a customer's per-server vs. per-OSI licensing decision, and this list is not exhaustive.

Additional licensing considerations

Additional licensing considerations that may affect your planning include the following:

- **Licensing of embedded technologies and installed components**—CloudSystem includes multiple embedded technologies and separately installed components, some of which are also offered by HPE as independent products. These technologies are provided collectively, as a single CloudSystem software solution, and the CloudSystem software license agreement covers the use of all included technology components.

- **Licensing of HPE OneView**—The rights to use HPE OneView are **not** granted by the CloudSystem Foundation or Enterprise software licenses.

- **Licensing of CloudSystem Enterprise Add-on Platform Applications and Analytics Suite**— The right to use this software is not granted by the CloudSystem Enterprise software license.

- **Licensing of Helion OpenStack and Helion Development Platform**—Customers purchase licenses for CloudSystem Foundation and CloudSystem Enterprise, not for the individual components or underlying platforms. Thus, the right to use Helion OpenStack and Helion Development Platform is included with the CloudSystem Foundation license. Both Helion OpenStack and Helion Development Platform versions that are included with the CloudSystem solution remain the same throughout the life of the solution.

License keys provided with purchase

License keys are provided for CloudSystem software components, as needed, to facilitate customer use of the software. Table 13-1 shows CloudSystem components that require license keys and the associated license type.

Table 13-1 License keys provided with purchase.

Purchased software	Components requiring license	License type provided
CloudSystem Foundation	None	N/A
CloudSystem Enterprise	CSA	OSI
	Matrix OE	Server
	Insight Control	Server
	iLO	Server

The license keys that the customer receives may or may not reflect the type of the licensing model chosen. For example, a customer may receive both OSI-based and server-based keys. Regardless of the number and type of keys provided, customer rights to use CloudSystem are defined in their license and purchase agreement.

License key management

License keys are required to enable the Enterprise components of the purchased CloudSystem software. To add license keys:

1. Activate the license on the HPE licensing site.

 Note

To visit the licensing homepage, scan this QR code or right-click it to open the hyperlink.

http://enterpriselicense.hpe.com/

---

2. Use Infrastructure Administrator privileges to add each license key to the appropriate management console. For example, add the Enterprise license key to the Cloud Service Management Console in the Enterprise appliance, or add the Matrix OE license to the Central Management Server (CMS). Add license keys to enable:

  • CloudSystem Enterprise appliance

  • Matrix OE

  • Insight Control

3. Purchase and add license keys as needed.

 Note

During the free trial period of 90 days, if a customer has not yet added a license key, CSA limits the number of new instances that can be created. License compliance is the responsibility of the customer and is subject to HPE audit at any time.

 Note

Additional licensing information can be found in the *Helion CloudSystem 9.0 Administrator Guide*, under "Manage CloudSystem software licensing and license keys" (section 10).

Predeployment planning task checklist

During your planning phase, use the checklist of predeployment tasks shown in Figure 13-14, which is available in the *Helion CloudSystem 9.0 Network Planning Guide*, under "Network planning" (Section 6).

Table 21 Checklist of pre-deployment tasks

☑	Pre-deployment task
	Read the Network Planning Guide.
	Evaluate network options (CVR vs DVR, VLAN vs VxLAN)
	Understand the purpose of CloudSystem networks.
	Determine if firewall rules are required and create the rules, if necessary.
	Plan static routes for the Data Center Management Network.
	Gather network service information for NTP servers, SMTP, DNS servers, and Active Directory.
	Verify that the management cluster has the required number and type of physical NICs.
	Verify that compute nodes have the required number and type of physical NICs.
	Add FQDN entries in the DNS server for the Data Center Management Network and the Consumer Access Network.
	Fill out the Network Planning worksheet and calculate the correct subnet size for each network.
	Assign unique customer VLAN IDs and CIDR ranges for each network.
	Create L3/L2 VLANs.
	Configure the TOR switch ports (trunks) and patch cables.
	Verify L3 network connectivity by pinging gateways from the management cluster.
	Verify DNS server (forward and reverse).
	Verify firewall rules.
	Install HP Helion CloudSystem 9.0 using the two CloudSystem installers.

Figure 13-14 Predeployment planning task checklist

Installation planning worksheet—Activity

Table 22 CloudSystem Installation Planning Worksheet

Installation Preparation	
Images	
Source	release package unpacked and added to your staging environment
Names	
Disk Format	Thin Provision (recommended)
Installation script (csstartgui.bat)	
Source	release package unpacked and added to your staging environment
Target	
CloudSystem Management Appliance Installer	
Management hypervisor type	

Figure 13-15 Planning worksheet—Activity

Use the CloudSystem Installation Planning Worksheet (Figure 13-15—Table 22) in the *Helion CloudSystem 9.0 Network Planning Guide* to answer these questions.

 Note

To access the guide, scan this QR code or enter the URL into your browser.

http://www.hp.com/go/CloudSystem/docs

1. What is the recommended disk format for the hypervisor?

2. For the CloudSystem Management Appliance Installer, are the CS key and CS certificate required or optional?

3. Which networks are configured under the Networks Settings: Management Trunk section?

4. What are the required network settings for the Enterprise appliance?

5. What is the recommended Glance disk size?

Installation planning worksheet activity—answers

1. What is the recommended disk format? **Thin provision**

2. For the CloudSystem Management Appliance Installer, are the CS key and CS certificate required or optional? **Optional**

3. Which networks are configured under the Networks Settings: Management Trunk section?

   - **Data Center Management Network**

   - **Consumer Access Network**

   - **Cloud Management Network**

   - **External network**

4. What are the required network settings for the Enterprise appliance?

   - **Data Center Management FQDN**

   - **Data Center Management Virtual IP address**

   - **Consumer Access Network Virtual IP address**

5. What is the recommended Glance disk size? **512 GB**

HPE support and other resources

Ensuring that you and your customers have access to HPE support and other resources is a key aspect of a satisfying customer relationship.

What to collect before contacting HPE

Before contacting HPE support, collect this information:

- Software product name

- Hardware product model number

- Operating system type and version

- Applicable error message

- Third-party hardware and software

- Technical support registration number (if applicable)

HPE support site

Visit the HPE support site to access product documentation, community forums, HPE specialists, and other resources. You can also click **Contact Us** in the top navigation bar to see different ways to contact HPE.

 Note

To access this site, scan this QR code or enter the URL into your browser. This webpage is for the United States. You can change the region using the top navigation bar.

https://www.hpe.com/us/en/support.html

How to register for software technical support and update service

Helion CloudSystem includes one year of 24×7 HPE Software Technical Support and Update Service. This service provides access to HPE technical resources for assistance in resolving software implementation or operations problems.

The service also provides access to software updates and reference manuals, either in electronic form or on physical media as they are made available from HPE. Customers who purchase an electronic license are eligible for electronic updates only.

With this service, Helion CloudSystem customers benefit from expedited problem resolution as well as proactive notification and delivery of software updates. Registration for this service takes place following online redemption of the license certificate.

 Note

For more information about this service, scan this QR code or enter the URL into your browser.

http://www.hp.com/services/insight

Learning check

1. What are the licensing models supported by Helion CloudSystem?

    _____

    _____

    _____

    _____

    _____

2. Describe a customer environment that should be considered for per-OSI licensing.

    _____

    _____

    _____

    _____

    _____

3. What information should be collected before contacting HPE support?

    _____

    _____

    _____

    _____

    _____

Learning check—answers

1. What are the licensing models supported by Helion CloudSystem?

    - **Per-server licensing**
    - **Per-OSI licensing**

2. Describe a customer environment that should be considered for per-OSI licensing

    - **Frequent movement of instances between private and public clouds**
    - **Low VM densities per server**
    - **Easier tracking of instances running on different servers**
    - **High server refresh rates**
    - **Desire to deploy Matrix OE on multiple servers**

3. What information should be collected before contacting HPE support?

- **Software product name**

- **Hardware product model number**

- **Operating system type and version**

- **Applicable error message**

- **Third-party hardware and software**

- **Technical support registration number (if applicable)**

Final activity: Fictional customer scenarios

Choose one of the following fictional customer scenarios and:

- Review and familiarize yourself with the scenario.

- Define company challenges and business and IT goals. Note that not all elements may be present in the scenario. You may need to identify missing elements and make some assumptions. This is typical of real-life customer circumstances.

- Create a solution that meets the customer's business and IT goals. Ensure that your solution meets both aspects, taking into account the four areas of systematic approach to preparing a solution proposal. Clearly identify any assumptions you made.

Scenario 1: Mid-Sized OEM

Bygg Bättre Prylar AB manufacturing company

Bygg Bättre Prylar AB (BBP) is a medium-sized manufacturing company located in Svängsta, Sweden. BBP manufactures parts for the aerospace industry as well as world-leading generators for water and wind applications.

In the company's latest annual report, executives attributed the company's recent success to research and development (R&D). The R&D Department has created several innovative solutions that improved the designing and manufacturing processes. However, leadership also predicts that the current company structure and IT infrastructure will inhibit momentum going forward.

Two R&D facilities are located in Svängsta. There is one primary manufacturing and prototyping facility located in Shenzhen, China. There are two additional manufacturing sites located throughout China. The designers work 40 hours, Monday through Friday during regular Swedish working hours (Central European Time). At the end of every day, all updated CAD files are bundled and transferred to the manufacturing and prototyping site in Shenzhen. When the transfer is complete, the CAD files are loaded into the manufacturing systems and verified. Currently, 20% of the files have some kind of problem. One of the most common problems is that the designers have very powerful workstations, and all CAD files are stored locally. Each evening, the files are loaded into a centrally located file share, which causes version control issues.

The senior executive vice president (EVP) of R&D and the chief information officer (CIO) from BBP met Antonio Neri, EVP and general manager at HPE, during an HPE executive IT event in Europe. They discussed solutions for distributed compute, storage, security, and networking. The BBP executives were interested in potential HPE solutions. They want you to review some of their business goals, including:

- Implementing a direct tie between R&D and the manufacturing plant for efficient prototyping with full orchestration and automation/management

- Using cloud storage to centralize CAD files and reduce version control and file-transfer challenges

- Centralizing R&D resources to remove the requirement for local workstations and storage

- Maintaining a secure environment

- Automating many of the management tasks

The senior EVP and CIO also want you to analyze their corporate enterprise resource planning (ERP) solution, which is currently a custom system with an Oracle Real Application Clusters (RAC) database. They have had extensive discussions with SAP about moving their ERP system to an SAP environment, and they are interested in moving forward.

Additionally, the manufacturing team is looking to modernize and virtualize their existing manufacturing execution systems to increase potential availability and improve disaster recovery. The engineering teams use rack-mounted servers for scale-up potential and implement all-flash storage for optimized performance.

Scenario 2: Large shipping company

Port on Demand Inc.

Port on Demand Inc. (PoD) is a large terminal operator and stevedoring company with 47 terminals and 73 locations across the United States, including offices and storage facilities. It is a major player in the global shipping industry. The corporate headquarters is located in Los Angeles, California, and the company's two main data centers are located in Phoenix, Arizona, and Elizabeth, New Jersey. Management believes the company's current IT model is their weakest link. They asked your company to make a recommendation on next steps and future strategy.

Some of the company's 47 terminals, including the Los Angeles location, have large IT requirements, and others only have one or two servers. Most locations have dedicated IT staff, and some locations can share staff due to geographical proximity. In addition to these terminals, the company currently operates seven regional data centers.

During discussions with the leadership team, you realize that upgrading and modernizing legacy IT using an all-at-once approach is not realistic; a modular, phased approach is more practical due to funding limitations. You suggest replacing legacy IT and developing new IT environments at the two main data centers in Phoenix and Elizabeth. Implementing these environments would include cloud and software-defined infrastructure strategies from HPE. PoD leadership is also interested in refreshing the IT at their terminals in line with the main data center's initiatives.

PoD requested:

- A modular approach with a self-serve infrastructure model

- Remote management capabilities only requiring low-cost resources for certain IT tasks

- No (or minimal) customization, with an interest in solutions that have been proven to help other companies reach business goals

- A fully integrated and tested IT solution delivered to terminals (branch locations)

- Better business response times for various initiatives, as well as increased cost control

- Improved balance of capital and operational expenditures

- Compatibility with Microsoft Skype for corporate communication

Scenario 1: Mid-Sized OEM example—potential solution

This solution is just one potential proposal for BBP. This solution combines cloud capabilities with traditional IT, which is appealing for many customers. It also covers many data center improvements.

The HPE solution could include the following:

- **Distributed cloud model spanning R&D in Sweden and manufacturing in China**—HPE Cloud Service Automation (CSA) with HPE Operations Orchestration (OO) to manage traditional IT along with public and private cloud instances.

- **Sharing of data over multiple regions (cloud storage)**—HPE Helion Content Depot, including installations at the two R&D facilities in Svängsta and at the main facility in Shenzhen.

  - Preference for an onsite solution, given the sensitive nature of data, connected with either a virtual private network or private dedicated suggested connection.

- **R&D**—HPE Moonshot or HPE ConvergedSystem 700 with workstation blades, using a Citrix virtual desktop infrastructure (VDI)/hosted desktop infrastructure (HDI) solution.

  - Consider key factors such as the intensity of the graphics workload and number of systems to deploy.

  - Establish a proof of concept (POC) to dive deeper.

- **Corporate**—SAP ERP system using HPE ProLiant DL560 Gen9 servers for the database layer with a pair of HPE 3PAR StoreServ 7450 for storage—application layer uses a number of HPE ProLiant DL180 Gen9 servers (specific amount to be determined).

- **MES**—HPE ConvergedSystem 250-HC StoreVirtual with two- or four-node implementations at the manufacturing sites in China, depending on compute requirements and a ConvergedSystem 700 single-rack configuration for the site in Shenzhen.

  - Integrated with Helion Content Depot and CSA for management and replication.

Deployment phases

Phase 1

- **HPE Cloud team**—Land Helion Content Depot in Sweden and in China for cloud storage.

- **HPE converged data center integration (CDI) team**—Work with the EMEA HPE Intel Solution Innovation Center to test a customer workstation implementation on Moonshot and workstation blades to determine the best path forward.

Phase 2

- **HPE CDI team**—Implement an HDI solution for the R&D team.

- **HPE Cloud team**—Start with a CSA and OO proposal for implementation within three months. Initiate an onsite POC to evaluate current processes and automation requirements.

Phase 3

- **HPE CDI team**—Deliver pilot site MES ConvergedSystem 250-HC StoreVirtual and ConvergedSystem 700 with HPE consulting services to assist with migration.

- **HPE Cloud team**—Finalize the CSA and OO implementation, integrating all the designed systems.

Phase 4

- **HPE servers/storage team**—Partner with SAP to understand the customer workload and characteristics for deploying the new ERP platform.

Scenario 2: Large shipping company—potential solution

Multiple strategies have the potential to meet the customer's requirements. The basic HPE solution strategy should show that:

- The focus is on retiring legacy IT over time rather than migrating all at once.

- The new IT environments are modular, and each module is an HPE Helion CloudSystem 9 of a specified size.

- Terminals will have CloudSystem 9, smaller converged systems, or hyper-scale servers.

- HPE Cloud Service Automation (CSA)/HPE Operations Orchestration (OO)/HPE Server Automation (SA) will be incorporated into the solutions.

- Software-defined networking (SDN) with HPE servers and storage is a key part of this solution.

The HPE solution design could include:

- **Primary data centers**—The company's two data centers will receive synchronous CSA implementations with OO and SA, built on HPE ConvergedSystem 700x with HPE Helion CloudSystem 9.0 software and Microsoft Hyper-V.

- **Terminals**—For the terminals, the company will use a hyper-converged design with HPE ConvergedSystem 250-HC StoreVirtual. This building-block approach is right-sized for the workload at each location. Larger sites can transition to single-rack ConvergedSystem 700. The terminals will be connected to the CSA automation environment within the main data centers.

- **SDN**—Across all IT environments, the company will implement the HPE Virtual Application Networks SDN Controller and configure it to control the switch gear within the ConvergedSystem 700x environments. Skype for Business should be an early adoption application for the brand-new IT implementations.

Deployment phases

Phase 1

- Install a ConvergedSystem 700x in one of the two main data centers to begin the project. Implement the VAN SDN Controller to manage the new environment.

- After the ConvergedSystem 700x is installed and running, set up Helion CloudSystem Enterprise software to deploy the Skype for Business platform. Move Skype for Business production traffic to the new platform. Identify other early adoption workloads to move into the new environment.

Phase 2

- Perform a virtualization evaluation within a few terminal environments to size the workload requirements for the modular systems. Design a small, medium, and large bundle for each site to include rack, power, and networking. Pilot the new systems in the terminal location for remote management and process evaluation.

- In the primary sites, configure the Helion CloudSystem Enterprise software to manage and monitor the terminal equipment.

Phase 3

- Define an implementation and migration schedule with PoD for the terminal locations and expansion of the primary sites moving forward.

Summary

- Preparing a solution to meet customer requirements requires a systematic four-step approach:
  - View the big picture.
  - Review customer requirements.
  - Architect the proposed solution.
  - Prepare the formal proposal.
- There are several licensing options and considerations available for specific IT implementations. It is important to consider the options, including Helion CloudSystem 9.0 licensing, hybrid cloud licensing, preinstallation preparation of the environment, and additional factors. Discuss these considerations with your customer to make sure they understand how it will affect the solution strategy.
- Ensuring that you and your customers have access to HPE support and other resources is a key aspect of a satisfying customer relationship. Make sure to collect all relevant information about the product or the solution before contacting HPE.

14 Practice Test

INTRODUCTION

The HPE ASE Designing HPE Data Center and Cloud Solutions certification exam is designed to test your skills to identify, describe, position, demonstrate, and specify the correct HPE Helion Cloud solution based on identified customer needs. Knowledge of how to architect, install, manage, and administer the HPE Helion Cloud solution, based on CloudSystem 9.0 Foundation or Enterprise, is tested in this exam. In-depth OpenStack knowledge is required to successfully pass the exam.

The intent of this book is to set expectations about the context of the exam and to help candidates prepare for it. Recommended training to prepare for this exam can be found at the HPE Certification and Learning website (**http://certification-learning.hpe.com**), as well as in books like this one. It is important to note that although training is recommended for exam preparation, successful completion of the training alone does not guarantee that you will pass the exam. In addition to training, exam items are based on knowledge gained from on-the-job experience and application, as well as other supplemental reference material that may be specified in this guide.

WHO SHOULD TAKE THIS EXAM?

This certification exam is designed for candidates with at least 2 years' experience in a presales consultative role who want to acquire the HPE ASE—Data Center and Cloud Architect V3 certification.

Alternatively, if eligible, you can take the HPE0-D35 delta exam instead of the HPE0-D34 (Designing HPE Data Center and Cloud Solutions).

EXAM DETAILS

The following are details about the exam:

- **Exam ID:** HPE0-D34
- **Number of items:** 65
- **Item types:** Multiple choice (single-response), multiple choice (multiple-response), point and click
- **Exam time:** 1 hour 45 minutes
- **Passing score:** 68%
- **Reference material:** No online or hard copy reference material will be allowed at the testing site.

HP0-D34 Testing Objectives

11%—Fundamental Converged Infrastructure and Cloud architectures and technologies

- Explain cloud resources and the tools used to manage cloud resources, service catalog, and service lifecycle
- Compare and contrast the types of cloud delivery models
- Describe and differentiate cloud enabling technologies and components
- Explain and position business benefits/value and risks/costs associated with cloud implementations

18%—HPE Converged Infrastructure and HPE Helion cloud products, solutions, and warranty/service offerings

- Describe and differentiate the functionality of HPE Helion CloudSystem software
- Describe the HPE strategy for cloud management

8%—Competitive positioning (recognizing opportunity)

- Recognize initial and additional opportunities
- Describe and analyze the cloud solution marketplaces
- Describe and illustrate how cloud solutions provide competitive advantage and add business value
- Demonstrate business acumen (TCO, ROI, IRR, NPV, TCA, CapEx, OpEx, and so forth)
- Identify potential sale and upsale opportunities and refer the customer to the appropriate contact (consulting, services, product, and so forth).

25%—Solution planning and design

- Discover and define the cloud solution opportunity
- Plan and design a cloud solution
- Review and validate a solution proposal
- Prepare and deliver customer presentation

15%—Solution implementation (install, configure, setup, customize, and integrate)

- Perform the predelivery tasks
- Implement the delivery tasks
- Implement postdelivery tasks.

9%—Solution enhancement (performance-tune, optimize, and upgrade)

- Compare the existing solution design to the customer requirements and document differences
- Upgrade or expand the solution.

14%—Solution management (administrative and operational tasks, POC)

- Maintain the management infrastructure
- Monitor and manage the HPE Helion Cloud resources (physical and virtual)
- Create, maintain, and perform backup and recovery processes, as applicable to HPE Converged Infrastructure and HPE Helion cloud

Test preparation questions and answers

The following questions will help you measure your understanding of the material presented in this book. Read all of the choices carefully, as there may be more than one correct answer. Choose all correct answers for each question.

Questions

1. What is the first step cloud administrators should take when installing Helion OpenStack 2.0 for the first time?

    a. Bootstrap initial server

    b. Install Linux

    c. Set up the deployment network

    d. Verify the cloud using Tempest tests

2. In the Helion OpenStack block architecture, what serves as the remote web software repository used for updates?

    a. Keystone

    b. Sherpa

    c. Heat

    d. Monasca

3. What type of servers power Helion Rack?

    a. HPE Cloudline servers

    b. HPE StoreEasy servers

    c. HPE Apollo servers

    d. HPE ProLiant Gen9 servers

4. Which Helion CloudSystem appliance contains the majority of OpenStack services used in CloudSystem?

    a. Management appliance

    b. Monitoring appliance

    c. Update appliance

    d. Cloud controller appliance

5. Which appliance in the Helion CloudSystem 9.0 architecture is optional and only used with VxLAN?

    a. Management appliance

    b. SDN appliance

    c. Monitoring appliance

    d. Update appliance

6. Which Helion CloudSystem 9.0 networks are part of the cloud data trunk network? (Select two.)

    a. Tenant networks

    b. Consumer Access Network

    c. Tenant Underlay Network

    d. Object storage network

    e. Object proxy network

7. In a CloudSystem environment, which appliance contains both admin and member roles?

    a. OpenStack Keystone service

    b. Management appliance (Operations Console)

    c. OpenStack user portal (Horizon)

    d. Key pairs

8. What should existing Helion CloudSystem 8.1 customers do before upgrading to version 9.x? (Select three.)

    a. Stop the CloudSystem 8.1 management appliances

    b. Update hardware and software as needed

    c. Contact Helion Professional Services

    d. Prepare the hypervisors for CloudSystem 9.x deployment

    e. Retire the 8.1 appliances

    f. Add a temporary second CLM network to vCenter

9. Which CloudSystem appliance task requires this command? (Select two.)

   ```
   ssh cloudadmin@[internal_name] sudo shutdown -h now
   ```

   a. Shut down a single CloudSystem virtual appliance

   b. Cluster appliances in a trio configuration

   c. List the status of CloudSystem appliances

   d. Shut down all CloudSystem appliances

10. What types of compute nodes can CloudSystem manage simultaneously? (Select two.)

   a. Deactivated compute nodes

   b. ESXi clusters

   c. Hyper-V compute nodes

   d. OVSvApp

11. Before virtual machine instances can be provisioned in the cloud, what is the minimum number of Provider or Tenant networks that must be created?

   a. Two

   b. Zero (there is no minimum)

   c. One

   d. Three

12. Which open-source project does CloudSystem use to provide Monitoring-as-a-Service?

   a. Zaqar

   b. Elasticsearch

   c. Kibana

   d. Monasca

13. What does Helion CloudSystem Enterprise use to provide physical server provisioning?

   a. Helion Eucalyptus

   b. Embedded MySQL

   c. CloudSystem Foundation instances

   d. HPE OneView and ICsp integration

14. Which console is used to perform all management tasks on CloudSystem Enterprise appliances?

    a. Operations Console

    b. Horizon Portal

    c. Marketplace Portal

    d. OneView

15. Which tool can be used in the CSA Topology Designer to provision physical servers in CloudSystem Enterprise?

    a. Helion Eucalyptus

    b. OneView

    c. Kibana

    d. OpenStack Sahara

16. What is missing from this graphic showing the flow of service provisioning in CSA?

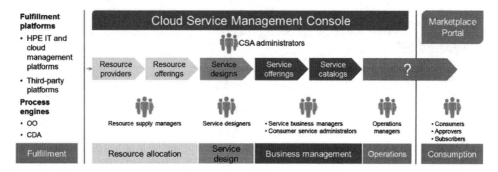

    a. Third-party platforms

    b. HPE IT and cloud platforms

    c. Subscriptions and service instances

17. Your customer asks you to show them how to provision HPE StoreServ storage with HPE OneView. If you do not have access to your own demo environment, what can you use instead?

    a. HPE SPOCK

    b. HPE Partner Ready Portal

    c. HPE Simplified Configuration Environment

    d. HPE Solution Demonstration Portal

18. Why are business leaders increasingly involved in IT purchasing decisions? (Select two.)

    a. They recognize the role technology plays in their success

    b. HPE requires c-suite involvement

    c. They want IT to be aligned with the company's goals

    d. They recognize that acquiring technology is a simple process

19. Which business financial metric has these characteristics?

    - Measures the economics of IT assets over their service life

    - Includes the direct cost of purchase, services, maintenance, and other indirect costs

    - Is part of project-specific financial measurements

    a. Internal rate of return (IIR)

    b. Return on investment (ROI)

    c. Total cost of ownership (TCO)

    d. Net book value (NBV)

20. What is the projected number of application releases that will be developed per year, per application by 2020?

    a. 36

    b. 120

    c. 55

    d. 220

21. A customer is considering cloud solutions from HPE and VMware. What should you say to highlight the benefits from HPE? (Select two.)

    a. For VMware, virtual infrastructure is the core of its offering, and cloud is just an extension.

    b. VMware does not have experience in on-premises virtualization.

    c. HPE and VMware both have partnerships with major players across cloud landscape.

    d. VMware lacks core application platform and development capability.

22. Which financial and operating key performance indicators can benefit from Helion solutions? (Select two.)

    a. Revenue growth

    b. Net Promoter Score (NPS)

    c. Customer loyalty

    d. Time to market

23. Which questions specific for a cloud customer should you consider when building a solution? (Select two.)

    a. Is the customer global?

    b. Does the customer need to use both private and public cloud services?

    c. Is the customer willing to purchase new hardware?

    d. Does the customer need IaaS, PaaS, SaaS, or XaaS?

24. Your customer needs help choosing a CloudSystem licensing model. Which model should you recommend under these circumstances?

    - The customer is taking initial steps into the cloud

    - VM densities per server are expected to increase

    - Server refresh rate is low

    a. CSA instance model

    b. Matrix OE model

    c. Per-server license model

    d. Per-OSI license model

25. New Helion CloudSystem customers receive 24x7 HPE Software Technical Support and Update Service for what length of time?

    a. Six months

    b. One year

    c. Two years

    d. Three years

Answers

1. ☑ **C** is correct. The deployment network is used to deploy Helion OpenStack for network booting of the servers to install Linux and then for Ansible access via SSH.

   ☒ **A, B,** and **D** are incorrect. Bootstrap initial server is the second step. Install Linux is the fifth step. Verify the cloud using Tempest tests is the last step.

   For more information, see Chapter 7.

2. ☑ **B** is correct. The Sherpa service provides a link to a remote web catalog of software available for purchase and download and for updating the Helion environment.

   ☒ **A, C,** and **D** are incorrect. Keystone provides identity (authentication and authorization) services. Heat provides orchestration services. Monasca provides monitoring services.

   For more information, see Chapter 7.

3. ☑ **D** is correct. ProLiant DL360 and DL380 Gen9 servers are provided with Helion Rack.

   ☒ **A, B,** and **C** are incorrect. HPE Cloudline, StoreEasy, and Apollo servers are not provided with Helion Rack.

   For more information, see Chapter 7.

4. ☑ **D** is correct. The cloud controller appliance contains Nova, Glance, Neutron, Cinder, Heat, and Swift services.

   ☒ **A, B,** and **C** are incorrect. The management appliance contains the Keystone service. The monitoring appliance contains the Monasca service. The update appliance contains the Sherpa service.

   For more information, see Chapter 8.

5. ☑ **B** is correct. The SDN appliance is only deployed in environments configured to support VxLAN for Tenant and Provider networks.

   ☒ **A, C,** and **D** are incorrect. The management appliance is not optional. It is responsible for standing up and managing CloudSystem virtual appliances. The monitoring appliance is not optional. It contains the monitoring services that are used to monitor the performance and health of CloudSystem virtual appliances and compute nodes. The update appliance is not optional. It manages patches and upgrades to the CloudSystem environment.

   For more information, see Chapter 8.

6. ☑ **A** and **C** are correct. Tenant networks are part of the cloud data trunk and are restricted, meaning that they can be accessed only by VM instances assigned to that network. Tenant Underlay Network is part of the cloud data trunk and is a single network VLAN that encapsulates and carries Tenant and Provider networks as VxLANs.

☒ **B, D,** and **E** are incorrect. Consumer Access Network is part of the management trunk. Object storage and object proxy networks are part of the storage trunk.

For more information, see Chapter 8.

7. ☑ **C** is correct. The OpenStack user portal (Horizon) contains admin and member roles.

☒ **A, B,** and **D** are incorrect. OpenStack Keystone is a service, not an appliance. Management appliance (Operations Console) contains only the admin role. Key pairs is a function of the Keystone service and is not an appliance.

For more information, see Chapter 8.

8. ☑ **B, D,** and **F** are correct. Before initiating the migration process, hardware and software should be upgraded as needed, hypervisors should be prepared for CloudSystem 9.x deployment, and a temporary second CLM network should be added to vCenter.

☒ **A, C,** and **E** are incorrect. Stop the CloudSystem 8.1 management appliance is performed after the migration has been completed. Contact Helion Professional Services is not a required action. Retire the 8.1 appliances is performed after the migration has been completed.

For more information, see Chapter 8.

9. ☑ **A** and **D** are correct. SSH and the cloudadmin credentials are used to connect to the first management appliance in the trio. From the first management appliance, SSH is used to connect to the desired appliance, and the shutdown command is issued. To shut down all appliances, SSH is used to log in to the first management appliance in the trio, ma1. All other appliances are then shut down in order, finishing with ma1.

☒ **B** and **C** are incorrect. Appliance clustering is configured during first time installation. The shutdown command is not used to list appliance status.

For more information, see Chapter 9.

10. ☑ **B** and **C** are correct. CloudSystem can simultaneously manage ESXi clusters, Hyper-V compute nodes, and KVM compute nodes.

☒ **A** and **D** are incorrect. Compute nodes must be activated before CloudSystem can manage them. OVSvApp is an appliance that runs on compute nodes.

For more information, see Chapter 9.

11. ☑ **C** is correct. Before virtual machine instances can be provisioned in the cloud, you must create at least one Provider or Tenant network.

☒ **A, B,** and **D** are incorrect. Before virtual machine instances can be provisioned in the cloud, you must create at least one Provider or Tenant network.

For more information, see Chapter 9.

12. ☑ **D** is correct. CloudSystem provides Monitoring-as-a-Service using the open-source project Monasca.

    ☒ **A, B,** and **C** are incorrect. Zaqar is a multi-tenant cloud messaging service for web and mobile developers. Elasticsearch is used by the CloudSystem logging service for database searching and indexing. Kibana is used by the CloudSystem logging service for data visualization.

For more information, see Chapter 9.

13. ☑ **D** is correct. CloudSystem Enterprise uses HPE OneView and ICsp to provision physical servers.

    ☒ **A, B,** and **C** are incorrect. Helion Eucalyptus is not used to provide physical server provisioning. It is an open solution for building private clouds that are compatible with Amazon Web Services (AWS). Embedded MySQL is not used to provide physical server provisioning. Virtual appliances are delivered in a three-node high-availability (HA) management cluster with embedded MySQL. CloudSystem Foundation instances are not used to provide physical server provisioning.

For more information, see Chapter 10.

14. ☑ **A** is correct. All CloudSystem Enterprise appliance management tasks are performed through the CloudSystem Operations Console.

    ☒ **B, C,** and **D** are incorrect. The Horizon Portal is not used to manage CloudSystem Enterprise appliances. It is used to manage VM instances, tenant networks, virtual routers, and so on. The Marketplace Portal is not used to manage CloudSystem Enterprise appliances. It is used by cloud consumers to manage service subscriptions. OneView is not used to manage CloudSystem Enterprise appliances. It is used to manage the servers, storage, and networks that comprise the Converged Infrastructure.

For more information, see Chapter 10.

15. ☑ **B** is correct. In CloudSystem Enterprise, users can use the CSA Topology Designer to provision physical servers using HPE OneView profiles.

    ☒ **A, C,** and **D** are incorrect. Helion Eucalyptus is not used to provision physical servers. It is an open solution for building private clouds that are compatible with Amazon Web Services (AWS). Kibana is not used to provision physical servers. It is used by the CloudSystem logging service for data visualization. OpenStack Sahara is not used to provision physical servers. It is used to create a data-intensive application cluster (Hadoop or Spark) on top of OpenStack.

For more information, see Chapter 10.

16. ☑ **C** is correct. Managing subscriptions and service instances is the last of the high-level steps for provisioning through CSA.

    ☒ **A** and **B** are incorrect. Third-party platforms and HPE IT and cloud platforms are not high-level steps for provisioning through CSA.

For more information, see Chapter 10.

17. ☑ **D** is correct. HPE Solution Demonstration Portal is correct. It provides a central location for all demonstrations, webinars, and supporting collateral that showcase HPE technologies.

    ☒ **A, B,** and **C** are incorrect. HPE SPOCK does not include product demos. It is used to obtain detailed information about supported HP Storage product configurations. HPE Partner Ready Portal does not include product demos. It provides access to sales tools and resources for sales engagement, training, demand generation, and business management. HPE Simplified Configuration Environment does not include product demos. It is used to gather customer requirements and to map them to a set of products or service options.

    For more information, see Chapter 11.

18. ☑ **A** and **C** are correct. The majority of today's business leaders recognize the role technology plays in their success and that this presents both an opportunity and a challenge. Business leaders need their investments in IT to deliver real business value.

    ☒ **B** and **D** are incorrect. HPE requiring c-suite involvement is desirable but is not a requirement. Acquiring technology can be a very complex and time-consuming process.

    For more information, see Chapter 11.

19. ☑ **C** is correct. TCO matches all of the characteristics.

    ☒ **A, B,** and **D** are incorrect. Internal rate of return (IIR) is a method of calculating rate of return. Return on investment (ROI) is a numerical representation (expressed as a percentage) of the earning power of a company's assets. Net book value (NBV) is an accounting tool for gradually reducing the recorded cost of a fixed asset.

    For more information, see Chapter 11.

20. ☑ **B** is correct. It is projected that by 2020, there will be 120 releases per year, per application.

    ☒ **A, C,** and **D** are incorrect. It is projected that by 2020, there will be 120 releases per year, per application.

    For more information, see Chapter 12.

21. ☑ **A** and **D** are correct. Cloud is at the core of HPE offerings, while for VMware, virtual infrastructure is at the core, with cloud as an extension. Unlike HPE, VMware lacks core application platform and development capability.

    ☒ **B** and **C** are incorrect. On-premises virtualization is one of VMware's strengths. HPE and VMware both have partnerships with major players across cloud landscape may be true, but does not highlight the benefits of working with HPE.

    For more information, see Chapter 12.

22. ☑ **A** and **D** are correct. Helion solutions can have a positive impact on revenue growth and can reduce time to market.

    ☒ **B** and **C** are incorrect. Net Promoter Score (NPS) is incorrect. It is a tool that can be used to measure the strength of a company's customer relationships. Customer loyalty is important, but it is not a financial or operating KPI.

    For more information, see Chapter 12.

23. ☑ **B** and **D** are correct. Building only on private or using hybrid resources is a key consideration for cloud customers. The required service delivery models can have a significant impact on solution design and must be considered.

    ☒ **A** and **C** are incorrect. Is the customer global? and Is the customer willing to purchase new hardware? are not specific to cloud customers, they apply to all customers.

    For more information, see Chapter 13.

24. ☑ **C** is correct. The Per-server license model satisfies all of the requirements.

    ☒ **A, B,** and **D** are incorrect. CSA instance model and Matrix OE model are not valid CloudSystem licensing models. The Per-OSI license model is recommended when server refresh rate is high.

25. ☑ **B** is correct. Helion CloudSystem includes one year of 24×7 HPE Software Technical Support and Update Service.

    ☒ **A, C,** and **D** are incorrect. Helion CloudSystem includes one year of 24×7 HPE Software Technical Support and Update Service.

Glossary of HPE Converged Infrastructure and Cloud Terms

Appendix 1

A

ADM	See *HPE Application Deployment Manager*.
Agentless data collection	The collection of inventory and performance data from managed systems without requiring installation or configuration of agents on the managed systems.
Alert	Represents one or more events or possible problems with a resource; usually has a severity, status, and description associated with it.
AP4SaaS	See *HPE Aggregation Platform for SaaS*.
Apache Hadoop	A distributed, open-source, Linux-based platform for data storage and processing; massively scalable and highly fault tolerant.
API	See *Application programming interface*.
Application lifecycle management	The process of planning, provisioning, customizing, and configuring applications and their resources, patching, and ultimately returning them to the pool of free resources.
Application programming interface	A set of programming language libraries used to communicate with a particular service/application/server.
Array	A physical storage device or an aggregate set of devices containing one or more storage pools that serves storage to clients and provides an API or a management console to allow an administrator to manage the storage provided by the device.

B

Bare metal server	A server that is not booted with a production operating system (could be a new server with no operating system); a server not yet known to the management software.
Baseline	A timeless demand profile used to generate demand profiles in forecasting.
BfS	See *Boot from SAN*.

Big data	A collection of data sets that, because of their size and complexity, are difficult to process using relational database management tools and desktop statistics and visualization packages.
Bill of materials	A list of parts needed to create a unit; a list of components or subcomponents, such as part numbers, of an equipment or a hardware configuration.
Blob store	A region of memory that is accessible by both the HPE iLO and the embedded software environment. Management software can communicate with the embedded environment through the iLO by reading files from and writing them to the blob store.
Block storage	Storage solution based on presenting a disk or a virtual disk to the operating system as if it were a local disk drive. Typically the operating system accesses the device directly, or builds a local file system on top of that device. The system typically assumes exclusive access to the presented device.
BoM	See *Bill of Materials.*
Boot from SAN	Booting of server operating system from a boot device (logical drive) located on a storage area network (SAN) as opposed to from an internal storage device.
BOOTP	See *Bootstrap protocol.*
Bootstrap protocol	Network protocol used by a network client to obtain an IP address from a configuration server. Used during the bootstrap process when starting a computer.
Bridging	A technique of providing a service sourced from multiple providers.
Bring Your Own Device	Policies and technologies that allow employees to bring personally owned mobile devices (laptops, tablets, and smart phones) to their workplace and use those devices to access privileged company information and applications.
Brokering	Providing access to services from one or more resource pools or service providers.
BURA	Backup, Recovery, and Archiving
Bursting	A technique used within HPE CloudSystem Matrix to provide additional resources on an as-needed basis to easily manage uneven service demands. Can be done by using local resources (local bursting) or by using public cloud resources. Also called *cloud bursting.*
Business continuity	Planning and activities performed by an organization to ensure that critical business functions continue being available to customers, suppliers, regulators, and other entities on a daily basis (not only during a time of disaster).
BYOD	See *Bring Your Own Device.*
C	
Capacity planning	The analysis of and planning for resource usage by workloads on a system or across a set of systems.
Capacity-planning simulation	The process of combining workload demand profiles, as prescribed by a scenario, to estimate the demand profiles of the systems that contain the workloads. Statistics gathered from the simulation can be summarized in reports.

CEE	See *Converged Enhanced Ethernet.*
Ceilometer	An OpenStack Telemetry service that provides a single point of contact for billing systems, supplying all counters they need to establish customer billing across all current and future OpenStack components.
CI	See *Converged infrastructure.*
CI-CM	See *HPE Converged Infrastructure Capability Model.*
CIDR	See *Classless Internet Domain Routing.*
Cinder	An OpenStack Block Storage component that provides persistent block-level storage devices for use with OpenStack compute instances. The block storage system manages the creation, attaching, and detaching of block devices to servers.
City 2.0	HPE vision for a sustainable future; a sustainable model for cities that takes full advantage of the technologies of the Information Age.
Classless Internet Domain Routing	A method for allocating IP addresses and routing Internet Protocol packets; designed to replace the previous addressing architecture of classful network design in the Internet to slow the growth of routing tables and the rapid exhaustion of IPv4 addresses.
CLI	See *command line interface.*
Client virtualization	See *desktop virtualization.*
Cloud auditor	Third-party individual or organization that examines a cloud provider's services from a security, privacy impact, and performance perspective.
Cloud broker	Organization acting as an intermediary between a cloud consumer and one or more cloud providers.
Cloud Cruiser	A heterogeneous cloud financial management solution that enables enterprises and service providers to optimize costs and increase profitability in public, private, and hybrid cloud environments.
Cloud deployment models	Define cloud infrastructure location, its ownership, management, method of access, and access population. Include public, private, and hybrid clouds.
Cloud Maps	HPE templates, best practices, workflows, sizers, deployment scripts, and white papers for creating service catalogs in HPE CloudSystem Matrix. Free and fee-based.
Cloud Service Automation	An HPE solution composed of automation software tools for brokering and managing application and infrastructure services in private and hybrid cloud environments. Available as part of HPE CloudSystem or as a stand-alone software solution.
Cloud service models	Categories of cloud services offered by the cloud providers to cloud service consumers. Include SaaS, PaaS, and IaaS.
Cloud service provider	Organization that makes cloud computing services available to others, especially to cloud consumers and cloud brokers. Also called *cloud providers.*
Cluster	A parallel or distributed computing system comprising many discrete systems that form a single, unified computing resource.

CNA	See *converged network adapter*.
Command line interface	A means of interacting with a computer device, a program, or an operating system where the user issues text-based line commands.
Commodity IT	Services that are required by or used by a large percentage of the users or departments in a business.
Connection template	A Virtual Connect FlexFabric construction that defines general connection requirements.
Consistency group	A group of LUNs that need to be treated the same from the perspective of data consistency.
Converged Enhanced Ethernet	Converges variety of data center traffic by stacking host protocols and consolidating network interfaces. Enables Ethernet to deliver a "lossless" transport technology with congestion management and flow control features. Also called *data center bridging*.
Converged infrastructure	An IT environment where server, storage, and networking resources are standardized, virtualized, organized into, and managed as resource pools.
Converged Network Adapter	A network adapter that combines the functionality of a host bus adapter (HBA) with a network interface controller (NIC).
Core	The actual data-processing engine within a processor. A single processor might have multiple cores, and a core might support multiple execution threads.
Create, read, update, and delete	Basic HTTP methods in the REST API.
Cross-technology logical servers	Logical servers that can be moved back and forth between being VC hosted (hosted by a physical server) or VM hosted (hosted by a virtual machine).
CRUD	See *create, read, update, and delete*.
CSA	See *HPE Cloud Service Automation*.
CSP	See *cloud service provider*.

D

Data center bridging	See *Converged Enhanced Ethernet*.
Deduplication	Technology similar to audio and video compression that compresses data files so that a smaller storage volume can hold the data. File and backup sets are examined with redundant information (white space) represented rather than transmitted whole.
Demand profile	A set of resource-demand readings made at regular intervals for a period of time. The demand profile of a workload, a system, or a complex is used when doing capacity planning. Demand profiles can be based on historical data or computed as part of a forecast.
De-provisioning	The process of returning a no-longer needed resource to the pool of available resources. May include scrubbing of sensitive data and generation of a final invoice.

Desktop virtualization	A VM running a desktop operating system that users access from multiple locations using a variety of thick and thin devices. Also called *client virtualization*.
Disaster recovery	A set of processes, policies, and procedures necessary to prepare for recovery or continuation of technology infrastructure critical to an organization after a natural or human-caused disaster. A subset of business continuity.
Discovery	A feature within a management application that finds and identifies network objects.
DMA	See *HPE Database and Middleware Automation*.
Domain	A collection of management elements that can be interconnected and communicate in a deployed environment. Domains can be managed through the same manageability network (directly or indirectly).

E

East/west traffic	The network traffic that goes between servers in a given data center (in the context of the data center LAN), or traffic that goes between servers in different data centers (in the context of the WAN). Can also refer to client-to-client traffic in collaboration tools.
EBIPA	See *Enclosure Bay IP Addressing*.
EFI	See *Extensible Firmware Interface*.
EG	See *Enclosure Group*.
Elasticity	Ability to dynamically expand or contract a computing resource based on demand.
Enclosure	A chassis that contains multiple server blades and interconnect devices.
Enclosure Bay IP Addressing	Automatically assigns IP addresses for the HPE iLO processors that are bridged through the HPE BladeSystem Onboard Administrator and to the interconnect module management ports.
Enclosure Group	Defines the desired physical configuration of a physical enclosure resource within the HPE OneView appliance. An enclosure must belong to a single enclosure group.
End User License Agreement	A contract between the software licensor and the purchaser, establishing the purchaser's right to use the software.
End-of-row network design	A network design where each server in the individual racks is connected to a common EoR aggregation switch directly, without connecting to individual ToR switches in each rack.
Endpoint	Within a Fibre Channel SAN, devices providing or consuming storage are known as endpoints and are identified by a World Wide Name (WWN).
	An endpoint is a URL that can be used to access services within OpenStack
EoR network design	See *end-of-row network design*.
EULA	See *End User License Agreement*.

Extensible Firmware Interface	Specification that defines the software interface between an operating system and the platform firmware. Unified EFI (UEFI) supersedes the original EFI specification by Intel and replaces the BIOS firmware interface.

F

Fabric	Fibre Channel network composed of one or more interconnected Fibre Channel switches.
Failover	The operation that takes place when a primary service (network, storage, or CPU) fails and the application continues operation on a secondary unit.
FCoE	See *Fibre Channel over Ethernet*.
Fibre Channel	A network technology primarily used for storage networks.
Fibre Channel over Ethernet	An encapsulation of Fibre Channel frames over Ethernet networks. This allows Fibre Channel to use 10 Gbps or higher-speed Ethernet networks while preserving the Fibre Channel protocol.
File storage	File storage enables users to access files and folders through a hierarchal formatted file system, such as NTFS. Files are stored based on name and file extension.
Firmware baseline	A specification of the firmware set to be used for the server profile or an environment.
FlexBranch	Technology that converges network functionality with services, enabling branch office employees to enjoy the same fast and reliable access to data and applications as employees at the main office.
FlexCampus	An infrastructure for mid-size or enterprise campus LANs; unifies wired and wireless networks to deliver media-optimized, secure, and identity-based access.
FlexFabric	Solution that provides the network infrastructure for the data center; simplifies data center infrastructure with converged network, computer, and storage resources across both virtual and physical environments to accommodate cloud computing models.
FlexManagement	A layer across the other three building blocks of the HPE FlexNetwork (see *FlexNetwork*) that enables management of all network segments through a single-console management application, HPE Intelligent Management Center (HPE IMC).
FlexNetwork	The fabric of the HPE Converged Infrastructure, seamlessly connecting users, applications, and data, delivering simplicity with a unified, consistent, standards-based architecture from applications to users; consists of FlexFabric, FlexCampus, FlexBranch, and FlexManagement.
FlexNIC	An abstraction of a portion of Virtual Connect Flex-10 and FlexFabric adapter connections, which are presented to the operating system as standard network interface cards (NICs). Each FlexNIC has a unique MAC address and supports port aggregation, failover, and VLAN tagging.
Forecast	A prediction of system utilization and workload demand profiles for a future time.
Forecast data range	A time interval specifying the set of historical data to use for generating a forecast.

Forecast model	A combination of a forecast data range and a set of annual projected growth rates that are used to estimate future utilization.
G	
Glance	An OpenStack Image service that provides discovery, registration, and delivery services for disk and server images. Stored images can be used as templates or to store and catalog an unlimited number of backups.
Guest OS	The operating system that is running on a virtual machine.
H	
HA	See *high availability*.
HBA	See *host bus adapter*.
Headroom	In general, the amount of a computing resource that is available on a system after all requirements for applications on the system are accounted for.
Heat	An OpenStack Orchestration service that orchestrates multiple composite cloud applications using templates.
High availability	System design approach and associated service implementation that ensures that a prearranged level of operational performance will be met during a contractual measurement period.
Horizon	An OpenStack Dashboard component that provides administrators and users a graphical interface to access, provision, and automate cloud-based resources. It allows for third-party products and services, such as billing, monitoring, and additional tools.
Host	A system or a partition that is running an instance of an operating system.
Host bus adapter	A circuit board or an integrated circuit adapter that provides input/output processing and physical connectivity between a server and a storage device.
Host name	The name of a system or a partition that is running an instance of an operating system.
Host OS	The operating system that is running on the host machine.
Hosted private cloud	Single-tenant, private cloud infrastructure created and operated by a third-party cloud provider and located on the provider's premises.
Hosting	The process of making services available via a network.
HPE StoreServ Persistent Ports	A tier 1 resiliency feature available on all HPE StoreServ systems; provides nondisruptive online software upgrades without relying on multipathing software to initiate failover.
HPE StoreServ	A family of modern storage systems from HPE that are autonomic, efficient, and federated.
HPE ActiveAnswers	An online portal that provides various tools to help you configure, size, use, and analyze performance of HPE products and solutions.
HPE ADM	See *HPE Application Deployment Manager*.

HPE Aggregation Platform for SaaS	An HPE Software offering that serves as the single point of access for applications (SaaS); enables operators to create a SaaS marketplace portal where customers discover SaaS products and bundles, run trials, subscribe to services, and consume them.
HPE AP4SaaS	See *HPE Aggregation Platform for SaaS.*
HPE Application Deployment Manager	An HPE Software offering that enables full application stack management and full application lifecycle management.
HPE Autonomy	An HPE market-leading software company that helps organizations understand the meaning buried in their information. It has developed a unique meaning-based technology that makes sense of and processes unstructured human information.
HPE CI-CM	See *HPE Converged Infrastructure Capability Model.*
HPE CloudSystem	A complete, packaged, integrated, open platform from HPE for private, public, and hybrid clouds. Part of HPE Converged Systems.
HPE CloudSystem Enterprise	An HPE CloudSystem offering for private or hybrid clouds using IaaS, PaaS, or SaaS service models. Supports third-party hardware and provides full lifecycle management and advanced application provisioning.
HPE CloudSystem Foundation	A foundational cloud offering from HPE that provides the essentials to deliver Infrastructure-as-a-Service (IaaS) while leveraging HPE Cloud OS and includes these OpenStack service offerings: Nova, Cinder, Neutron, Keystone, Glance, and Horizon.
HPE Converged Infrastructure Capability Model	A method for organizations to develop their own roadmap for transitioning to a converged infrastructure and to cloud; uses a realistic and structured approach based on business and IT objectives and quantitative metrics.
HPE ConvergedSystems	HPE turnkey systems that integrate hardware, software, and services. Tested, integrated, optimized, and scalable for virtualization, cloud, and next-generation applications.
HPE Data Center Smart Grid	Technology within an HPE Converged Infrastructure that collects and communicates thousands of measurements across IT systems to make the data center less complex, more flexible, and cost-effective. Includes a central component called *HPE Thermal Logic.*
HPE Database and Middleware Automation	An HPE Software offering that automates manual, repetitive administrative tasks with databases and application servers.
HPE DMA	See *HPE Database and Middleware Automation.*
HPE EcoPOD	An HPE high-efficiency, turnkey data center solution that can scale to almost any capacity in a small space.
HPE IMC	See *HPE Intelligent Management Center.*
HPE Insight Control	System management software capable of managing a wide variety of systems, including HPE servers, clusters, desktops, workstations, and portables.

HPE Insight Online	A web-based, personalized IT and support dashboard; provides the information you need to monitor HPE devices in your IT environment from anywhere, at any time, and at no additional cost.
HPE integrated Lights-Out	A management processor that virtualizes system controls, enables remote management, simplifies server setup, and engages in health monitoring and power and thermal control of HPE servers.
HPE Intelligent Management Center	A centralized network management platform from HPE that allows you to manage both physical and virtual networks; supports both HPE and third-party network devices.
HPE Intelligent Provisioning	HPE technology that configures HPE ProLiant Gen8 and later servers and deploys "off-the-shelf" and HPE branded versions of leading server operating systems; replaces HPE SmartStart.
HPE Network Automation	An HPE Software offering that automates the complete operational lifecycle of network devices, from provisioning to policy-based change management, compliance, and security administration.
HPE OneView	A converged infrastructure, software-defined management platform providing a single integrated view of the IT infrastructure, a software-based approach to lifecycle management, and an open development environment.
HPE OO	See *HPE Operations Orchestration*.
HPE Operations Orchestration	An HPE software solution that automates data center tasks and processes using workflows. Part of HPE CloudSystem Enterprise
HPE SA	See *HPE Server Automation*.
HPE SAVA	See *HPE Server Automation Virtual Appliance*.
HPE Server Automation	HPE software that provides complete lifecycle management for physical and virtual servers and applications—from establishing a baseline to provisioning, patching, configuration management, and compliance assurance.
HPE Server Automation Virtual Appliance	Part of the HPE Cloud Automation Solutions, HPE Server Automation Standard, also called *HPE Server Automation Virtual Appliance (SAVA)*, is a single HPE Server Automation core that is packaged as a virtual machine.
HPE SiteScope	An agentless HPE solution that helps ensure the availability and performance of distributed IT infrastructure and application components.
HPE Smart Update Manager	A technology included in many HPE systems software maintenance and management products for installing and updating firmware and system software components on HPE ProLiant and HPE Integrity servers, enclosures, and options.
HPE SUM	See *HPE Smart Update Manager*.
HPE Thermal Logic	HPE technology that enables administrators to dynamically track and control power limits based on workload demands within the chassis, and reclaim overprovisioned power and cooling capacity without impacting performance.

HPE UCMDB	See *HPE Universal Configuration Management Database*.
HPE Universal Configuration Management Database	An HPE Software offering that stores, controls, and manages software and infrastructure components along with associated relationships and dependencies.
HPE Virtual Connect	HPE hardware virtualization product, primarily for server blades.
HPE Virtual Connect Enterprise Manager	Centralizes network connection management and workload mobility for HPE CloudSystem Matrix servers that use Virtual Connect to access local area networks (LANs), storage area networks (SANs), and converged network environments.
HPE Virtual Connect Manager	An embedded software utility, used for smaller Virtual Connect configurations, that defines and manages server connection profiles for each server blade installed in the enclosure.
HPE ConvergedSystem for Virtualization	A complete, packaged, and integrated solution from HPE that is tuned for virtual application environments and provides simplified deployment, management, and security of physical and virtual resources. Part of HPE ConvergedSystems.
Hybrid cloud	The cloud infrastructure contains two or more clouds (private, community, or public) that remain separate but are bound together by technology that enables data and application portability (that is, bursting in CloudSystem Matrix).
Hybrid cloud delivery	A cloud delivery model where the normal computing demands are met using an in-house private cloud and demand spikes are met by using resources from public cloud providers.
Hyper-threading	Intel technology that enables certain processors to create a second virtual core that allows additional efficiencies of processing. This is not a true multicore processor, but it may add performance benefits.
Hypervisor	A virtualization software layer that distributes the physical server resources among multiple virtual machines.

I

IaaS	See *Infrastructure as a Service*.
iLO	See *HPE integrated Lights-Out*.
IMC	See *HPE Intelligent Management Center*.
Infrastructure as a Service	Cloud service model providing processing, storage, network, or other fundamental computing resources to the consumer.
Infrastructure layer	A layer within the HPE Cloud Functional Reference Architecture that provides the physical resources, such as servers, storage, networking, power, and cooling.
Infrastructure service	A running configuration of infrastructure resources that is designed to run a business application such as a multitier web application. Infrastructure resources include server blades, virtual machines, SAN disks, networks, and IP addresses.
Intelligent PDU	See *Intelligent Power Distribution Unit*.

Intelligent power discovery	HPE technology that enables customers to capture power data across racks and rows of servers to identify and eliminate areas of waste, extend the life of the data center, and ensure uptime by eliminating human error during power planning and provisioning.
Intelligent Power Distribution Unit	A PDU that can be automatically discovered and controlled.
Intelligent Provisioning	See *HPE Intelligent Provisioning*.
Intelligent Resilient Framework	An HPE virtualization technology for HPE Comware switches that transforms multiple physical network devices into a single logical device.
Interconnect module	A hardware module that enables communication between the server hardware in the enclosure and the data center networks.
Internet Group Management Protocol	Communications protocol used on IP networks to establish multicast group memberships.
iPDU	See *Intelligent Power Distribution Unit*.
IRF	See *Intelligent Resilient Framework*.
iSCSI	See *Internet Small Computer System Interface*.
J	
K	
Keystone	An OpenStack Identity service that provides a central directory of users mapped to the OpenStack services they access. It acts as a common authentication system across the cloud operating system and can integrate with directory services such as LDAP.
L	
LACP	See *Link Aggregation Control Protocol*.
LAMP stack	A solution stack that is composed entirely of free and open-source software, suitable for building highly available dynamic web sites; LAMP = Linux (operating system), Apache HTTP Server (web server), MySQL (database), and PHP/Perl/Python (scripting).
LAN on motherboard	Typically, a chip or a chipset embedded on a PC system board to handle network connections; less expensive than the traditional network interface cards; also frees up a PCI slot.
Lease period	The duration, or lifetime, of an infrastructure service. It is set or changed by the user.
LI	See *logical interconnect*.
LIG	See *logical interconnect group*.
Link Aggregation Control Protocol	A method to control the bundling of several physical Ethernet ports together to form a single logical channel; allows a network device to negotiate an automatic bundling of links by sending LACP packets to the peer.
Linux Preboot Environment	A Preboot eXecution Environment (PXE) that uses a Linux image.
LinuxPE	See *Linux Preboot Environment*.

Logical disk	A partition that may contain a file system or a database or be used by a volume manager or a hypervisor to present higher-order volumes. Volumes are backed up by capacity that has been created from lower-level capacity.
Logical interconnect	An HPE OneView construct that represents available networks, uplink sets, and stacking links for a set of physical interconnects in a single enclosure.
Logical interconnect group	Within HPE OneView, acts as a recipe for creating a logical interconnect representing available networks, uplink sets, stacking links, and interconnect settings for a set of physical interconnects in a single enclosure.
Logical resource template	A logical resource offering, describing a service that can be obtained from the resource self-service interface.
Logical unit	The entity within a SCSI target that executes I/O commands. Each logical unit exported by an array controller corresponds to a volume.
Logical unit number	An address used in the SCSI protocol to access an array within a target; refers to a volume that has been presented to one or more initiators.
Logical volume	The storage medium that is associated with a logical disk. It typically resides on one or more hard drives.
LOM	See *LAN on motherboard*.
LUN	See *logical unit number*.
LV	See *logical volume*.
M	
MAC address	See *Media Access Control address*.
Management LAN	A LAN dedicated to the communications necessary for managing systems. Typically of moderate bandwidth (10/100 Mbps) and secured.
Management network	See *management LAN*.
Management protocol	A set of protocols, such as WBEM, HTTP, or SNMP, used to establish communication with discovered systems.
Media Access Control address	A unique identifier assigned to network interfaces for communications on the physical network segment.
Monitoring	The process of collecting server, storage, network, and application availability and performance data across the physical and virtual infrastructure.
MPIO	See *Multipath I/O*.
Multipath I/O	A fault-tolerance and performance enhancement technique whereby there is more than one physical path between the CPU in a computer system and its mass storage devices through the buses, controllers, switches, and bridge devices connecting them.
Multi-tenancy	Software architecture where a single instance of the software runs on a server and serves multiple clients (organizations or tenants); each client organization receives its own customized virtual application instance.

Multithreading	The ability of an application and operating system to allow parallel computing by dividing processing between multiple processors or cores.

N

Network	Medium for interconnecting a specific set of server links; typically relates to a single L2 broadcast domain; may map to a specific Ethernet VLAN, a multitenant network, or a specific IP subnet.
Network set	A specific set of networks within a single domain that can be used together.
Network Time Protocol	Network protocol for time synchronization between devices.
Network virtualization	Takes physical network devices and adds an abstraction layer that brings these devices into a single network resource pool from which virtual networks can be provisioned.
Neutron	The OpenStack Networking component that is a system for managing networks and IP addresses. It ensures that the network will not be the bottleneck or the limiting factor in cloud deployments and provides users real self-service.
North-south traffic	The client server network traffic that goes between users in a branch office and the data center that hosts the application that they are accessing.
NTP	See *Network Time Protocol*.

O

OA	See *Onboard Administrator*.
Object storage	Object storage enables users to store and retrieve files that have been stored in namespaces. The storage itself is unorganized, but still enables users to retrieve content that has been stored.
Onboard Administrator	Central point (single management processor) for controlling an entire BladeSystem enclosure; offers configuration, power, and administrative control over the rack, its associated blades, iLOs, network switches, and storage components.
On-site hosted private cloud	Private cloud infrastructure owned by the organization and residing on the organization's premises, but managed by a third-party cloud provider.
On-site private cloud	Private cloud infrastructure owned and operated by an organization and located on the organization's premises.
OO	See *HPE Operations Orchestration*.
Open network	A Fibre Channel network that does not have zones.
OpenFlow	A protocol that is a developing standard that structures communication between the control and data planes (or forwarding, infrastructure layers) of supported network devices; centralizes control of those devices, and simplifies network management.
OpenSSH	A set of network connectivity tools providing encrypted communication sessions over a computer network using SSH.
OpenStack	An open source cloud platform project founded by IaaS provider Rackspace and NASA. Now supported by technology companies such as HPE, Intel, Cisco, AT&T, and Dell.

Orchestration	The execution of predefined workflows used during provisioning and management of cloud services.
Overprovisioning	Overallocation of resources to have enough capacity for peak demands.

P

PaaS	See *Platform as a Service*.
Partition	1. A subset of server hardware that includes core, memory, and I/O resources on which an operating system can be run. 2. A resource partition that runs within a single operating system.
PDR	See *Power Distribution Rack*.
PDU	See *Power Distribution Unit*.
Peak	The highest utilization value in the selected time interval.
Peer Motion	StoreServ software that allows you to balance workloads across multiple arrays in the same location or across geographies, and to shift workloads between systems dynamically without affecting application performance or availability.
Physical capacity	The maximum amount of storage available on an array or a pool, as constrained by the physical structure of the device.
Physical resource group	A collection of similar physical resources that allows for indicating common desired settings across the members of the group. Each group has group settings to describe the common desired settings.
Physical server	A single instance of a physical server that is physically attached to one or more interconnect modules and may have a server profile assigned to it.
Physical server type	Defines the unique physical configuration of physical server resources within the HPE OneView appliance; includes details such as the base server model (SKU), LOMs, mezzanine cards and their location, storage, memory, and CPU configuration.
Platform as a Service	Cloud service model providing a development platform for consumer-developed or acquired applications created with programming languages or tools from the provider.
PoC	See *Proof of Concept*.
Policy	A collection of rules and settings that control compute resources.
Power distribution rack	A data center power distribution rack that improves power management by moving power distribution to the row level. It decentralizes power, improves cable management, decreases diagnostic time for problems, and saves installation costs.
Power Distribution Unit	Rack device that distributes conditioned AC or DC power within a rack.
PR group	See *physical resource group*.
Preboot eXecution Environment	An environment to boot computers using a network interface independently of data storage devices (such as hard disks) or installed operating systems.
Private cloud	A cloud infrastructure that is operated solely for a single organization. Can be owned or operated by that organization or by a third party. Can be on or off premises.

Private multi-tenant cloud	See *virtual private cloud*.
Proof of Concept	Realization of a certain method or the ideas to demonstrate the feasibility of the method; a demonstration in principle, whose purpose is to verify that a concept or a theory has the potential of being used. A PoC is usually small and may or may not be complete.
Provisioning	The process of deploying an operating system, an application, or a service from a template.
Public cloud	A cloud infrastructure that is made available to the general public or a large industry and is owned and operated by the organization selling cloud services.
PXE	See *Preboot eXecution Environment*.

Q

QoS	See *Quality of Service*.
Quality of Service	A combination of qualitative and quantitative factors such as uptime, response time, and available bandwidth, that collectively describe how well a system performs.

R

Rack	A set of components cabled together to communicate between themselves. A rack is a container for an enclosure.
RAID	See *redundant array of independent disks*.
RBAC	See *role-based access control*.
RBSU	See *ROM-Based Setup Utility*.
Redundant Array of Independent Disks	A RAID volume consists of more than one drive but appears to the operating system to be a single logical disk. RAID improves performance by disk striping, which involves partitioning each drive's storage space into units and placing those on multiple disks.
Redundant SAN	The duplication of components to prevent failure of a SAN solution.
Reference architecture	Provides a template solution, common vocabulary, structures and elements, and relationships for an architecture for a particular domain.
Relative headroom	The percentage by which the demand on a resource can grow before the utilization limits set for the resource are exceeded.
Representational State Transfer	An architectural style consisting of a coordinated set of constraints applied to components, connectors, and data elements within a distributed hypermedia system.
Resource manager	A generic term representing an architectural pattern used in the CI management architecture. Components implementing the resource manager pattern have the responsibility of implementing services from a group of resources.
REST	See *Representational State Transfer*.
Return on investment	A measure of how much a company earns for each dollar (or monetary unit) of investment it makes.
ROI	See *return on investment*.

Role-based access control	An approach of restricting system access to authorized users where an administrator allocates specific levels of access to people in different roles.
ROM-Based Setup Utility	A configuration utility embedded in the system ROM of HPE ProLiant Gen2 and later servers. Pressing F9 from the startup sequence starts this utility and allows configuration of system parameters.

S

SA	See *HPE Server Automation*.
SaaS	See *Software as a Service*.
SAN	See *storage area network*.
SAVA	See *HPE Server Automation Virtual Appliance*.
Scalability	Ability to expand a computing resource to match the current workload.
SCMB	See *State Change Message Bus*.
SDG	See *Service Delivery Guide*.
SDN	See *Software Defined Networking*.
SDP	See *Solution Demo Portal*.
Secure HTTP	An extension to the HTTP protocol that supports sending data securely over the web.
Secure Shell	A program to log into another system over a network and execute commands on that system. It also enables you to move files from one system to another, and provides authentication and secure communication over insecure channels.
Secure Sockets Layer	A standard protocol layer that lies between HTTP and TCP and provides privacy and message integrity between a client and a server. A common usage is to provide authentication of the server, so clients can be assured of communication with that server.
Self-service	Ability of cloud consumers to peruse, configure, and provision, via self-service portals, their own IT resources without any interaction with the cloud provider.
Server blade	Typically a very dense server system, containing microprocessor, memory, and network connections, that can be easily inserted into a rack-mountable enclosure to share power supplies, fans, switches, and other components with other server blades.
Server hardware type	Defines the physical configuration for server hardware, and the settings that are available to server profiles to be assigned to that type of server hardware.
Server profile	A description of a server including its identity (serial number and UUID), hardware requirements (CPU, memory, and so on), configuration settings, and connections to data center networks and fabrics in a manner independent of the underlying hardware.
Server profile template	A type of server profile that cannot be instantiated directly but can be used to generate an instantiable server profile.
Server template	See *server profile template*.

Server virtualization	A single physical server hosts a number of VMs, each with an operating system, and includes virtualized devices such as hard drives, NICs, optical drives, and printers.
Service aggregation	A cloud broker combines two or more cloud provider services into a new service that better satisfies a customer need.
Service catalog	A list of services that the cloud provider offers to the client.
Service Delivery Guide	An HPE Technical Services (TS) document that defines the scope, steps, components, and tools for successful delivery of services.
Service intermediation	A cloud broker injects a value-added service between the cloud provider and the cloud consumer.
Service-level agreement	Contract between a consumer of computing resources and the provider of those resources. Includes the minimum level of service, particularly in terms of availability.
Service Pack for ProLiant	A comprehensive collection of firmware and system software components, all tested together as a single solution stack that includes drivers, agents, utilities, and firmware packages for HPE ProLiant servers, controllers, storage, server blades, and enclosures.
Service template	A design blueprint specifying the requirements for an infrastructure service with servers, storage, and networks. Contains customization points that use workflows during execution of a request.
Serviceguard	Specialized software for protecting mission-critical applications from a variety of hardware and software failures. With Serviceguard, multiple servers and server partitions are organized into an enterprise cluster that delivers highly available application services.
Serviceguard cluster	A networked grouping of HPE Integrity servers having sufficient redundancy of software and hardware that a single point of failure will not significantly disrupt service.
Simple Network Management Protocol	One management protocol supported by HP Systems Insight Manager (HP SIM). Traditional management protocol used extensively by networking systems and most servers.
Simple Object Access Protocol	A lightweight protocol for exchange of information in a decentralized, distributed environment.
Simulation	A mock situation (scenario) that allows you to experiment with various capacity-planning solutions.
Single Point of Connectivity Knowledge	An HPE portal for obtaining detailed information about supported HPE storage product configurations. SPOCK is located at: http://h20272.www2.hp.com
Single Sign-On	Ability to log in once with a single set of credentials and gain access to multiple systems without having to log in to each individually.
SLA	See *service-level agreement*.
SNMP	See *Simple Network Management Protocol*.
SNMP trap	An asynchronous event generated by an SNMP agent that the system uses to communicate a fault.

SOAP	See *Simple Object Access Protocol*.
Software as a Service	Cloud service model providing access to applications hosted on the cloud infrastructure.
Software Defined Networking	An approach to networking where the network administrators manage network services through abstraction of lower-level functionality; abstracts the control plane, presenting you with a single, simple interface for interacting with the network dynamically.
Solution Demo Portal	A portal from HPE that provides a central location for demonstrations, webinars, and supporting collateral that showcase how HPE technologies lead, innovate, and transform enterprise business; includes live and prerecorded demonstrations.
SPOCK	See *Single Point of Connectivity Knowledge*.
SPP	See *Service Pack for ProLiant*.
SSH	See *Secure Shell*.
SSL	See *Secure Sockets Layer*.
SSO	See *Single Sign-On*.
State Change Message Bus	An interface that uses asynchronous messaging to notify subscribers of changes to managed resources, both logical and physical.
Storage area network	A network (or subnetwork) that connects data storage devices with associated data servers. A SAN is typically part of an overall network of computing resources.
Storage federation	Technology that enables online, nondisruptive movement of storage volumes between arrays.
Storage virtualization	For physical storage devices, adds an abstraction layer that brings these devices into a single storage resource pool from which virtual disks are provisioned.
Subscribed capacity	The amount of storage within a storage pool that has been promised to clients. Thinly provisioned volumes may contribute more to the subscribed capacity of the pool of which they are a member than they contribute to the committed capacity.
Supply layer	A layer within the HPE Cloud Functional Reference Architecture where the resources (typically virtualized) are defined, created, configured, and monitored.
Swift	An OpenStack Object Storage component that is a scalable redundant storage system. Objects and files are written to multiple disks throughout servers in the data center, with the OpenStack software responsible for data replication and integrity.
Switch group	A collection of interconnect modules that may have similar attributes, such as network membership, uplink configuration, or basic module settings.
T	
Tenant array	Servers hosting users' application workloads, as opposed to servers running monitoring and/or management software.
Thermal Logic	See *HPE Thermal Logic*.

Time-to-value	A measure of how long it takes a resource to begin providing a direct benefit to a business.
Top-of-Rack network design	A network design where a network switch is placed on every rack and all computing devices within the rack connect to it. These network switches can be connected to aggregation switches using a few cables.
ToR network design	See *Top-of-Rack network design*.
U	
UC&C	See *Unified Communication and Collaboration*.
UCMDB	See *HPE Universal Configuration Management Database*.
UI	See *User Interface*.
Unified Communication and Collaboration	Merging voice, video, messaging, presence, and data across a single, unified network; integrates telephony, audio conferencing, email and voice messaging, web and video collaboration, and other communications into a seamless communications platform.
Uniform Resource Identifier	A string of characters used to identify a name of a web resource; a particular view of a physical or logical resource or some metadata.
URI	See *Uniform Resource Identifier*.
User interface	A system through which users interact with a computer.
Utilization limits	The limits set on the usage of system resources such as CPU, memory, or network I/O by an application. Utilization limits are expressed as a percent of the system capacity and the amount of time an application is allowed to exceed this limit.
V	
VAN	See *Virtual Application Network*.
VC	See *HPE Virtual Connect*.
VCEM	See *HPE Virtual Connect Enterprise Manager*.
vCenter	See *VMware Virtual Center*.
VCM	See *HPE Virtual Connect Manager*.
vCPU	Virtual CPU. A single-core virtual processor in a virtual machine.
VCSU	See *Virtual Connect Support Utility*.
VDI	See *Virtual Desktop Infrastructure*.
Vdisk	See *virtual disk*.
Virtual Application Network	A framework that enables network agility and accelerates application and service delivery in the data center; automates network operations using Software Defined Networking (SDN) technology.
Virtual Connect Support Utility	Windows-based command line utility that enables administrators to upgrade Virtual Connect-Ethernet and Virtual Connect-Fibre Channel firmware, and to perform other maintenance tasks remotely on both the BladeSystem c7000 and c3000 enclosures.

Virtual Desktop Infrastructure	An infrastructure that places user desktops in the data center and replaces user PCs with thin clients; reduces the need for desktop support resources and lowers desktop PC capital expenditures.
Virtual device	An emulation of a physical device. This emulation, used as a device by an Integrity VM virtual machine, effectively maps a virtual device to an entity on the VM host.
Virtual disk	A virtual logical disk or volume in storage virtualization applications.
Virtual local area network	A standard that enables network administrators to group end users by logical function rather than by physical location. Created on switches to segment networks into smaller broadcast domains, enhance network security, and simplify network management.
Virtual machine	A software simulation of a fully operational computer that can have its own operating system, storage, and applications.
Virtual private cloud	Single-tenant, private cloud infrastructure hosted in a multitenant, public cloud environmnent. Also called *private multitenant cloud*.
Virtual resource pools	Within the HPE Converged Infrastructure, provide a common modular infrastructure of virtualized compute, memory, storage, and network resources.
VLAN	See *virtual local area network*.
VM	See *virtual machine*.
VM host	A server running HPE Integrity Virtual Machines, VMware ESX, VMware ESXi, or Microsoft Hyper-V that provides multiple virtual machines, each running its own instance of an operating system.
VM hosted logical server	A logical server running on a virtual machine under the control of a hypervisor.
VMware Virtual Center	VMware enterprise-level virtualization management product.
Volume	A logical drive provided by a storage pool that can be presented to a host system.

W

WAIK	See *Windows Automation Installation Kit*.
WBEM	See *Web-based enterprise management*.
Web-based enterprise management	An industry initiative providing management of systems, networks, users, and applications across multivendor environments. It simplifies system management and provides uniform access to data. A Distributed Management Task Force (DMTF) standard.
What-if scenario	A configuration of systems and workloads that is different from the current configuration. Capacity-planning simulations are run using what-if scenarios as experiments before making any actual configuration changes.
Windows Automation Installation Kit	A collection of tools and technologies from Microsoft designed to help deploy Microsoft Windows operating system images to target computers or virtual hard disks.
Windows Management Instrumentation	An API in the Windows operating system that enables management and control of systems in a network.

Wizard	A sequential series of pages that transforms a complex task into simple steps and guides you through them. The wizard makes sure that you provide all required information and do not omit any steps.
WMI	See *Windows Management Instrumentation*.
Workflow	A set of actions, such as approvals, manual steps, or notifications, that execute alongside the service templates.
Workload	A collection of processes in a stand-alone server, or virtual machine compartment.
World Wide ID or World Wide Name	A unique identifier assigned to a Fibre Channel device.
WWID	See *World Wide ID* or *World Wide Name*.
WWN	See *World Wide ID* or *World Wide Name*.

X

Y

Z

Zone	Within a Fibre Channel SAN, a zone is a subset of the endpoints on a network. Endpoints may communicate only with other endpoints that are in the same zone. Endpoints may be in multiple zones.
Zoned network	Fibre Channel network that has zones.

Index